T0262855

# Cyber-Physical Systems

## Integrated Computing and Engineering Design

# Cyber-Physical Systems

## Integrated Computing and Engineering Design

# Fei Hu

## CRC Press
Taylor & Francis Group
Boca Raton   London   New York

CRC Press is an imprint of the
Taylor & Francis Group, an **informa** business

CRC Press
Taylor & Francis Group
6000 Broken Sound Parkway NW, Suite 300
Boca Raton, FL 33487-2742

© 2014 by Taylor & Francis Group, LLC
CRC Press is an imprint of Taylor & Francis Group, an Informa business

No claim to original U.S. Government works

Version Date: 20130819

International Standard Book Number-13: 978-1-4665-7700-8 (Hardback)

**Library of Congress Cataloging-in-Publication Data**

Hu, Fei, 1972-
  Cyber-physical systems : integrated computing and engineering design / Fei Hu.
    pages cm
  Summary: "Many things need to be done to realize the benefits of cyber-physical systems. With these goals for CPS improvement, the book discusses the many challenges that must first be overcome, and provides a roadmap on how to do it. Accounting for random events that can occur in a real environment can be troublesome for a system when trying to ensure safety, security, and predictability"-- Provided by publisher.
    Includes bibliographical references.
    ISBN 978-1-4665-7700-8 (hardback)
    1. Automatic control. 2. Cybernetics. 3. Systems engineering. I. Title.

TJ213.H698 2013
629.8'9--dc23                                                                                                      2013014922

**Visit the Taylor & Francis Web site at**
**http://www.taylorandfrancis.com**

**and the CRC Press Web site at**
**http://www.crcpress.com**

To Fang Yang, Gloria Yang Hu,
Edward and Edwin (twins) Yang Hu.

# Contents

# Preface

Cyber-physical systems (CPSs) have become one of the hottest computer applications today. A CPS has the tight integration of cyber and physical objects. Here, the term *cyber objects* refers to any computing hardware/software resources that can achieve computation, communication, and control functions in a discrete, logical, and switched environment. Also, *physical objects* refers to any natural or human-made systems that are governed by the laws of physics and operate in continuous time. It is believed that CPSs will transform how we interact with the physical world, just like the Internet transformed how we interact with one another. A CPS could be a system at multiple scales, from big smart bridges with fluctuation detection and responding functions, to autonomous cars, to tiny implanted medical devices. As a matter of fact, the ultimate purpose of using cyber infrastructure (including sensing, computing, and communication hardware/software) is to intelligently *monitor* (from physical to cyber) and *control* (from cyber to physical) our physical world. Figure 1 illustrates such a concept.

This book will comprehensively cover the principles and design of CPSs. It is not a simple collection of different experts' opinions. Instead, it is written in a systematic way as follows: first, it discusses the basic concept of CPS and some challenging design issues. Second, it provides the most important design theories and modeling methods for a practical CPS. Third, it discusses

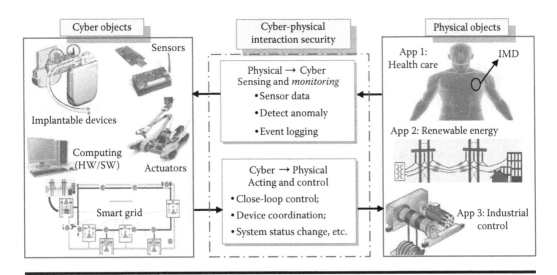

**Figure 1  Cyber-physical systems (CPSs): Examples and concept.**

sensor-based CPS, which uses embedded sensors and actuators to interact with the physical world. Fourth, it discusses concrete CPS designs for popular civilian applications including building and energy management. Finally, because of the importance of human health care in society, it provides some CPS examples in rehabilitation applications such as virtual reality-based disability recovery platforms.

**Features of the Book**: Compared to other computing books, this book has the following special features:

1. *Emphasizes the tight integration of cyber computing and physical objects control*: Unlike traditional computing books that describe pure computing knowledge, this book emphasizes the cyber-to-physical (control) and physical-to-cyber (sensing) through the understanding of physical laws as well as the impacts of computing hardware/software on the physical characteristics. As an example, for implanted medical device (IMD) applications, conventional computing discussions focus on the internal design of an IMD from a circuit and software viewpoint. However, we will discuss how an IMD can monitor the human tissue status and then control the IMD outputs (such as electrical pulses) to change the organ/tissue's status. Such a sensing/control model should be built in order to understand how physical sensing triggers different physical control outputs.

2. *Uses concrete case studies to explain challenging CPS design*: In Sections IV and V we will use several important civilian and health care applications to illustrate the detailed CPS design process. For example, we will show how energy can be significantly saved by using smart sensors and air conditioning controllers. We will also use a virtual reality-based rehabilitation system to explain how a patient can interact with a virtual world to train his or her body flexibility.

3. *Covers important CPS theory foundations and models*: Some CPS researchers need to understand the theory foundations and modeling skills before they can design a practical CPS. Therefore, in Section II, we will explain how some important theories (such as networked control) can be used to build a CPS control strategy. We will also explain how we can use software to model a CPS in order to understand large-scale CPS design. Such a model is important because it is not realistic to always build a large-scale CPS.

**Targeted Audiences**: This book is suitable for the following types of readers:

1. *College students*: This book can serve as the textbook or reference book for college courses on CPS. CPS courses could be offered in computer science, electrical and computer engineering, information technology and science, or other departments.

2. *Researchers*: Because we explain CPS from concepts to theory foundations, the contents are very useful to researchers (e.g., graduate students and professors) who are interested in CPS design.

3. *Computer scientists*: We have provided many computing algorithms on CPS sensing and control models in this book. Thus, computer scientists could refer to those principles in their own design.

4. *Engineers*: We have also provided many useful CPS design principles. Thus, company engineers could use those principles in their product design. The last eight chapters especially provide concrete CPS case studies.

**Book Architecture**: This book uses five sections to cover system, hardware, software, and security issues in CPS.

*Section I: Basics.* This section describes the basic concepts of CPS from the architecture and design challenges viewpoint. We will explain the differences between CPS and general computing systems.

*Section II: Design Principles.* This section describes the design principles of CPS including the important theories on signal processing, machine learning, and modeling methods.

*Section III: Sensor-Based Cyber-Physical Systems.* This section focuses on the sensor–actuator interactions in a CPS. We will explain how sensors can collect the physical world status and how we can use the sensing data to control physical objects.

*Section IV: Civilian Cyber-Physical System Applications.* We will discuss concrete CPS applications in building energy management, smart grid, and transportation systems.

*Section V: Health Care Cyber-Physical System Applications.* This section will detail some exciting health care applications that use CPS architecture. We will describe rehabilitation systems based on virtual reality and robots.

# Disclaimer

We sincerely thank all authors who have published CPS materials and directly/indirectly contributed to this book through our citations. We have tried our best to provide credits to all cited publications in this book. Because of the time limit, this book could have some errors or missing contents. If you have questions on the contents of this book, please contact the author (Fei Hu: fei@eng.ua.edu), and we will correct the errors and thus improve this book in the future editions.

MATLAB® is a registered trademark of The MathWorks, Inc. For product information, please contact:

The MathWorks, Inc.
3 Apple Hill Drive
Natick, MA 01760-2098 USA
Tel: 508 647 7000
Fax: 508-647-7001
E-mail: info@mathworks.com
Web: www.mathworks.com

# Author

 **Dr. Fei Hu** is currently an associate professor in the Department of Electrical and Computer Engineering at the University of Alabama (main campus), Tuscaloosa, Alabama, USA. He obtained his PhD degrees at Tongji University (Shanghai, China) in the field of signal processing (in 1999) and at Clarkson University (New York, USA) in the field of electrical and computer engineering (in 2002). He has published over 160 journal/conference papers and books.

Dr. Hu's research has been supported by the US National Science Foundation, Cisco, Sprint, and other sources. His research expertise can be summarized as *3S: Security, Signals, Sensors*:

(1) Security, which focuses on how to overcome different cyber attacks in a complex wireless or wired network. Recently, he focused on cyber-physical system security and medical security issues.

(2) Signals, which mainly refers to *intelligent signal processing*, that is, using machine learning algorithms to process sensing signals in a smart way in order to extract patterns (i.e., achieve pattern recognition).

(3) Sensors, which includes microsensor design and wireless sensor networking issues.

# Contributors

This book is mainly written by Fei Hu, with the help from some of his students and colleagues. We especially thank the following people:

**Jaber Abu-Qahouq**
Department of Electrical and Computer
  Engineering
University of Alabama–Tuscaloosa
Tuscaloosa, Alabama

**Ahmed Alsadah**
Department of Electrical and Computer
  Engineering
University of Alabama–Tuscaloosa
Tuscaloosa, Alabama

**Preston Arnett**
Department of Electrical and Computer
  Engineering
University of Alabama–Tuscaloosa
Tuscaloosa, Alabama

**Matthew Bell**
Department of Electrical and Computer
  Engineering
University of Alabama–Tuscaloosa
Tuscaloosa, Alabama

**Brock Bennett**
Department of Electrical and Computer
  Engineering
University of Alabama–Tuscaloosa
Tuscaloosa, Alabama

**Trenton Bennett**
Department of Electrical and Computer
  Engineering
University of Alabama–Tuscaloosa
Tuscaloosa, Alabama

**Sarah Betzig**
Department of Electrical and Computer
  Engineering
University of Alabama–Tuscaloosa
Tuscaloosa, Alabama

**Erica Boyle**
Department of Electrical and Computer
  Engineering
University of Alabama–Tuscaloosa
Tuscaloosa, Alabama

**David Brown**
Department of Physical Therapy
University of Alabama–Birmingham
Birmingham, Alabama

**Chad Buckallew**
Department of Electrical and Computer
  Engineering
University of Alabama–Tuscaloosa
Tuscaloosa, Alabama

**Xiaojun Cao**
Department of Computer Science
Georgia State University
Atlanta, Georgia

**Christopher Chadwick**
Department of Electrical and Computer
   Engineering
University of Alabama–Tuscaloosa
Tuscaloosa, Alabama

**Derek Chandler**
Department of Electrical and Computer
   Engineering
University of Alabama–Tuscaloosa
Tuscaloosa, Alabama

**Sarah Duncan**
Department of Electrical and Computer
   Engineering
University of Alabama–Tuscaloosa
Tuscaloosa, Alabama

**Bryant Grace**
Department of Electrical and Computer
   Engineering
University of Alabama–Tuscaloosa
Tuscaloosa, Alabama

**John Grace**
Department of Electrical and Computer
   Engineering
University of Alabama–Tuscaloosa
Tuscaloosa, Alabama

**Mengcheng Guo**
Department of Electrical and Computer
   Engineering
University of Alabama–Tuscaloosa
Tuscaloosa, Alabama

**Steven Guy**
Department of Electrical and Computer
   Engineering
University of Alabama–Tuscaloosa
Tuscaloosa, Alabama

**Qi Hao**
Department of Electrical and Computer
   Engineering
University of Alabama–Tuscaloosa
Tuscaloosa, Alabama

**Yang-ki Hong**
Department of Electrical and Computer
   Engineering
University of Alabama–Tuscaloosa
Tuscaloosa, Alabama

**Walter Hudgens**
Department of Electrical and Computer
   Engineering
University of Alabama–Tuscaloosa
Tuscaloosa, Alabama

**Tony Huynh**
Department of Electrical and Computer
   Engineering
University of Alabama–Tuscaloosa
Tuscaloosa, Alabama

**Michael Johnson**
Department of Electrical and Computer
   Engineering
University of Alabama–Tuscaloosa
Tuscaloosa, Alabama

**Rebecca Landrum**
Department of Electrical and Computer
   Engineering
University of Alabama–Tuscaloosa
Tuscaloosa, Alabama

**Kassie McCarley**
Department of Electrical and Computer
   Engineering
University of Alabama–Tuscaloosa
Tuscaloosa, Alabama

**Loilin Muirhead**
Department of Electrical and Computer
Engineering
University of Alabama–Tuscaloosa
Tuscaloosa, Alabama

**Sarah Pace**
Department of Electrical and Computer
Engineering
University of Alabama–Tuscaloosa
Tuscaloosa, Alabama

**Cameron Patterson**
Department of Electrical and Computer
Engineering
University of Alabama–Tuscaloosa
Tuscaloosa, Alabama

**Joseph Pierson**
Department of Electrical and Computer
Engineering
University of Alabama–Tuscaloosa
Tuscaloosa, Alabama

**Jonathan Pittman**
Department of Electrical and Computer
Engineering
University of Alabama–Tuscaloosa
Tuscaloosa, Alabama

**Tony Randolph**
Department of Electrical and Computer
Engineering
University of Alabama–Tuscaloosa
Tuscaloosa, Alabama

**Kenneth Ricks**
Department of Electrical and Computer
Engineering
University of Alabama–Tuscaloosa
Tuscaloosa, Alabama

**Edward Sazonov**
Department of Electrical and Computer
Engineering
University of Alabama–Tuscaloosa
Tuscaloosa, Alabama

**Matt Smith**
Department of Electrical and Computer
Engineering
University of Alabama–Tuscaloosa
Tuscaloosa, Alabama

**Wenlong Tang**
Department of Electrical and Computer
Engineering
University of Alabama–Tuscaloosa
Tuscaloosa, Alabama

**Roger Vasquez**
Department of Electrical and Computer
Engineering
University of Alabama–Tuscaloosa
Tuscaloosa, Alabama

**Yufan Wang**
Department of Computer Science
Georgia State University
Atlanta, Georgia

**Ian Wolfe**
Department of Electrical and Computer
Engineering
University of Alabama–Tuscaloosa
Tuscaloosa, Alabama

**Hoyun Won**
Department of Electrical and Computer
Engineering
University of Alabama–Tuscaloosa
Tuscaloosa, Alabama

**Lv Wu**
Department of Electrical and Computer
Engineering
University of Alabama–Tuscaloosa
Tuscaloosa, Alabama

**Yeqing Wu**
Department of Electrical and Computer
Engineering
University of Alabama–Tuscaloosa
Tuscaloosa, Alabama

**Ting Zhang**
Department of Electrical and Computer
 Engineering
University of Alabama–Tuscaloosa
Tuscaloosa, Alabama

**Huiying Zhen**
Department of Electrical and Computer
 Engineering
University of Alabama–Tuscaloosa
Tuscaloosa, Alabama

# BASICS

1

## Chapter 1

# Cyber-Physical Systems Concepts

Christopher Chadwick, Sarah Betzig, and Fei Hu

## Contents

## 1.1 Introduction

Cyber-physical systems (CPSs) incorporate the close interactions between computing resources (such as computer hardware and software) and physical objects (such as a car, human body, etc.). CPS already exists today. For example, many cars on the market use a computer-based antilock brake system (see Figure 1.1). CPSs could revolutionize many areas from health care to manufacturing to transportation. However, the ability to realize these grand visions is limited by today's computing technology, particularly by the lack of system reliability and timing control [1]. A CPS typically includes a network of devices that receive and perform physical actions while simultaneously being controlled and monitored by computational and communication software [2].

Computational and digital technologies will soon be found in, and play an integral role in, many physical structures and devices. Just as the Internet has revolutionized how people interact

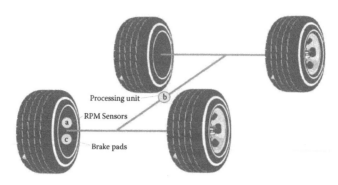

**Figure 1.1** **CPS example: antilock brake system. When the driver holds down the brake pedal, the CPS in the car pumps the brakes automatically. (a) Sensors near the tires collect information about the rate of rotation of each tire and send these data back to the (b) processing unit. This information is then used to determine the system's status. A series of commands to carry out the response is sent to the (c) brake pads, which activate or deactivate appropriately.**

with each other and allowed us to more easily connect and trade with one another, CPSs also have the same potential to change how we interact with the physical world around us. The implementation of the Internet is based on the integration of major advancements in network technology, applications, and infrastructure. Likewise, CPSs can be seen as the integration of embedded systems, sensors, and control systems.

There are a number of technological advancements that are opening the door for CPS improvement [3–7]. Devices are becoming cheaper as they get smaller and smaller. There are also breakthroughs in wireless communication, Internet bandwidth, and the constant increase in alternative energy sources and energy potential. Computer parts are becoming increasingly more high capacity at lower power consumption and smaller form factors. The CPS field is also becoming more and more demanding in the areas of aerospace, defense, the environment, health care, and automation.

A CPS can be thought of as the utilization of the logical and discrete properties of computers to control and oversee the continuous and dynamic properties of physical systems. Using precise calculations to control a seemingly unpredictable physical environment is a great challenge. The uncertainty and lag from a real-time physical system to discrete-time digital control is an obstacle that must be overcome. The failures or safety issues must be contained and dealt with in an efficient manner. Synchronization within a system and overcomplexity are also obstacles that must be taken into consideration in order for the CPS field to grow. Systems that are designed around trial-and-error methods should be replaced with more reliable and easily adjustable blueprints. Analysis and engineering are needed to make the way for more robust systems so that accidents do not occur [8]. Figure 1.2 shows that a CPS system can activate a security system with smart camera sensors.

In Table 1.1 we have listed the major differences between cyber resources and physical objects. Those differences make the CPS design very challenging.

With the architecture of software design in CPSs, two main challenges arise: first, the huge diversity of hardware platforms used in embedded applications and, second, the diversity of these applications themselves. Rather than focusing on execution times and utilizing resources, these applications will not only account for real-time requirements but will also concentrate on being safe, reliable, and more predictable. A greater diversity in hardware platforms will have a huge impact on various aspects of system software. However, most of today's systems concentrate only

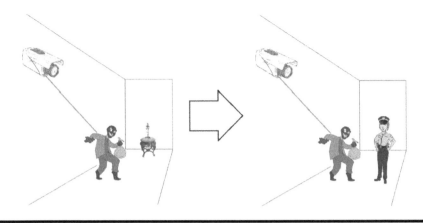

**Figure 1.2    The CPS reads the status of the environment through sensors, and from the sensing result, the computer formulates how the physical part of the system should respond. For example, if the system detects an imposter through the cameras, laser beams, infrared detection, motion detection, pressure pads, or other sensing devices, it can call the police automatically.**

on improving and extending already existing commercial systems. Given the diversity of applications and hardware features for most systems, one basic cookie-cutter system design may not work for other systems. Thus a base software architecture should be employed where various devices can be constructed or altered with ease for individual applications, while simultaneously being safe, reliable, and efficient [6].

CPS software system design must include a diverse pool of specific application tools that can automatically and independently perform a variety of functions according to the specific constraints of the system's hardware capabilities and application requirements. This avoids the situation of having one function that is more important than another and instead suggests an optimal system design where all components and applications are equally conditional upon each other. In traditional devices, memory allocation and cache footprints have been hindered by extraneous services that slow down the system as well as by using code that is much more complex than it needs to be. Such problems would infringe on the safety and dependability of these systems, making them less efficient.

Most software systems today are for the applications that require a trusted kernel through a specific interface. These interfaces are typically assigned at a level where the basic needs are met; however, they are not extensive enough to know the specific task as seen from the lower application levels. In the future, a CPS should be able to program at the specific level that is suited to any given application. If applications can specifically designate their own service and task, the overall system

**Table 1.1    Comparison of Cyber and Physical Properties of CPS**

|  | *Cyber* | *Physical* |
|---|---|---|
| Method of ensuring proper order | Sequences | Real time |
| Event synchronization | Synchronous | Asynchronous |
| Time properties | Discrete | Continuous |
| Structure | Computing abstractions | Physical laws |

will have an easy task of organizing and managing the applications as a whole. For example, an advanced application programmer interface (API) may allow an avionics control system to regulate its speed and altitude, while its lower-level applications that may control the engine or wing rudders can operate independently according to other requirements and specifications.

In order to meet the above-mentioned software architecture specifications, the system has to be constructed as a collection of several different components that are independent from other applications through code or hardware. Otherwise, they would be compiled together in a singular address point in accordance with distinct specifications. For example, if there is a memory management unit in the lower level of the processing hardware, then a whole array of services can be mapped to other hardware domains automatically. This degree of isolation induces inter-service communication lag; hence the system can be self-regulated with more dynamic communication channels in order to balance properly between speed and confinement properties. Also, the ability to self-predict and adjust accordingly with other devices can be achieved as the system adapts to the optimal configuration. We have to ensure that the CPS can acquire the configuration that suitably adapts to a complex hierarchy of descriptions defined by the programmers in a hardware-independent manner. This makes the proper software modules run in the hardware automatically.

## 1.2 CPS Application Examples

### 1.2.1 Claytronics

Here, we provide a few CPS application examples. One potential application of the CPS is claytronics, a technology that could essentially give humans the ability to program matter itself to meet various needs. Currently, most technological systems consist of a computer core that can be reprogrammed and some physical parts that interact with the world. For example, a robotic arm has a computer chip that allows users to produce various commands in order to grip some objects. While the computer part of this technology is very easy to repurpose, the physical part is difficult to control: after a physical system is in place, it can only be effectively used for the purpose that its builders had in mind when designing it. Claytronics would make the matter composing technology programmable. Thus, the physical properties of a claytronic system, such as the shape and color, could be changed with relative ease by using the adaptability of software to hardware [5].

This ability to reprogram the world to fit human needs can derive from the structure of claytronics technologies. The building blocks of claytronic systems are catoms (claytronic atoms), tiny programmable devices that can join together in a way that their macroscopic properties resemble those of regular matter. Each catom needs to be self-sufficient and therefore contains at minimum the following components: a microprocessor, energy storage, networking circuit, sensors, visual output device, transportation system, and an efficient means of binding to other catoms. Just as thousands of tiny pixels are needed to form an image that is clear to a human's eye, millions of catoms on the order of 1 to 0.1 mm are necessary to create claytronic systems.

In order for claytronic technologies to be accessible to society, they should be mass-producible. Mass production of tiny cyber-physical devices is already in place. For instance, photolithography is used to mass produce computer chips. The assembly of tiny three-dimensional objects through two-dimensional processes is already in practice as well. With present-day manufacturing technologies, spheres on the order of 1 mm could be produced. Each sphere contains electrostatic actuators to transport the sphere. It also has 256 kB of memory and an 8086 equivalent microprocessor and is able to deliver enough energy to power the whole assembly. Therefore claytronics is a concept that may be feasible in the near future.

Programming claytronic systems is another challenge. It has two basic levels: the algorithmic level and the control level. The algorithmic level addresses how the system as a whole will accomplish a goal, and the control level deals with how each catom must behave in order to contribute to the accomplishment of the goal of the claytronic system. A certain large-scale behavior is produced from numerous individuals following a simple set of rules. Over the past half-century, emergent systems have been studied, and while much of this research has focused on predicting large-scale behavior based on small-scale rules, recent research suggests that the reverse may be possible; that is, knowing the desired large-scale behavior, one could determine a set of small-scale rules necessary to produce this behavior. For claytronics this means that a user could input a command for the large-scale object, and the system would be able to translate this command into actions for individual catoms to perform. This process is essential for any claytronic system.

Uncertainty is an additional obstacle for claytronics. By giving each catom its own processing unit we may help the system to reduce the uncertainty level. By doing so, the system could easily recover from the failure of one individual catom because its individual contribution to the overall task would be quite small. Still, it is necessary to be able to detect, measure, and deal with uncertainties, including catom failures, communication delays, and environmental changes. Presently, systems typically either ignore or have quite specific ways of dealing with particular types of uncertainty. Uncertainty detection and adaptation is an area of claytronics that requires further research. Simulation is an extremely valuable tool that will allow researchers to antagonize claytronics systems with various types of uncertainty, observe the systems' responses, and fine-tune the systems' correction algorithms.

## 1.2.2 Virtual Reality System

There are a few limitations that must first be resolved before the next major step toward CPS advancement can be made. First, designing a system that is highly flexible must be achieved. A CPS must be developed to the point where adjustments and modifications to an already existing system can be made with relative ease. Second, CPSs as a whole have not yet progressed beyond the stage of experimentation into a more practical, useful tool. The practice, in general, needs to be safe and dependable, as well as accurate enough to operate in real time. Finally, we need to build a CPS that is user friendly. A system that is accessible and easily understood by a wider audience must be achieved in order to promote growth and sustainability [4].

If these limitations can be overcome, we may build a CPS with exciting virtual reality effects. A user would be able to observe and interact with a large array of different virtual environments. Each would be fully equipped with its own sensors, sounds, behaviors, and so on. One application, in particular, that is being proposed is to replicate a CPS similar to that of a large multiplayer video game, thus creating a synthetic world with numerous cyber-physical objects that can also be easily accessed to a large number of people on a shared network simultaneously. However, one of the biggest constraints to such an application would be ease of use. A CPS must become not only an instrument for the programmers and engineers who design it but also a practical tool for the public to promote educational and touristic purposes.

A possible solution to this dilemma is to design a more generic, toolkit-oriented layout. This would provide modules for input data with visualizations and also multiple cross-linked modules to accommodate human communications. Additionally, this system could offer multiple CPS interfaces based on various modeling and code languages that can be integrated and applied uniformly to the cyber-physical environment. Thus, this toolkit could offer an in-depth approach to rendering and analyzing data on varying levels of study.

## 1.3 Some CPS Design Challenges

### 1.3.1 Reliability and Uncertainty

Consumers expect CPSs to be reliable and consistent. Indeed, in many applications, such as medical systems, it is crucial that CPSs perform requested tasks on time and predictably. Unfortunately, current technologies leave much to be explored in this aspect. The behavior of electronic components is not perfectly consistent or predictable.

A disconnection often lies between program execution and physical requirements. A program has, essentially, 100% reliability in the sense that it will go through the exact same set of commands in exactly the same order every time it is run. However, physical systems rely not only on function but also on timing, and computer programs can be imperfect in this regard. This mismatch causes much uncertainty and unreliability and becomes a problem for CPSs.

There are essentially two ways to deal with uncertainty in computer systems: either reduce the uncertainty in each part as much as possible so that the reliability is very close to 100% or implement algorithms to correct errors caused by imperfect reliability in the electronic components. This problem is further complicated by the fact that, in general, the smaller the electronic component, the greater the uncertainty in its operation. Generally, engineers opt both to use parts with a high reliability and to implement error control systems to adapt to uncertainty.

Though some basic CPSs have been successfully made to address the lack of precise timing, advanced CPSs that engineers envision for the future cannot function with the current level of timing inconsistencies. Networking is potentially a key part of the future of CPSs. While current systems can be tested thoroughly and fine-tuned until they are consistent, this will not be possible for future, networked CPSs. Networking means the software will no longer be self-contained, and opening a system to the environment in this way means exposing it to additional unpredictability and to the various complex scenarios under which it will be expected to operate. This makes the ability to adapt to uncertainties even more essential.

To make CPS behavior predictable, some artificial intelligence or machine learning schemes can be used to predict the next-time system state. For example, we may build a regression model or hidden Markov model to describe the CPS time evolution dynamics. Thus, we can predict the next-step system state based on the history state evolutions.

Timing is also crucial in CPSs. Even a 1-s delay could cause a catastrophe. For example, imagine what will happen if the time it takes a self-driven car to respond to the command "stop" varies each time the command is given. To achieve a real-time physical control, the cyber computing parts should avoid any complex algorithms or dead loop. The network delay among sensors and controllers should be minimized.

Much of the problem stems from the fact that most programming languages show a disregard for timing. However, CPSs rely heavily on timing. Another obstacle is that concurrent threads are not often used in programming as they can be confusing to programmers. It may be possible to salvage current computation and programming techniques, but a serious overhaul will be required to accommodate the delicate timing needs of CPSs. Some programs, such as Split-C and Clik, support simultaneous threads, and others, like Guava, limit current languages to be more readily interfaced with CPSs.

Despite this massive disconnection between current computing capabilities and what CPSs require, networks have the possibility to maintain a clock shared by all devices, and the clock's precision can be controlled. Digital circuits themselves have a high degree of timing precision. It may be possible to introduce time sensitivity into current computing to make CPSs a reality.

One possible way to avoid this conflict and speed up the development of CPSs is to introduce a system that would translate physical specifications into software and embedded systems requirements. This would enable physical systems engineers to more easily see the consequences of their physical design choices on the cyber part of the system, and therefore more clearly discern feasible solutions for the system as a whole. This would greatly ease communication between the cyber teams and the physical technicians and speed up the development process of the new CPS. Such a tool would need to have strong simulation capabilities and a wide variety of hardware prototyping options.

### 1.3.2 Levels of Abstraction

Embedded systems, which CPSs fundamentally rely on, have different levels of abstraction (see Figure 1.3). This is helpful in that it allows someone to work on one part of the system without having to understand or alter the rest of the system. For example, a programmer can change his or her code without having to worry about the actual electronic components that will execute the program's commands because the codes have been designed with certain abstraction; that is, they are suitable to any machines as long as the machine understands the code abstraction. However, the way in which this separation is currently implemented causes several orders of magnitude of timing precision to be lost. CPSs can be extremely time sensitive, and precise timing is very important. It is, however, quite encouraging that high precision is available on the most basic level digital circuits. This opens the possibility for more precise CPSs in the future.

The disconnection between different layers of abstraction can be a serious problem for CPSs. The teams designing CPSs are often divided into sections, with each subgroup handling a different layer of abstraction of the project. A common example is that a group of physical system engineers will work on the physical (e.g., mechanical, electrical, chemical, and biological) part of the project; they clearly define the precise actions that must be carried out by the physical parts of the system. A group of computer system engineers may design the software, including memory real-time systems, and networking. Finally, embedded system engineers work on the electronic hardware, handling tasks such as power distribution, system testing, and mechanical packaging. The requirements determined by the physical team are passed to the teams working on software and embedded systems, and they use these specifications as a goal for their work. Each layer of

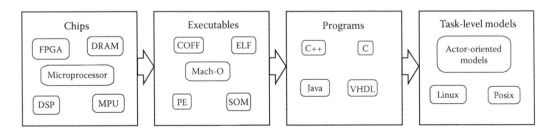

**Figure 1.3  Layers of abstraction in computing.**

abstraction is crucial to the realization of a CPS, so this kind of teamwork is essential to the successful construction of the CPS. However, important information may be lost in communication between the groups. Such a team structure results in software and embedded system engineers relaying problems back to hardware engineers, who must change their designs in an attempt to fix the problems the other groups are having, resulting in a cyclical pattern of information flow. This system of passing requirements and results back and forth often has the side effect of keeping each team from understanding what impact small changes will have on the operation of the system as a whole. The software specialists may choose a certain programming architecture that seems appropriate from their perspective, but the physical side of the system may require a completely different software structure to function properly. Such miscommunication impedes the progress of CPS production [7].

A group of senior engineering students from Rensselaer Polytechnic Institute (RPI) were able to build a CPS (specifically a human transportation vehicle) [7]. They were aided by visual-based software that can efficiently simulate physical system models. Using the simulation results, the students were able to build a scale model, test it, and construct a working full-scale model. Ref. [7] highlights a key factor of successful CPS construction: graphical simulation software. Other studies have confirmed that the visual nature of the software is crucial. Users more easily adopt the software if it has a wide variety of commands available. This tool would help bridge the gap between development in the different layers of abstraction, greatly speeding up the development of CPSs.

### 1.3.3 Cyber-Physical Mismatch

Today, many frequently used programming languages do not have commands dealing with timing and synchronization. In many electronic products today, a delicate balance is reached between hardware and software, and even a slight change in either one can throw off this balance, thus causing errors. For example, airlines stock up on parts when installing a new hardware system because if part of the system is replaced with a nonidentical part, the software will have to be retested and possibly changed. This is not only economically cumbersome, but it also keeps these consumers from having up-to-date technology. Thus, they are stuck using the old model until they use up all their stockpiled parts and can afford to catch up with the technology. So dividing a system into layers of abstraction, if not managed properly, may actually impede progress toward successful CPSs.

In a CPS today, the interaction and coordination between the physical elements and the cyber elements of a system are key aspects. In the physical world, one of the most dominant characteristics is its dynamics or the state of the system constantly changes over time. Alternatively, in the cyber world, these dynamics are more appropriately defined as a series of sequences that do not have temporal semantics. One of the greatest problems that engineers and researchers are faced with today is the point at which these two corresponding subsystems intersect one another. There are two basic approaches to analyzing this problem: *cyberizing the physical* (CtP), which is where cyber interfaces and properties are imposed on a physical system; and *physicalizing the cyber* (PtC), which is when software and cyber components are represented dynamically in real time [3].

Before these two approaches are further discussed, we should first understand the basic layout of a CPS. A CPS is the integration of computer processing with physical input/output. Embedded computers that are networked together utilize a series of feedback loops to oversee and control a variety of physical processes. A typical CPS has a basic layout consisting of interconnected sensors and actuators as shown in Figure 1.4. An actuator is simply a type of motor that moves or controls

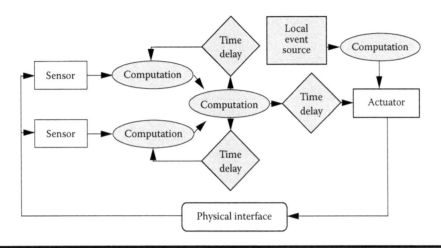

**Figure 1.4  Simplified layout of a CPS.**

another mechanism. The sensors provide data taken from a physical object, which is, in turn, utilized by the actuator to perform a function. After data are collected by the sensors, a number of algorithms are performed and looped over again until a proper command is calculated and sent to the actuator. It is important to note that time delays must be accounted for when running computations.

## 1.3.4 Superdense Timing

One of the greatest challenges that have been brought to the attention of many researchers is the notion of a uniform concept of time across all parts of a system operating simultaneously when correlating the cyber object to the physical object. There have been many problems in the dynamic systems with control processes that relate to synchronization, flocking, and formation control. Within the network of cyber components (such as sensors), a uniform concept of time cannot be realized. Systems that utilize synchronization and time-triggered networks can approximate a time model, but imperfections must be included to accurately model the dynamics. The dynamics and natural uncertainties of the physical world are difficult to accurately model. The concept of time just does not mix well with the cyber world. Time is continuous in the real world but must become discrete in the cyber world.

A model of time that is used to support a time continuum as well as discrete, untimed sequences is known as superdense time. Superdense time is an ordered list of elements denoted as $(t,n)$. The integer $t$ is some real number that represents discrete time, and $n$ represents ordered sequences that are independent of time. This model has the form

$$x : R \times N \to R$$

In this model, $t$ is a real integer at a specific time and $x$ could be multiple values with a specific order. This model can be altered to the form

$$x : R \times N \to R \cup \{\varepsilon\}$$

**Table 1.2 Summary of Challenges and Their Potential Solutions**

| Potential CPS Applications | Major Challenges | Potential Solutions |
|---|---|---|
| Claytronics | Disconnect between layers of abstraction | Visual-based design software |
| Security and environmental applications | Inadequate consistency | Time-sensitive programming languages |
| Medical and transportation applications | Lack of accurate timing | Networking and superdense time |

where $\varepsilon$ represents the absence of a value. This model allows discrete events to be ordered together with continuous signals so that a physical system can be accurately portrayed. The superdense model is commonly utilized for pairing synchronous models with discrete times.

When the physical world is controlled by cyber objects, a major problem that arises is the fact that time must have a semantic property. CPSs rely on the timing of the physical actions that are taken rather than the actual speed of computing. In many cases, cyber objects have difficulty in monitoring the physical world when they must regulate and control highly unstable and often unpredictable physical processes. In order for CPSs to overcome this obstacle they must become extremely reliable and predictable. They have to be more intricately networked with adaptive behaviors that can evolve and learn without being susceptible to failure as a result of minor altercations within the system framework. Near-perfect timing is essential for CPS processes to progress further. Precision must supersede quality. The ability to generate accurate and timely reactions with minimum latency is a must in order to achieve cyber-physical abstractions, and such a goal may be achieved by utilizing the base concept of superdense time [3].

Table 1.2 lists the typical challenges in CPS design and their potential solutions.

## 1.4 Conclusions

In order for CPSs to be improved upon, major advancements in the general perception and recognition of CPS technology need to be developed further. To do so, there must first be a deeper understanding of the integration of real-time processing with embedded wireless networks that work in harmony with a varying array of sensors across a dynamically physical environment. Also, a deep investigation of the interactions of the cyber world and mechanically manufactured devices needs to be understood. We need effective sensing and controlling of the physical systems through a robust software architecture.

With this new focused discipline, system hierarchies, protocols, language tools, and analytical procedures will define a path to improved CPS. Better programming concepts and hardware design should be utilized. A better grasp of the natural limitations, physics, and chemistry of data that are detected by cyber systems must be obtained to ensure stability, safety, and quality. Also, system models that can structurally and systematically describe physical system behaviors must be done in a coordinated fashion. We should also be aware of the constraints and quantitative limitations of a design that are governed by corresponding components and human errors.

There are many challenges to overcome. Accounting for random events that can occur in a real environment can be difficult for a CPS when trying to ensure safety, security, and predictability. With the notion of predictability, we may achieve precise timing and synchronization. Ensuring that a CPS is real time in all operations can be very difficult when taking into account the delay of data being processed by the sensors until action is taken by the actuators. Remodeling or designing a system and remaining within a given set of cost constraints is also very important. System architectures must also remain consistent for large-scale designs and protocols. Improved model-based development, CPS certification and verification, network autonomy, and embedded system scalability are all issues that will have to be addressed.

CPSs will one day reshape how humans intermingle and connect with the physical world. Energy-efficient power sources, highly accelerated agriculture, advanced health care systems, and secure systems will all be realized based on CPS design principles.

# References

1. E. A. Lee. 2008. Cyber physical systems: Design challenges. *International Symposium on Object/Component/Service-Oriented Real-Time Distributed Computing (ISORC)*, Orlando, FL, pp. 363–369.
2. E. A. Lee. 2006. Cyber-physical systems–Are computing foundations adequate? NSF Workshop on Cyber-Physical Systems: Research Motivation, Techniques and Roadmap, Austin, TX.
3. E. A. Lee. 2010. CPS Foundations. In *Proceedings of the 47th Design Automation Conference (DAC 2010), Association for Computing Machinery*, Anaheim, CA, pp. 737–742.
4. R. H. Campbell, G. Garnett, and R. E. McGrath. 2006. Cyber-physical systems: Position paper—CPS environments. Cyber-Physical Systems, National Science Foundation, Austin, TX. Available at http://varma.ece.cmu.edu/cps/Position-Papers/Roy-Campbell.pdf.
5. J. Campbell, S. Goldstein, and Todd Mowry. 2006. Cyber-physical systems. Cyber-Physical Systems, National Science Foundation, Austin, TX. Available at http://varma.ece.cmu.edu/cps/Position-Papers/Goldstein-Mowry-Campbell.pdf.
6. R. West and G. Parmer. 2006. A software architecture for next-generation cyber-physical systems. Cyber-Physical Systems, National Science Foundation, Austin, TX. Available at http://varma.ece.cmu.edu/cps/Position-Papers/richard-west.pdf.
7. J. Kornerup. 2006. A vision for overcoming the challenges of building cyber-physical systems. Cyber-Physical Systems, National Science Foundation, Austin, TX. Available at http://varma.ece.cmu.edu/cps/Position-Papers/Jacob-Kornerup.pdf.
8. R. Rajkumar, I. Lee, L. Sha, and J. Stankkovic. 2010. Cyber-physical systems: The next computing revolution. *Proceedings of the 47th Design Automation Conference (DAC 2010)*, Anaheim, CA.

# Chapter 2

# Cyber-Physical Systems: Design Challenges

Cameron Patterson, Roger Vasquez, and Fei Hu

## Contents

## 2.1 Introduction

When looking at cyber-physical systems (CPSs), two main components must occur. First, there must be a physical process. After this physical process is identified, sensors can be used to measure different aspects. The computer system ("cyber") will then process these data to perform the desired task. Many CPSs today have wireless connections of sensors and actuators. There are many challenge design issues in wireless CPSs. Section 2.2 will examine these challenges. We start with looking at the challenge of bringing multiple disciplines together to create such systems. We then discuss some necessary metrics and software tools. The next challenge is how to correctly represent sensor readings. We will go through two different algorithms to solve such an issue. The last challenge is the use of network coding for CPS design.

In Section 2.3 we review case studies performed on some wireless CPSs. The first case study deals with containers being monitored remotely using a wireless CPS. Another wireless CPS is then used to monitor algae cleanup in Lake Tai in China. The final case study is on the use of the CPS to control a smart home.

In Section 2.4, we provide examples that deal with a multilayer heterogeneous system used in some CPSs. The first is an example of using mobile networking for CPS infrastructure. The next example uses multilayering protocols to perform agent-based recovery of heterogeneous data.

In Section 2.5, we explain how these systems can increase energy efficiency. The first example creates a wireless joint sleep schedule for a radio. The next example looks into examining thermal activity when performing task scheduling.

In Section 2.6, we list other important factors when designing a practical wireless CPS. These include adaptation, unification, dependability, and consistency. In Section 2.7, we conclude with some thoughts on wireless CPSs as a whole by considering the topics discussed in the previous sections.

## 2.2 Challenges for Wireless CPS

The first step to understanding a topic is to understand its challenges and possible proposed solutions. In this section, some general challenges in wireless CPSs are discussed.

### 2.2.1 Multidisciplinary Knowledge Requirements

The study of networked CPSs is a complex field that requires the effort of multiple disciplines to advance and solve the challenges that arise. Such research needs to involve different technology communities to work together. Typically, a new system design requires a new calculus, new metrics, new software tool sets, new network control and middleware, and new foundations for education [1]. These topics all have challenges associated with them, along with proposed solutions. The challenge for calculus is to merge time-based occurrences with event-based systems. The challenge for metrics is to create a method that will chart the network-oriented metrics in accordance with the performance of the systems. The next necessary component is software tool sets. A challenge of creating new software is to handle the complexity of the cyber and physical interactions. Network control and middleware aims to increase the reliability and accuracy of the systems while also keeping in mind the time sensitivity. Another challenge is to increase the number of scientists and engineers who study networking.

For the new calculus, research for a new theory to solve the issues of the future cyber-physical world will need to be performed. To keep the unity of this broad topic, the new ideas that are created will be more useful if they are all based on a common metric. To make the systems more efficient, a milestone for software development is to fabricate and check the CPS. With all these advances, a milestone for the networked CPS control is to have the software available and useable for industries. In training young engineers and scientists, there must be an emphasis on basics of networking studies. With the increased studies and interest in the various fields of CPS networking, new engineers and scientists will begin to create newer and better advancements for future wireless CPSs.

## 2.2.2 Challenge Example 1: Sensor Readings Representation

The previous section discussed some general challenges, whereas in this section we provide an example of a more specific challenge. Some applications in airplanes and cars require correct representations of sensor readings. There are two algorithms that can overcome this problem [2]. To implement these algorithms, a dominance-based medium access control (MAC) protocol can be used, and the wireless form of this protocol is known as WiDom [3]. In this MAC protocol, there are recessive and dominant bits. The logic "1" is recessive, and the logic "0" is dominant. The nodes with recessive bits must take the time to sense the medium before sending data. However, the dominant bits can be freely sent across the medium. This protocol helps to achieve the following requirements: the majority of sensor nodes are used to formulate the sensor readings for communication through the broadcasting media and low delay of the computation. Along with this protocol, the interpolation is also a key tool. In essence, the interpolation is to estimate new data points that are missing based on the current sensor readings.

The two proposed interpolation algorithms are known as the basic algorithm and differential interpolation algorithm [2]. For the basic algorithm, the interpolation value is set to zero on all spots in the beginning. Then the interpolation values are determined. Such values are compared with the actual readings from the sensor for interpolation error correction purpose. This process is repeated for some rounds as desired by the programmer. This algorithm has some flaws. The differential algorithm handles some of its flaws such as relating the physical world better in the system. The past interpolation processes are used to predict future processes. This algorithm uses the differential values of the points from the interpolation calculations. Equations 2.1 and 2.2 are used to calculate the interpolation in the differential algorithm. Iterations are performed using these equations. Here the function $f(x, y)$ represents the sensor readings in an area. The control points $q_i$ are in the set $S$. Each of these control points uses the following three variables as attributes: $x_i$, $y_i$, and $s_i$. The weights are defined by the variable $w_i(x, y)$.

$$f(x,y) = \begin{cases} 0 & \text{if } S = \varnothing \\ s_i(j) & \text{if } \exists q_i \in S; \ x_i = x, y_i = y \\ \dfrac{\displaystyle\sum_{j=1}^{j} s_i(j)w_i(x,y)}{\displaystyle\sum_{i \in S} w_i(x,y)} & \text{otherwise} \end{cases} \tag{2.1}$$

$$s_i(j) = s_i + (j - i)g_i \tag{2.2}$$

**Table 2.1   Percent of Average Error for Algorithms**

| Algorithm | Different Increase of up to 4% | Increase of 4% | Scaling of 1% | Number of Times Repeated |
|---|---|---|---|---|
| Basic algorithm | 18.23 | 38.78 | 4.75 | 20 |
| Differential algorithm 1A | 6.19 | 8.99 | 4.96 | 20 |
| Differential algorithm 2 | 5.70 | 9.01 | 4.74 | 20 |
| Basic algorithm | 9.23 | 15.82 | 7.10 | 10 |
| Differential algorithm 1A | 9.75 | 10.96 | 7.91 | 10 |
| Differential algorithm 2 | 8.49 | 10.36 | 7.94 | 10 |

Their simulation results show that the basic algorithm has certain efficiency. For example, when looking at smooth signals that have slow changes, the algorithm works properly. However, the basic algorithm does not perform well when the changes are abrupt and during interpolation. The differential algorithm handles these situations much better according to the simulation data shown in Table 2.1 [2]. Through this research the challenge of developing a more efficient algorithm to represent sensor readings was accomplished.

## 2.2.3 Challenge Example 2: Network Coding

In this section, we discuss network coding challenges. A form of computer coding has been used in network communication. It is referred to as COPE [4]. COPE addresses the challenges in wireless network's limited throughput. Researchers first try to create smaller network models for easier testing. After the smaller size testing is complete, it is time to transfer it to a real functioning network. One problem lies in that it does not perform well when converted to the larger size. Another problem is that some in-between nodes end up becoming a bottleneck. This also slows down the network. The rest of the section will discuss these problems in depth.

In a typical system a node stores the packets until it is ready to send. Then the node passes the packets along to the next node in the routing chain and finally to its destination. The nodes use an XOR function to combine packets going to the same destination in order to save time and cost. This scheme works well when packets arrive at nodes at the same time; however, when the packets arrive at the node at different times, combinations of the packets cannot be achieved. There are three ways to solve this problem [4]. First, a program has been created from COPE called encoding scores. The program is able to calculate how much time a system can save by combining certain packets for transmission. The encoding score is determined using the size gain ratio and distance gain ratio. Next, an internode contention is used to assign priority on packets with higher scores from the first step. These high-priority packets will be transmitted first. Finally, a packet is sent to more than one node on the way to the destination. This gives the packet more chances to be combined with other packets.

$$\Delta l = (l_1 + l_2 + \ldots + l_j) - \max \{l_1, l_2, \ldots, l_j\} \tag{2.3}$$

$$\text{size gain ratio } (C) = \frac{\Delta l}{\sum_{i=1}^{j} l_i}, 0 \leq \text{size gain ratio } (C) \leq 1 \tag{2.4}$$

$$\text{distance gain } (x_i) = \left\{ \frac{d(A,z) - d(N_i,z)}{r_{\text{radius}}} \right\} \tag{2.5}$$

$$\text{distance gain ratio } (C) = \frac{\sum_{m=1}^{j} \text{distance gain } (x_m)}{j}, \quad 0 \leq \text{distance gain ratio } (C) \leq 1 \tag{2.6}$$

$$\text{encoding score } (C) = \text{size gain ratio } (C) \times \text{distance gain ratio } (C) \tag{2.7}$$

These equations are used to determine the encoding score. Equation 2.3 is used to determine the size coding gain, $\Delta l$, where $l_1$ through $l_j$ are the packets' different sizes. Then Equation 2.4 uses this information to determine the size gain ratio that will be a value between zero and one. The variable $d$ in Equation 2.5 represents a distance between the nodes specified in the parentheses with $z$ being the destination and $N_i$ being the next hop. The original node is represented by $A$. The radius of the transmission range is represented by $r_{\text{radius}}$. The distance gain and number of packets, $j$, are then used to calculate the distance gain ratio, which is also a value between one and zero. The encoding score is found by multiplying the size gain ratio and the distance gain ratio of the combination $C$.

Research has been done in related fields for COPE. Some studies target how COPE could have intersecting nodes in order for packets to be able to be combined. COPE is able to operate on both the physical layer and network layer for packet transmission. Some researchers proposed a new system based on COPE called enhanced distributed coordination function (EDCF). For the EDFC system to work correctly, it uses COPE characteristics for the combined packets that can be broken down into their original transmitted properties. It also places priority on packets. This ensures that the packets that are more important get transmitted when the lines are free. Not only do packets in the same node compete to be sent, but also nodes in the whole system compete to transmit the packets. The node that is allowed to transmit is again based on the priority score: The higher the score, the sooner the transmission. When packets are sent to different nodes, the combined scores are determined before sending the packet. The other nodes stop trying to send the packets. The reason systems like this work so well over systems that do not use networking code is because if a system does not have some way of determining paths, it could use a short path. The COPE-based system has seen up to 40% greater efficiency over a system with no code [4]. Coding efficiency can be determined by dividing the encoding transmission amount by the total transmission amount.

### 2.2.4 Challenge Example 3: Mobile Cyber-Physical Applications with Internet Integration

Some current challenges that people face today are due to the increase in mobile Internet devices and developing the programs that run these devices. Smart phones today such as the Apple iPhone or the Motorola Droid have everything from a camera to a global positioning system (GPS) in them. These devices are able to trigger operations in the cyber-physical layer. With the help of the Internet, these devices are able to interact with people and things around the globe. However, with the advancing of these technologies come several challenges that researchers are facing.

Even though researchers are becoming able to use these mobile Internet devices to build a cyber-physical layer, the researchers face several challenges. The first of these problems is battery life. As cell phone processors have got bigger, so have the codes these processors carry out. Older models would allow the users to go longer periods of time between battery charges. With the new technology, some users are unable to make it run for a whole day without charging their device. The problem researchers have run into in fixing this problem is that there is really no way for them to predict the amount of power a specific user will use just based on the program the phone or other device is going to run. They must first find a way to develop the software and find users willing to test it and report their battery life. One solution to the problem is model-driven power consumption analysis (MDE) [5]. With MDE, the researchers are not only able to predict power consumption of the programs, but they are also able to condense the things that are using unnecessary power.

Another factor facing researchers is integrating mobile Internet devices with the sensors that surround us in everyday life. Sensors are everywhere from street corners to traffic lights. The researchers are working on service-oriented device architecture (SODA), which is able to bridge the gap between the mobile Internet devices and the conventional sensors by creating a form of firewall to protect the sensors from being overwhelmed by the advanced mobile devices programs. This allows users to access and control a wide variety of things that they could not previously do.

## 2.3 Case Studies

The section goes into detail about a few applications of the wireless networking in CPSs. After understanding the basis and some challenges others have faced, researchers can begin to apply this knowledge to real-life situations.

### 2.3.1 Remote Container Monitoring

This section discusses an application that addresses the tracking and monitoring of containers on ships. Because terrorism and smuggling are struggles that the nation faces on a daily basis, researchers are attempting to develop a way to track and monitor cargo whether at port or at sea. Another factor in this research is how items are shipped globally on a regular basis. The main task for researchers is being able to perform these tasks remotely. In order to be able to accomplish the task, the researchers would have to use both cellular and satellite mobile networks. Several applications such as ZigBee, sensors, and radio frequency identification (RFID) would also have to be incorporated for the system to work efficiently. With all these things working together, users can track containers whenever they want.

There are two different methods used to track and monitor containers today. One is cable transmission. It uses four different cable and sockets placed strategically in the container. These sensors monitor the compression of the container and temperature by feeding back a Boolean value. The other is carrier transmission. This method uses modems located in the refrigerated cars and sends back frequencies to help the user monitor the state of the container. These methods are not the most cost efficient and are time consuming with all the work to get them set up. The solution the researchers found is wireless communication, which will be able to monitor the containers at a greater efficiency.

The proposed system has local communication and monitoring and also remote communication and monitoring, which is shown in Figure 2.1. It also has a place so that the users can tap in and monitor their containers. The local monitoring is implemented using both passive and active

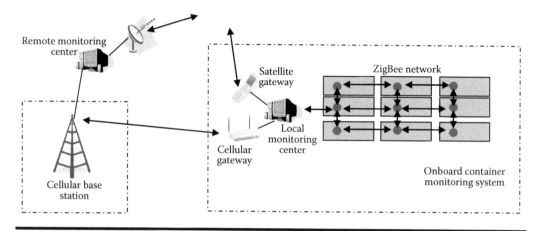

**Figure 2.1    Remote container monitoring system.**

RFID attached to the containers. Users are able to enter the date that goes along with the container. When the container is scanned by passive RFID reading equipment, it instantly lists the items and specifications for the container. When the user employs the active RFID, this allows for the container to constantly transmit data. The local communication uses the ZigBee application. This is more for intercontainer networks. ZigBee is able to send and receive packets at a low energy level for optimal energy saving. Crewmen on the ship use the ZigBee networks that are established in the local communication to monitor the containers. Researchers provided two different options for people on land to monitor cargo: cellular gateway and satellite gateway [6]. The crew decides between the two methods by the Always Best Connected (ABC) principle. This principle looks at different variables such as price or bandwidth. ABC is performed with various algorithms. Cellular is used when the ships are closer to land, and satellite is used when the ships are at sea. The remote monitoring function allows the whole logistics process to be watched. All this is taking place while the user is able to tap into the client inquiry terminal and oversee the whole process.

With the new system implemented it allows for customers to track their products or merchandise. Worldwide coverage is also achievable when using this method.

## 2.3.2  CPS for Algae Cleanup and Monitoring

The purpose of the research discussed in this section is that it addresses a problem that occurs for roughly 1 million Chinese people. Lake Tai is one of the places that battle blue-green algae. When the algae take over, it causes the water to be contaminated for 10 days. With the problem growing, scientists have devised a system using a sensor device and an algorithm to command the sensor to do the following tasks: detect algae, track it on the water, and send a response team quickly and efficiently. Right now the sensor only detects specific algae; this is because it takes a long time to develop a system and algorithm for detection. The goal of the project is detect the blue-green algae and determine the degree of the outbreak. There will then be an automated system that tells the treatment boats where to go after the threat is determined. Research has allowed the use of applications to determine the location of the algae, the station where the detecting takes place, and the technology for the salvage boats. The system is implemented using different sensors that transmit on a 3G network. Sensors use different algorithms to determine what needs to be sent across the network. After the information is sent, it is stored at the information center. The center analyzes the features of the data. If the data are determined to be a threat, the location of the algae is transmitted to the salvage boats that will clean up the algae.

The system consists of four parts: algae estimation device and algorithm, sensing station, cleanup subsystem, and geographic information system (GIS) based management Web site [7]. Figure 2.2 displays a basic overview of the system design. The system uses a camera to take pictures of the water periodically. The pictures are then examined by a control board. An algorithm is used to determine the amount of algae in a picture based on the shade of the color as well as the amount of space the algae takes up in the picture. Other pictures are taken from the shore by cameras. Those pictures are sent over the 3G network to be analyzed. After the threat is determined, the next step is to notify the boat. This step is mainly manual. The directions are sent via the GPS on the boat. The directions also can be sent by voice with a turn-by-turn approach. After the boat is on site, it handles the cleaning and storing of the algae. After the algae are collected, it is then taken to the factories for dehydration and proper disposal. Simultaneously, all the data including the boat's location and amount of algae can be monitored on the Web site. This helps in the algae cleanup while not wasting resources.

There are several variables that have to be assigned to the different steps of the operation in order for the process to work smoothly. Certain items must be specified, such as a GPS location, the area of algae on surface, water depth, and degree of algae danger. The variables assigned to these items are $L_w$, $A_w$, $D_w$, and $S_w$, respectively [7]. The boat variables are location from the GPS device, sail speed of the boat, how much room is left to hold algae, collecting speed, cost per hour including both labor and energy, and the Boolean variable that tells if a boat is available or not. The variables used for these parameters include $L_b$, $V_b$, $C_b$, $S_b$, $P_b$, and $B_b$, respectively. When the algae is harvested, it is then transported to factories. In order for the factories to work efficiently they must determine the location of the factory, dehydrating speed of that particular factory, and how much room the factory has left for boats to bring in algae. The variables assigned to these factors are $L_F$, $V_F$, and $C_F$, respectively.

The variables described above are used in what is known as the agile sensing actuator control (ASAC) mechanism. This mechanism has three layers. The first layer is known as the device layer,

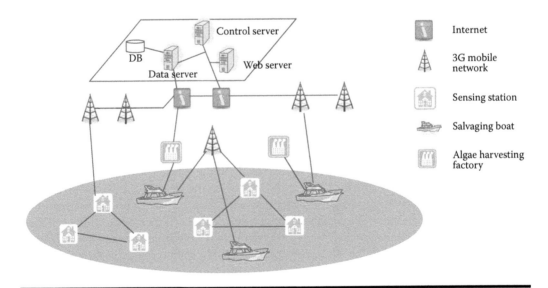

**Figure 2.2   Proposed system.**

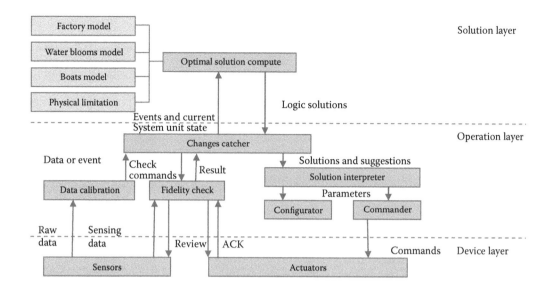

**Figure 2.3  Agile sensing actuator control mechanism layers.**

which consists of sensors and actuators. The second layer is the operations layer, which performs more detailed tasks such as checking actual data and changes in the lake water. The third layer is called the solution layer. In this layer, the correct step is chosen to be carried out by the system. Linear programming is used to determine this solution. A detailed overview of the ASAC is shown in Figure 2.3.

By using the applications discussed earlier in this section, workers are able to locate the blue-green algae. The system is able to determine the threat level and send out the salvage boats to harvest the algae. The final step involves the transportation of the algae to dehydration facilities. These tasks are accomplished through wireless capabilities of CPSs.

## 2.3.3 Smart Community CPSs

This section addresses an advanced form of CPS and its applications. CPSs are made up of several different physical devices. These devices are managed and run by software systems. The devices then perform real-world tasks in a smart home. Smart homes have trouble creating a way to bring all remote monitoring together. This dilemma may be solved by using a system known as smart community cyber-physical system (SCCPS) [8]. The system uses ZigBee and a technology called global packet radio service (GPRS). This is all designed in a combination of the mixed mode of B/S and C/S. This model is supposed to allow the user to have a smart home without bugs.

A key to the success of the smart community CPS is the setting it which it is built. It uses both B/S and C/S. B/S is for the user interface. It allows easier interactions between the user and the interface. The C/S addresses more of the productivity side. It allows the system to work faster while still being secure. When the two modes are combined, it allows the user to have an easier user interface without sacrificing speed or reliability. On the contrary, when addressing the cost, the B/S works on an embedded platform and uses TCP. This will work well enough, but it is not without a complication. When using this embedded platform, every time an appliance is moved

**Table 2.2  Package Structures**

| Package Type | Length (Bytes) | Package Structure | Command | Sender | Receiver |
|---|---|---|---|---|---|
| Request for registration | 16 | Head and check code | 0x01 | AMS | Alarm |
| Registration | 32 | Head, data, and check code | 0x11 | Alarm | AMS |
| Heartbeat | 16 | Head and check code | 0x1F | Alarm | AMS |
| Alert | 32 | Head, data, and check code | 0x1A | Alarm | AMS |

in a home, it is necessary to rewire the network. The researchers' solution to this rewiring is a wireless network. They propose the use of both ZigBee and GPRS.

The network will allow the proposed smart home to manage and run the applications of the home through the combination of ZigBee and GPRS technology. The appliances of the house and other smart items are connected through ZigBee in a star shape. This shape allows the user to add and subtract appliances without disturbing the rest of the system. The center of the star is the GPRS technology. It is what the whole system is built around. The system uses an embedded Web server to communicate. The domestic controller acts like a medium in the system. It receives a request from the browser, answers, and sends back. It also acts as a medium between the alarm and the smart appliances. When an appliance recognizes something is wrong, a connection is made between the appliance and the domestic controller. The domestic controller then sends a packet to the alarm telling it to go off.

As with every system, a precaution must be set to help prevent false alarms. The system uses advanced management system (AMS) technology to manage the packets. The details about the different types of packets such as length and structure are shown in Table 2.2. AMS uses TCP protocols to ensure that the packets are received for the alarm. Then when the packet for an alarm is received, AMS uses a message-driven mechanism to ensure that the alarm really does need to alert the user. Another threat is someone being able to gain access outside of the system. To prevent this, the smart community cyber-physical system includes an authority management module. When the alarm is over and it has been sent to the user, the GPRS resets the system, getting it ready to run to receive another notification if/when something is wrong.

## 2.4 Multilayer CPS Design

This section gives two examples on how multiple layers and heterogeneous properties are used in wireless CPSs. These methods can improve data recovery as well as allowing a system to better determine the most efficient wireless access network to use.

### 2.4.1 Heterogeneous Wireless Network for Mobile CPS

As technology advances, the way CPSs are formed must also keep up with new technology. CPSs allow a user to access the Internet through various forms. For example, one can check the weather using different devices including a laptop, cell phone, or other Internet-ready devices. To gain this access, a CPS uses a terminal that can gain access through several ways. These include a wireless local area access network (WLAN) and wireless wide area access network (WWAN). Even though WLAN's name has wireless in it, it also has the capability of being connected with wires. The problem is that this can only be done for so many homes because the throughput eventually becomes so constrained. In most cases the WLAN coverage is not very big and the WWAN sometimes exceeds 90% coverage. In order to create a CPS mobile network, a user must use heterogeneous multilayer wireless networking where a user may use both WLAN and WWAN when the WLAN does not offer enough coverage.

If a WLAN is not capable of sustaining a CPS alone, it is necessary to be able to establish a suitable way of determining how to gain access based on the location and available coverage. CPS terminals can only connect to a WLAN when they are in the WLAN's coverage area. While the CPS is in the WLAN's coverage area, it is also able to connect to the WWAN. The WWAN has much higher processing ability and speed, but it also comes at a higher cost. An algorithm was determined to implement the heterogeneous multilayer wireless networking. When the algorithm is implemented, it will be able to determine what is available and whether WLAN or WWAN meets the user's needs. The first step of the algorithm is to assign a WLAN bandwidth to the CPS terminals that will not exceed it. Next, the algorithm assigns WWAN connections to terminals that are either outside the WLAN coverage or where the WLAN throughput is not large enough.

The algorithm is carried out by the actual terminal that detects the quality of the signal. When the signal strength is determined, the terminal finds the channel quality indicator that is reported to the terminal so that it can choose the best possible network. There are only three possibilities for the algorithm to decide how to serve the terminal (see equations below). First, if the terminal is located inside the WLAN/WWAN and does not have a lot of mobility, it will be served by the WLAN. Second, when a terminal is located outside the WLAN, it will be served by the WWAN. The last option would be when the terminal is located inside the WLAN but will be served by WWAN because of the mobility of the terminal. With the terminal being able to determine the best network, a CPS can be used even when it is out of range of a WLAN.

$$\text{System} = \begin{cases} \text{WLAN}, & \text{if } \text{SINR}_{\text{WLAN}} \geq \text{SINR}_{\text{WWAN}} \\ \text{WWAN}, & \text{if } \text{SINR}_{\text{WLAN}} < \text{SINR}_{\text{WWAN}} \end{cases} \tag{2.8}$$

$$\text{SINR}'_{\text{WLAN}} = \text{SINR}_{\text{WLAN}} + W_{\text{Price}} \tag{2.9}$$

Terminal schedules must also be compared. In this experiment, there are two different schemes for the terminal scheduling [9]. The first scheme is a traditional one using Equation 2.8. The different signal-to-interference-plus-noise ratios (SINRs) were compared to see if WLAN or WWAN needs to be used. The option with the higher SINR was chosen. Scheme 2 is the proposed scheme that uses Equation 2.9. Equation 2.9 demonstrates the difference between Scheme 1 and Scheme 2. The proposed scheme ended up with an 8.7% gain over the traditional scheme [9]. This design proved that a CPS system could efficiently and successfully operate using both WLAN and WWAN under a heterogeneous multilayer networking.

### 2.4.2 Agent-Based Recovery of Data Heterogeneity

CPSs are used today to operate several different devices. One type is communication devices, which optimize heterogeneity in hardware. This chapter focuses more on the data aspect of the heterogeneity. Communication devices use several different forms of sensors today. One of these sensors used by cells phones is the GPS. These devices read data and communicate to the CPS on a real-time basis. The system conveys back real-time information. This chapter concentrates on the water industry and more specifically the water distribution network. It also explains how CPSs play an important role in allowing these systems to work efficiently.

An intelligent water distribution network utilizes several different physical components. This system is shown in Figure 2.4 [10]. Among the physical components are pipes, values, and reservoirs. To help maximize physical components, intelligent water allocation sensors are used. The goal of the intelligent water distribution network is to provide water when the user turns on the tap. Using this system, researchers are able to track water use. They are also able to predict where the majority of water will be consumed. Further, the system can send out maintenance workers to certain locations without the presence of an actual human.

The multilayer system of the intelligent water distribution works very efficiently. One layer is the actual water flow, such as a reservoir of a sink. This layer has sensors that communicate back to the higher level like computer devices on how much and when the water will be used. This allows the computers to be able to allocate water to where it will be needed at the correct times. As mentioned, this method also allows for monitoring of the maintenance side of water flow. It can do this by monitoring what amount is being used at a house and how much water is being sent to that section. If there is a leak, then more will be sent to that section than what is being used so then they will know that there is a leak or malfunction.

With the different layers working together, a faster and more dependable water supply is able to be delivered.

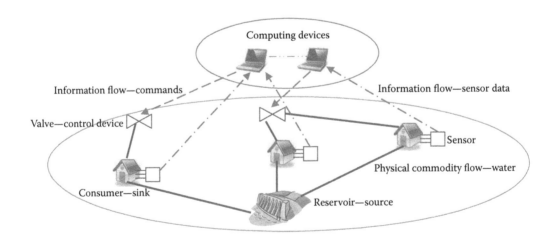

**Figure 2.4    Water distribution system.**

## 2.5 Energy Efficiency

The following section gives examples of how wireless CPSs help in the process of energy efficiency that is important in creating more cost-efficient systems. Less energy can be used by implementing a sleep mechanism or monitoring thermal awareness and cooling.

### 2.5.1 Wireless Radio Joint Sleep Schedule

A specific task in this aspect was to examine radio sleep scheduling using wireless nodes. The challenge of using a minimal amount of energy must be accomplished while also considering the timing of the system. The energy use will be reduced by decreasing the supply voltage of the processor and the clock frequency. To achieve these tasks, the proposed design of the precedence graph is a tree topology. All nodes need to be taken into account when looking at the energy reduction. The researchers' goal is to create this reduction for the whole network.

A proposed idea to reduce the power that a radio uses is to put radios in sleep mode when they are not in use. This brings up the problem of the nodes communicating properly because the nodes can only communicate if both radios are awake. The radios only need to wake up when communication needs to take place so that not much energy is wasted. The solution proposed is to not have a fixed sleep schedule. Alternatively, the decision was made that a joint sleep schedule with mode assignments would increase energy efficiency of the whole system [11]. Static energy management schemes for the network topology are also necessary for this solution to work correctly.

The three different models that are introduced are the following: task model, CPU model, and network model. For the task model, each task set has precedence constraints as well as a shared period. The acyclic graph represents the costs and constraints. The CPU power model compares the frequency and power of different modes. The network model for the test is based on a single-hop network where the sleep schedule is set up according to the mode assignments. A simple example of the sleep schedule for three nodes is shown in Figure 2.5. The symbol alpha shows the time the nodes are on or awake. This demonstrates how both nodes must be on for communication to take place.

When considering the problem, it is necessary to look at the input and output. The task graph is used as the input. For this situation the task graph is represented by $G = (V, E)$. The task set is $V$, and the dependence between each task is the set $E$. The output is a joint sleep schedule along with mode assignments. The output creates a minimized amount of energy consumed. The following equation shows the breakdown of the energy consumption: $E_i = e_{ki} + e_{ri}$. The first variable, $e_{ki}$,

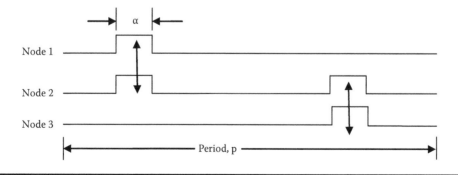

**Figure 2.5  Node sleep schedules.**

represents the CPU execution energy of node $i$ in this case. Then $e_{ri}$ represents radio energy of the same node. The goal is to minimize the summation of all the nodes' energy used in the task set $V$.

In developing algorithms for the joint sleeping schedule and mode assignment, the type chosen depends on the task graph. For a tree task graph, the *JointAssignTree* algorithm should be used. From this algorithm, the following theorem was developed: $X_i[j]$ ($1 \le i \le N$) is the system's minimum cost where the subtree ended on $u_i$ [11]. Next, when the task graph is a DAG, then *JointAssignDAG* algorithm should be used. To achieve a long execution period while also using the minimal energy, the execution node is set with the slowest operating mode. Equation 2.10 is used at each node $i$. It introduces a variable that divides the energy difference of mode changes by the total execution time. Whichever node has the highest value of $EW_i$ will be the node that goes up one level of energy consumption mode. This process is continually occurring for the JointAssignDAG algorithm.

$$EW_i = \frac{e_{ki} - e_{ji}}{\Delta t} \tag{2.10}$$

In the experimentation of this research, there were eight different DAG task graphs used for testing. To determine if the joint sleep schedule algorithm was indeed the best choice, the normalized energy costs of the joint assign algorithms was compared against other algorithms that had prefixed schedules. The other algorithms in both the tree and DAG situation had higher energy costs than the JointAssignTree and JointAssignDAG algorithms. Then, when comparing JointAssignTree versus JointAssignDAG, the tree algorithm shows lower energy costs. Therefore, the researchers successfully determined an algorithm to achieve lower energy costs over the whole network through using joint sleep scheduling and mode assignment using a tree task graph.

### 2.5.2 Thermal Awareness in Task Scheduling

In Ref. [12] the need for an energy-efficient, sustainable power source for data centers is addressed. The majority of that cost goes to cooling off the data centers. The only thing that is truly holding the data centers back is the lack of sustainable power. Right now data centers consume roughly 3% of the nation's total power. For the larger data centers, the price to operate can be millions of dollars. The goal of researchers is to shrink the need for cooling in data centers. When this need is minimized, then the overall energy efficiency will be increased greatly. Researchers plan to address this by thermal-aware task management [12].

One problem the researchers faced when it came to implementing the thermal-aware task management is air circulation. Servers produce hot air from their processors. It is the job of the cooling system to bring in cool air for their intake to keep the system cool. Another problem is when the hot air from the processors does not leave the data center and the intake takes in hot air. This causes hot spots. The easiest way to mediate this is to run air conditioners in the data center at a much lower temperature than the red line of the processors. This is so that even if the processors take in any hot air, the cold air in the room will still keep the hot air from causing the processor to cross over the red line. Clearly, this is not a very energy efficient response owing to the fact that you are running the air conditioners at a lower temperature. The researchers are looking at ways of mapping out the circulation of air at the data centers. They are using a method called Recirculation Heat Index [12]. This method looks at how much heat an individual processor would produce for a given task. It also looks at the location of a processor to determine if the processor's air output is directly next to another processor's input. Tasks are then assigned to processors. This method addresses the cooling cost problem.

Researchers have started using biomedical sensors to address the problem of overheating. These sensors allow for several things to take place. First, these sensors monitor the amount of work on the processors and the heat being used. Second, these sensors are able to tell computers to increase the air conditioning if needed for an immediate fix. If the sensor senses that the problem will not be fixed and the processor is working too hard, the computer will then move some of the work being placed on the processor to a different processor. This helps keep one processor from becoming overheated and also ensures that all the heat will not be generated in one location.

As companies' data centers continue to grow, and the processing and storage capacity keeps increasing, the need for solutions for overheating data systems likewise increases. For now, tracking the heating and the use of sensors has helped these centers become more energy efficient.

## 2.6 CPS Design Lessons

The following sections address four components of making a good design. They are adaptation, unification, dependability, and consistency. These concepts not only make better CPSs, but they also allow design processes to be better transferred to other applications.

### 2.6.1 Adaptation

A CPS can be used to implement intelligent adaptation for networks systems that are very complex. Researchers are now able to use CPSs to take both cyber- and real-world applications and make them interact seamlessly. The goal of this chapter is to show how an embedded system can be implemented to help take away the restrictions of what a CPS can do in real-world situations. Embedded systems typically have not been used in CPSs owing to the fact that these systems require so many different programs and software to work correctly. Now embedded systems are able to combine these different programs and software to work uniformly. When researchers use these embedded systems in CPSs, it allows for more intelligence capabilities on the part of the CPSs. When these intelligence capabilities are possible, it allows for the raw physical process to be able to work quicker and more efficiently. The raw physical systems are the sensors and devices that interact with the physical world and have little or no self-reliability. Without a CPS, a raw physical process would be unable to function. CPSs not only manage the raw physical processes' inputs, but they allow control of the outputs as well. With the CPSs controlling the raw physical processes, and the CPS being managed by a management station, it is possible to expand the boundaries of the intelligence capabilities of the system [13].

This idea is implemented by expanding the amount that the intelligence physical world covers. Researchers do this by assigning limited repair capability to the intelligence of the physical world. This allows for the comprehensive repair to be assigned to the CPS. This permits for the expansion of the overall system when it comes to intelligence issues. With the intelligent physical world handling some of the original information coming in, it allows the actual CPS to concentrate on other applications and programs. This style of technology lowers the overall operating cost. Researchers used an embedded system to better divide the workload and to create a more layered system. They divide this workload by using autonomic networks to generalize information coming into the system and to handle inconsistencies with the information.

As CPSs become more complex, it is essential for the systems to expand their technology boundaries and efficiency. This can be done by using embedded systems to divide the work load and to create a CPS that is made up of layers. Doing this reduces cost and increases efficiency. With the system being divided into layers, it allows the overall networks to integrate more easily.

## 2.6.2 Unification

Today CPSs do everything from monitoring remote locations to helping in the medical field. CPSs even go as far as to implement the distribution of power to a power grid. With all these different applications working on different domains, they present a problem when it comes to integration and unification. Researchers are trying to accomplish certain things when it comes to unification of different CPSs. They would like for the CPSs to be safe and reliable, have no need for human interaction, be able to adapt, and be very efficient in both energy and in time. When researchers are trying to link together these different domains, they have to gain a better understanding of what a CPS can be broken into. CPSs can be broken into cyber parts. These cyber parts can be represented by differential equations. They can also be broken down into several different forms of sensors and processors and even into actual physical components. The goal of this chapter is to find a way to relate these physical components to the differential equations of the cyber parts.

The first approach to unification that was introduced was the Kahn process [14]. The process targeted the ability for several different processes to be able to work together for a common goal. This system uses FIFO buffers, which take a sampling of different processes that the components are reading in. It then determines what is relevant for the overall goal of the system. The system will need to be able to run and manage several different processes simultaneously. The problem facing researchers is that for a CPS to be able to work uniformly it must be designed using models rather than processes. These different models led to the fragmentation of domains. Researchers are trying to be able to develop a way to make the models to be more homogenous. This type of model is still in the theoretical phase. The first step toward such a model is the hybrid data flow graphs. This breaks up a system so that processes are more easily managed. It uses both discrete and continuous semantics to distribute tasks and to manage the system's functionality. The hybrid data flow graph deals with more values than the actual layout of a system. It still lacks a way to be able to model the system. The solution to this problem is to be able to combine the technology of the hybrid data flow graphs with the technology of the Kalman process. The combination of these two processes is called unified process networks. This new method allows for the integration of the basic networks, the networks that deal with time, and it also uses the graphs from the hybrid data flow models to help monitor the system progress.

As cyber-physical layers become more advanced and have more layers, researchers are constantly looking for new ways to integrate these different process. Because the systems are ever-changing, it is hard to say whether or not the solution will ever be solved completely. Although with the combination of the Kalman process and the hybrid data flow graphs, the researchers are headed in the right direction toward creating a better solution.

## 2.6.3 Dependability

Dependability is an important quality for any CPS. Research has been done in the hopes of combining an intelligence adaptation with network systems that are complex. The framework of these systems is based on intelligent physical worlds (IPW) [15]. The IPW helps to add the adaptive behavior to the system. Allowing this to happen makes the system more dependable. Here raw physical processes (RPP) data are collected, and the system is controlled by an intelligent computational world. For ease of understanding, intelligent physical world and intelligent computational world will be denoted as $A_p$ and $A_c$, respectively. Together $A_p$ and $A_c$ create a management-oriented feedback. This will allow $A_c$ to see the behavior changes that $A_p$ makes. To clarify this, a real-world example of this would be a smart home and a security system that monitors the home remotely. The smart home would be $A_p$, and the security system would be $A_c$. Dependability is represented by $S$ and is based on three components: measurability, predictability, and adaptability.

To achieve these desired components, piecewise linearity and separability of functions must be implemented. This allows the IPW's operating regions to be identified. Then it can fix a problem if it knows the region where the problem is located. An example of this is when TCP window size is varied. When looking at $A_p$ as a black box system, there are three defining characteristics: programmability, observability, and computability. Using these characteristics, three important categories in dependability can be achieved: quality of service (QoS), fault tolerance, and timeliness. The QoS is a measurement of how well the systems resources are allocated, depending on which part of the system the user is currently using. Fault tolerance is seen as a percentage. It can be better understood by observing how a system is impacted by a failure. Timeliness and fault tolerance help better define QoS. These factors need to be met so that dependability can be achieved during times of failure.

The researchers hope to achieve the dependability desired by ensuring that the QoS requirements are met. By having a high QoS, the complex system should keep its dependability. We may observe how the intelligent physical world and intelligent computational world react and work together.

## 2.6.4 Consistency

This section will describe the needed consistency between architectural views and base architecture. One way to achieve consistency early on is to use model-based development (MBD). It uses computer models to find where systems might fail while they are in the development process. Having a consistent model is important to ensure the best system without wasting any time or money.

When designing a system, there is a choice between weak or strong structural consistency. For weak consistency, each component in the view is accounted for in the base architecture (BA), and every path of communication and physical connection between view elements is allowed in the BA by connectors [16]. This means that the system should know all the possible paths. If a wrong connection or assumption is made, it will not be in the BA. The strong consistency option has all the parts of a weak consistency plus some additional properties. All the elements in the BA have to be represented in the view [16]. This means that it is checking completeness instead of sufficiency.

Researchers are looking into achieving consistency through an architectural approach. This allows engineers who are verifying the models to recognize inconsistent parts between the components. Each individual system model must be validated with the topology of the system. This means that the behaviors must match up. Checking this is not easy because the behavior is hard to show on a topology of the system. However, if everything matches up with the BA, then consistency of design can be better retained.

To show this multiview consistency, additional tools are needed. AcmeStudio framework is used to help do this. It creates architecture design environments. Figure 2.6 shows the design flow used to check consistency. The BA and system view are compared for consistency in AcmeStudio.

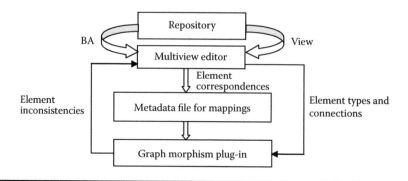

**Figure 2.6   Consistency checking design flow using AcmeStudio.**

The graph morphism plug-in checks the consistency of the system view. This is done by looking at the component-connector graph of both the view and the BA. If the elements are inconsistent, it will go back through the multiview editor to make corrections. However, the researchers are still trying to develop an algorithm to perform this action of making corrections. With all these methods to check for consistency between the view and the BA, researchers hope to prevent errors in the CPS while the CPS is still in development. If this structural consistency can be achieved easily within systems, it will, in turn, save developers time and money that might be wasted on future issues that may arise because of lack of system consistency.

## 2.7  Conclusions

Wireless CPSs combine the physical aspects of systems with a cyber system. After establishing and studying various challenges associated with these systems, one can better understand the basic concepts of CPSs. Another way to learn more about the wireless CPS is to observe different case studies that have implemented these systems (e.g., algae clean up in a lake in China). We have also looked into more detail, multilayered heterogeneous systems. These systems can perform tasks such as help to decide between using WWAN or WWLAN. Then, using all this information, an important use of wireless CPSs is to create more energy-efficient systems. Finally, this chapter reviews four important factors needed to design a CPS. They are adaptation, unification, dependability, and consistency.

## References

1. M. S. Branicky et al. 2006. Multi-disciplinary challenges and directions in networked cyber-physical systems. Cyber-Physical Systems, National Science Foundation, Austin, TX.
2. A. Ehyaei et al. 2011. Scalable data acquisition for densely instrumented cyber-physical systems. *ACM Second International Conference on Cyber-Physical Systems*, Chicago, IL. doi 10.1109/ICCPS.2011.15.
3. B. Andersson, N. Pereira, and E. Tovar. 2007. Wisdom: A dominance protocol for wireless medium access. *IEEE Trans. Ind. Inf.* 3:2.
4. C.-Y. Liu et al. 2011. Coding-based contention forwarding protocol for cyber-physical systems. *2011 Second International Conference on Networking and Computing*, National Cheng Kung University, Tainan, Taiwan, doi 10.1109/ICNC.2011.38.

5. J. White et al. 2010. R&D challenges and solutions for mobile cyber-physical applications and supporting Internet services. *J. Internet Serv. Appl.* 1:1, 45–56.

6. Y. Bai et al. 2010. Remote container monitoring with wireless networking and cyber-physical system. *Mobile Congress (GMC), 2010 Global*, IEEE, Shanghai. doi 10.1109/GMC.2010.5634569.

7. D. Li et al. 2011. A cyber physical networking system for monitoring and cleaning up blue-green algae blooms with agile sensor and actuator control mechanism on Lake Tai, *IEEE Conference on Computer Communications Workshops*, Shanghai. doi 10.1109/INFCOMW.2011.5928908.

8. L. Ma et al. 2010. Net-in-net: Interaction modeling for smart community cyber-physical system. *7th International Conference on Ubiquitous Intelligence and Computing and 7th International Conference on Autonomic and Trusted Computing*, IEEE, Xi'an. doi 10.1109/UIC-ATC.2010.15.

9. J. Shen et al. 2010. Heterogeneous multi-layer wireless networking for mobile CPS. *7th International Conference on Ubiquitous Intelligence and Computing and 7th International Conference on Autonomic and Trusted Computing*, IEEE, Xi'an. doi 10.1109/UIC-ATC.2010.30.

10. J. Lin et al. 2011. An agent-based approach to reconciling data heterogeneity in cyber-physical systems. *IEEE International Symposium on Parallel and Distributed Processing Workshops and PhD Forum*, Shanghai. doi 10.1109/IPDPS.2011.130.

11. C. J. Xue et al. 2009. Joint sleep scheduling and mode assignment in wireless cyber-physical systems. *29th IEEE International Conference on Distributed Computing Systems Workshops*, Washington, DC. doi 10.1109/ICDCSW.2009.13.

12. Q. Tang et al. 2008. Energy-efficient thermal-aware task scheduling for homogeneous high-performance computing data centers: A cyber-physical approach. *IEEE Trans. Parallel Distrib. Syst.* 19:11, doi 10.1109/TPDS.2008.111.

13. K. Ravindran and M. Rabby. 2011. Cyber-physical systems based modeling of adaptation intelligence in network systems. *Systems, Man, and Cybernetics*, IEEE, doi 10.1109/ICSMC.2011.6084087.

14. C. Grimm and J. Ou. 2011. Unifying process networks for design of cyber physical systems. *Electronic System Level Synthesis Conference*, San Diego, CA. doi 10.1109/ESLsyn.2011.5952280.

15. K. Ravindran. 2011. Cyber-physical systems based modeling of dependability of complex network systems. *Sixth International Conference on Availability, Reliability and Security*, Vienna. doi 10.1109/ARES.2011.90.

16. A. Bhave et al. 2011. View consistency in architecture for cyber-physical systems. *IEEE/ACM International Conference on Cyber-Physical Systems*, Chicago, IL. doi 10.1109/ICCPS.2011.17.

# Chapter 3

# Mobile Cyber-Physical Systems

Yeqing Wu and Fei Hu

## Contents

## 3.1 Introduction

In mobile cyber-physical systems (CPSs), mobile sensing and computing devices are tightly connected via communication networking systems that are coupled with the control of the physical world. Because of the inherent mobility of the physical system, a wireless network is the main communication technique used in mobile CPSs. The design of mobile CPSs confronts several unique challenges. In mobile CPSs, the computing devices (machines and devices) that monitor or control the physical objects via sensors must operate in a cooperative and real-time fashion. For instance, when the local available resources become insufficient for the current local tasks, linking the mobile device to other devices, with spare resources, enables complex computing tasks. That is, for both centralized and distributed mobile CPS comprising numerous computing machines or devices, efficient wireless communications among resources are essential.

There are many devices and machines that monitor the physical objects and must move with the physical objects they measure. In order to improve the efficiency of resource use in the system, these machines need to be controlled in a cooperative manner. This adds to the communication load on the wireless network connecting the computing resources in the system. The physical world has limitations, and some of its objects will be independent entities with different and sometimes competing needs. Sensors and actuators suffer from inaccuracies because of technique limitations in the physical world. When people are involved with the mobile CPS, the situation becomes more complex. As a result, some components in even well-designed mobile CPSs may not be perfectly controllable. In large mobile CPSs, a single, wide-area, high-speed wireless network is unlikely. Such systems need to provide a computational environment that consists of multilevel heterogeneous subsystems interacting with each other. These subsystems can have different processing capabilities, wireless communication interfaces, and time precisions in which physical objects are measured. The systems depend on a variety of smaller wireless provisioning points, which control the movement of devices along with concerns of power assumptions.

The capability of mobile CPSs also gives rise to another significant challenge: How do we protect the privacy of the people being monitored? There have been significant exposures of personal information we traditionally consider as private [1]. For instance, a college supervision system with wireless deployment tracks and gathers information about individuals. The public release of the supervising information about individual behaviors may raise privacy problems. Traditional notions of security, authentication, and authorization are unlikely to provide much traction in mobile CPSs [1].

### 3.1.1 Examples of Mobile CPS

Examples of mobile CPSs include various industrial monitor and control systems.

#### 3.1.1.1 Vehicular CPS

Vehicular communication systems are a basic element of both intravehicle and intervehicle CPSs as shown in Figure 3.1 [2]. Modern vehicles are complex systems of control, sensing, computing, and communication. The number of electronic control units (ECUs) is usually more than 70 in modern high-end vehicles. These ECUs process up to 2500 signals, including elementary information such as vehicle speed, and support up to 500 features, including brake by wire and active safety [2]. The increasing number of ECUs adds burden on the vehicular communication system. Traditionally, multiple wiring communication networks, such as controller area networks (CANs), are deployed within a single vehicle. This design adds significant weight to vehicles and thus reduces gas efficiency. More recently, embedded wireless sensor networks have been deployed as the basic communication systems of future automotive CPSs. They can provide intervehicle communication that enables communication and cooperation among vehicles on the road to improve traffic and avoid possible traffic accidents. It is expected that wireless vehicular CPSs will be ubiquitously served as both intravehicle and intervehicle CPSs. Mobile tracking systems that provide traffic accident alarms to the responders [3] are an example of a wireless vehicular CPS.

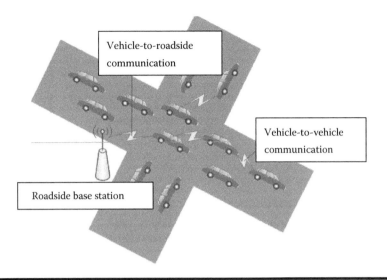

**Figure 3.1    Wireless intervehicle CPS.**

## 3.1.1.2 Mobile Supervision System

Mobile supervision systems often have a mobile surveillance unit (MSU), which is capable of remote monitoring. The MSU may consist of multiple sensors like movable cameras with pan, tilt, and zoom functions and local recording for easy review by the worker. The movable sensors can be located anywhere for maximum flexibility. All the sensing and computing resources communicate with each other or to a control center via a communication network. Such mobile CPSs are beneficial especially in remote locations in which the operating humans cannot monitor in person or visit on a daily basis. These systems alleviate the workers from daily in-person monitoring of targeted entities of the physical world and prevent them from possible danger that may occur in secluded locations. The mobile supervision systems are cost-effective solutions for clients with security needs. A reliable mobile supervision system covering multiple sites may cost a fraction of what the security workers would be paid to watch the sites. Standard mobile security systems are powered by electricity, and mobile CPS offers an alternative power unit equipped with solar panels or other types of new energy sources. Health care systems that monitor cardiac patients [4] and the overspeed supervision system (shown in Figure 3.2) are both examples of mobile supervision systems.

## 3.1.1.3 Smart Grid CPS

The smart grid, as the next-generation electricity network, is a platform to integrate information from the real world and operational technology applied to an electric grid. It provides sustainable options to the way the industry generates and delivers electric energy and guides the customer to manage energy services more flexibly. One of the key driving forces behind this is energy efficiency at both the local and global scale. In order to achieve this, fine-grained monitoring and management are needed. Thus, CPS is the central element of the smart grid. Mobility is a prominent option in smart grids. It is expected that a smart grid can gather information about energy consumption of any mobile devices. The future smart grid will heavily depend on mobile CPSs, because of their capability of monitoring, sharing, and managing information on the electricity business in a dynamic environment.

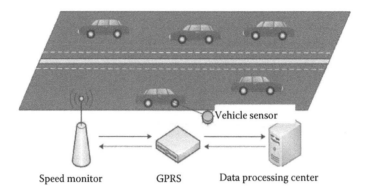

Figure 3.2   Overspeed supervision system. (P. Leijdekkers and V. Gay, Personal heart monitoring and rehabilitation system using smart phones. In *Proceedings of the International Conference on Mobile Business*, © 2006 IEEE.)

## 3.2  Cognitive Radio Network for Mobile CPS

Cognitive radio (CR) as the next-generation wireless communication technique has been recognized to provide efficient communications through dynamic spectrum allocation for communication in mobile CPSs [5]. In traditional wireless communication, a static spectrum allocation scheme divides the radio spectrum into licensed and unlicensed spectrum bands [6]. Unlicensed users are forbidden to access any of the licensed spectrum bands that have been purchased. Many popular wireless communication systems, including Wi-Fi [7] and Bluetooth [8], have been operating in unlicensed spectrum bands without incurring any spectrum cost. It was reported by FCC Spectrum Policy Task Force (2002) that there were about 70% white spaces in the licensed spectrum even in a crowded area [9]. Consequently, CR has been proposed to promote the utilization of the spectrum. Laptops, smartphones, and other consumer communication devices can access the CR network (CRN). In a CRN, the unlicensed users or secondary users (SUs) are allowed to dynamically access the unused parts of licensed spectrums that are owned by the primary users (PUs) or the licensed users. However, SUs should not cause any harmful interference to PUs. In order to achieve this goal, SUs must detect the existence of the PUs and find the spectrum holes. These spectrum holes are often referred to as white space. The white space may consist of unused spectrum bands in a given location and fragments of time. After this, SUs can select an available channel with good quality to transmit data. SUs need to estimate the channel states and monitor the return of PUs to avoid interfering PUs. When PUs return, SUs should vacate the selected channel and choose another one to utilize.

## 3.3  Communication Model

Within a mobile CPS, CR communication assumes complex spatial and temporal dynamics because the network topology constantly changes due to the mobility of the physical entities in the system. The unpredictable communication quality over time and across different applications also increases the complexity of the communication network. Furthermore, a wide variety of environmental factors, such as temperature and humidity, can affect the quality of wireless communications. Improving the overall utilization of radio spectrum throughput of the CRN and maintaining the required quality of service (QoS) for all the computing resources in the mobile CPS become critical and challenging.

There are two basic deployments of CRNs: centralized and distributed. Many CRNs are a combination of the two. In a centralized CRN, each network is composed of a single SU base station (BS) and SU hosts as shown in Figure 3.3. A model of centralized CRN that uses a special cognitive pilot channel for signalization and control was proposed [10]. The central signalization and control unit is in charge of the communication among heterogeneous wireless networks. The unit gathers channel information via pilot signals and announces the spectrum availability and the necessity of leaving a spectrum slot if that one is occupied or will be occupied by a PU. The control is performed by using frequency division and time division multiplexing techniques [10].

A cognitive wireless ad hoc network (CWAN) is a distributed CRN. It can be stationary or mobile through a dynamic mechanism to connect SUs using the provision of network relay functions. Figure 3.4 shows an example deployment of a distributed CRN. The SU could search for

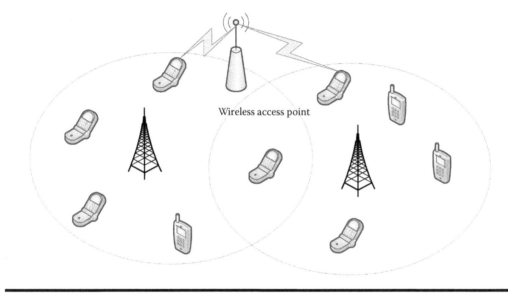

**Figure 3.3** **Example of centralized CRN deployment. (N. Bolívar, J. L. Marzo, and E. Rodriguez-Colina. Distributed control using cognitive pilot channels in a centralized cognitive radio network,** *Sixth Advanced International Conference in Telecommunications,* **pp. 30–34 © 2010 IEEE.)**

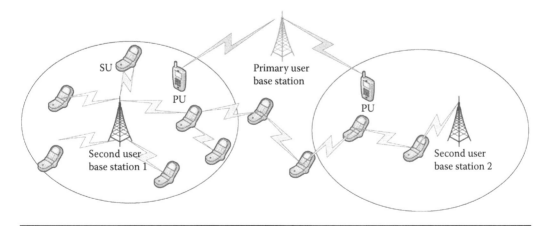

**Figure 3.4** **Example of distributed CRN deployment.**

unused licensed spectrum bands to use. This deployment is useful for telecom operators to extend wireless access among users that are outside the BS coverage [6]. Compared to a centralized CRN, the resource allocation and performance optimization are more challenging for a distributed CRN because of the need for increased coordination between users.

## 3.4 Cognition Cycle

The cognition cycle (CC) of a node is a state machine that shows the states of its cognitive process. The simple CC for a network node includes six states: observe, act, orient, plan, learn, and decide [11]. A CC starts with an observed state. The observed information is evaluated during the orient state. The node can enter the *act* state to react immediately, or enter the *decide* state to further evaluate its various options, or enter the *plan* state to plan for a longer term before entering decide and act states. Throughout the cycle, the radio uses the observations and decisions to achieve a better performance for the node as time goes by.

The notation of CC of a node does not fully capture the network elements of the cognitive process in CRN. In [11], the CC was expanded to frame the operations for a CRN. Figure 3.5 shows an example of CC in a network.

In the figure, the CC in the network consists of two distinct levels of cognitive processes: node level and network level. Each node in the CRN can make individual decisions about how to react to a given situation and collective decisions can be taken by multiple nodes. Two node-level CCs for nodes *x* and *y* are depicted. The processing depicted in the center of the figure is intended to capture this notion of the network-level cognitive processing. The node-level orient stage is concerned with the determination of whether the decision to be taken is individual or cooperative. Individual decisions can be taken by a node alone and affect only node-level cognitive processing. An example of such a unilateral decision is one about dropping a packet on the route. Cooperative decisions are those that affect network-level cognitive processing. An example is about the choice of routing protocol (network-level cognitive processing).

An example of the application of a network-level CC is the IEEE 802.22 WRAN [12]. In WRAN, each SU host is associated with one SU BS. The SU BS coordinates its SUs to make

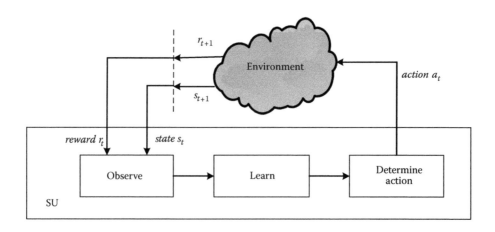

**Figure 3.5  Example of CC in CRN.**

network-level decisions, for example, to instruct its SUs to operate in available high-quality spectrum bands in order to maximize the overall network throughput and delay performance.

Dynamic spectrum access is a specific area of the CR field. A CRN dynamically identifies the white space and uses its high-quality portions. White space is the spectrum bands that are not being used by other systems. As spectrum availability changes over time, the CRN must respond and change the transmission characteristics quickly to avoid causing harmful interference to the licensed users or PUs. Thus, CRNs must quickly identify which spectrum bands are allowed for use and move to a new unused spectrum band if the current operating spectrum band becomes occupied by a licensed user.

CRN uses several methods to gather information with respect to spectrum availability from its spectral environment. In [5] the authors proposed an efficient method to discover the spectrum opportunities with medium access control (MAC) layer sensing in CRNs. In [13] the authors analyze the functionalities and role of the MAC layer in CRNs. The authors also summarize the classification of cognitive MAC protocols. Two common methods are geolocation/database and spectrum sensing [12]. The geolocation/database method exploits additional information that is related to the location of the CR devices and a database of licensed users to determine which spectrum bands are locally available for reuse by the CRN. The spectrum sensing method automatically observes the spectrum and identifies which spectrum bands are occupied by the licensed users.

## 3.5 Communication Protocols

The wireless communication standard called WirelessHART can be used for communication in CRNs. It is based on field-tested international standards, including the HART protocol (IEC 61158), EDDL (IEC 61804-3), and IEEE 802.15.4., which specify adaptive frequency hopping and time-division multiple access. In this section, we focus on a prominent example of CRN architecture, the IEEE 802.22 WRAN, which was designed to operate in the TV broadcast bands while assuring that no harmful interference is caused to the incumbent operation, that is, digital TV and analog TV broadcasting, and low power licensed devices such as wireless microphones [12].

The application for the IEEE 802.22 WRAN standard will be providing wireless broadband access to a rural area of typically 17–30 km and up to a maximum of 100 km in radius from a BS and serving up to 255 fixed units of customer premises equipment (CPE) with outdoor directional antennas located at nominally 10 m above ground level, similar to a typical VHF/UHF TV receiving installation [12]. IEEE 802.22 reference architecture consists of a physical (PHY) layer, PHY layer management entity (PLME), a MAC layer, PHY and MAC layer management entities (MLMEs), a station management entity (SME), and higher layers such as IP and asynchronous transfer mode (ATM).

In general, the PHY layer primarily performs three functions: the main data communications, the spectrum sensing function, and the geolocation function. The latter two functions are designed for the cognitive functionalities of the CRN. IEEE 802.22 defines a single time domain duplex mode and will define a frequency division duplex mode based on orthogonal frequency division multiple access to provide a reliable end-to-end link suitable for efficient communication [12]. IEEE 802.22 will not support multiple-antenna techniques because of the physical size of antennas at low spectrum. In order to support different TV transmission standards, the sampling frequency, carrier spacing, symbol duration, signal bandwidth, and data rates are scaled by the channel bandwidth used in particular standard for worldwide operation. The MAC layer primarily performs three functions: dynamic spectrum access (DSA), dynamic spectrum sharing (DSS), and dynamic spectrum management (DSM). DSA is in charge of accessing white spaces, detecting PU signals, and timely vacating the SU occupied channel if the reappearance of PU is detected in the channel. DSS is in charge of coordinating SUs for channel sharing. The DSM function enables multichannel data packet transmission through channel bonding. The PHY layer interfaces with the MAC layer through the PHY service access point (SAP). The MAC layer interfaces with the convergence sublayer bridge (e.g., 802.1d) with MAC SAP. The MLME, SME, and PLME interface each other with MLME-PLME SAP, SME-MLME SAP, and SME-PLME SAP.

## 3.6 Quality of Service Architecture

Mobile CPSs that use wireless networks as the carriers of mission-critical sensing and control information often have stringent requirements on predictable QoS such as reliability and latency. QoS is the ability to provide different levels of priority or guarantees different levels of performance for different users and applications. Such abilities are important for real-time applications, such as multimedia streaming, especially when the communication resources are insufficient. There are many QoS parameters that are required by different applications. Some common parameters include end-to-end delay, packet loss, throughput, and jitter. The end-to-end delay can be measured by the average time to transmit a data packet from the source to the destination. It can be either deterministic or probabilistic. Real-time applications often demand stringent and deterministic end-to-end delay. Packet loss occurs when one or more packets of data transmitted across a communication network fail to reach their destination. Communication quality in CRNs can be severely affected by the packet loss. The throughput measures the number of data units passing through a communication network per time period. Because of the limited network resources, effective schemes are required to manage the traffic flow such that the stringent requirements of throughput for certain users are guaranteed. Jitter is the packet delay variation, which measures the difference in deterministic end-to-end delay among selected packets. Large jitter can seriously degrade the transmission quality. In a demanding system, effective schemes are used to constrain the maximum jitter rates in the system for seamless transmission [15].

There are two categories of schemes for providing QoS for data transmissions over CRNs based on the design architecture: single-layered approach and cross-layered approach.

## 3.7 Methods to Enhance System Efficiency

In traditional policy-based management mechanisms, each device/user in the system adheres to a strict and static predefined set of rules. The major drawback of these mechanisms is that the underlying system cannot adapt to its changing environment dynamically. This drawback heavily affects the efficiency of mobile CPSs using wireless communication due to the complexity and dynamic operation in a wireless network. Intelligent schemes enable each device/user to observe, learn, and respond to its dynamic operating environment in an efficient manner and thus promote the overall efficiency of the systems. Intelligent management mechanisms for mobile CPSs may be implemented by exploiting a variety of machine learning and pattern recognition approaches. This section provides an overview of approaches for system efficiency enhancement, focusing on reinforcement learning (RL). In mobile CPSs, the efficiency of the communication networks is critical to achieve the quality of the whole systems because of limited communication resources in mobile CPSs. Methods that decrease the communication load while maintaining the required QoS are beneficial. Compressive sensing is a common method used to reduce the size of the sensed data while guaranteeing the quality of recovered information from the sensed data.

### 3.7.1 Intelligent Management Mechanisms

In a mobile CPS with intelligent management mechanisms, the device/user usually learns about its actions from the feedback, and evaluates and modifies its actions accordingly. The common intelligent approaches include Bayesian networks, neural networks, and RL.

Continuous learning is important for a system operating in a dynamic environment. RL has been used as an intelligent and continuous learning approach to achieve system-wide performance enhancement. It has been applied for resource management and dynamic path selection in a system with a communication network. $Q$ learning is the most widely used online RL approach. The $Q$ learning is associated with updating a $Q$ table, whose elements are the learned action value $Q(state, event, action)$, reward measurement of an action, and rules. Note that some elements are optional, for example, an event can be a NULL set. The size of the $Q$ table is $|(state, event)|$ * $|action|$, where $|.|$ is the cardinality function. The element $Q(state, event, action)$ is the reward or discounted reward of a certain action under the particular state–event pair. Let $s$ denote the state, $e$ denote the event, and $a$ denote the action. The $Q$ values can be updated using the previous $Q$ values and the reward of the action:

$$Q_{t+1}(s_t, e_t, a_t) = (1 - \partial_{t+1}(s_t, e_t, a_t))Q_t(s_t, e_t, a_t)$$
$$+ \partial_{t+1}(s_t, e_t, a_t)[r_{t+1}(s_t, e_t, a_t)] + \gamma \max_{a \in A} Q_t(s_{t+1}, e_{t+1}, a)] \tag{3.1}$$

where $Q_t(s_t, e_t, a_t)$ is the old $Q$ value, $\partial_{t+1}(s_t, e_t, a_t)$ is the learning rate at time $t + 1$ (it may be the same for all situations), and $0 \leq \partial \leq 1$. $r_{t+1}$ is the reward observed after taking action $a_t$ under the $(s_t, e_t)$ pair. $\gamma$ is the discount factor ($0 \leq \gamma \leq 1$). The algorithm ends when an absorbing state–event pair is achieved. The higher the value of $\gamma$, the less important its previous learned knowledge, giving a one-shot network behavior. The higher the value of $\gamma$, the greater the weight put on the future return, which is the maximum old $Q$ value in the state–event pair at the next time period. RL searches for optimal or suboptimal sequences of actions that maximize its accumulated reward through choosing the action with the maximum $Q$ value at each time instant.

In [14], an example of the usage of discounted reward in Equation 3.1 is stated. At a particular node, the states can be the upstream nodes, and the action is chosen in a certain upstream node to transmit the data. The immediate reward is the inverse of the immediate cost; it represents the time delay introduced by an upstream node. The discounted reward is the inverse of the cost, and it measures the end-to-end delay from an upstream node to a destination node in a multihop routing scheme. At any time instant, the agent chooses an upstream node such that the corresponding $Q$ value of the state–action pair at the upstream node is maximized.

Figure 3.6 shows the simplified framework of $Q$ learning. At time instant $t$, a host identifies the subset of actions in adherence to predefined rules that exclude actions that violate the network constraints. Next, it chooses an exploitation action with the maximum $Q$ value or a randomly chosen exploration action from the coherent action subset, which is used to increase the knowledge about the operating environment of the host. At time instant $t + 1$, the host updates its $Q$ table with the observed reward/cost of its previous action.

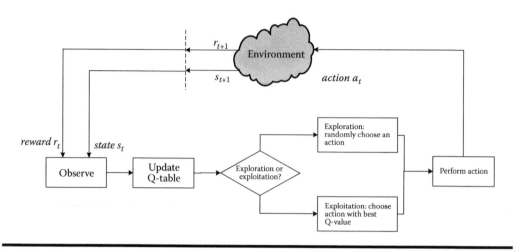

**Figure 3.6   Framework of Q learning.**

## 3.8 Challenges

Many factors in mobile CPSs make their system design complex. Such factors include the complex spatial and temporal dynamics in wireless communication, heterogeneous subsystems, or applications with different QoS requirements and INP methods, unpredictable network behaviors such as malicious attack, dynamic environment factors such as intrusive human behaviors, and weather conditions [16]. How to efficiently model the system and environmental dynamics, uncertainties, and their interactions is a challenge posed on the system design. In this section, we focus on the security and survivability of mobile CPSs.

### 3.8.1 Security

The popularity of mobile CPSs leads to the potential exposure of the details of our daily lives that are traditionally considered as private. The ability of mobile CPSs to partially control the physical world leverages the possibility of malicious attacks. The legal restrictions on keeping private information of the clients/users confidential have been set up. Although there are legal holes, cultural restrictions should alleviate the situation. We should build systems that respect people's privacy without unduly pursuing commercial benefits. Privacy is not absolute for most people. People willing to expose their personal information or behaviors may change along the changing context in which they find themselves. When the privacy preference of a person changes, the system must be able to track this change and adapt its behaviors accordingly immediately.

Another aspect of system security involves malicious attack. Malicious attacks can be categorized into two types: node capture and false data injection. Captured nodes can defeat authentication, which creates an insider threat that enables insider attacks. Injecting false data degrades the integrity of the system data. In general, security techniques to detect or avoid untrusted software and hardware can be categorized as either design-time techniques or run-time techniques [17]. Formal verification is a commonly used design-time method. Design-time techniques seek to verify that a computing system is flaw free before the implementation. Such techniques are time consuming and are extremely expensive. Furthermore, the specifications designed for the verification may not capture all vulnerabilities of the system. As a result, only a limited set of small-scale applications use design-time techniques. Run-time security techniques build run-time trusted components to provide guarantees for certain system components. Encryption and authentication modules are typical examples of run-time techniques. The trusted components can be changed and modified at design time. In large-scale mobile CPSs, establishing security contexts among system components before the deployment is unlikely owing to the large number of devices belonging to many different administrative control domains. One can utilize run-time security techniques to build an intelligent authentication module that can leverage everyday activities to help establish secure rules of device connections [17]. However, it is worth noting that run-time techniques cannot ensure security in all system components.

### 3.8.2 Survivability

In a wireless network, one cannot prevent nodes from traveling to the edge of connectivity. Environmental considerations such as bad weather cause significant variations affecting the stability of wireless bandwidth. Furthermore, behaviors of intelligent devices are still subject to human users. These users often have conflicting and competing needs. Cultural and legal restrictions will alleviate this status for the foreseeable future. However, misconfigured devices and malicious attackers will inevitably pollute the wireless spectrum. Furthermore, mobile CPSs have unique challenges in terms of intrusion detection due to the inherent mobility of the system, heterogeneity, and resource constraints. Distributed intrusion detection applications are usually used in mobile CPSs because they are fault tolerant.

In a mobile CPS, energy replenishment is often not possible and nodes may be compromised at times. Thus, the ability to withstand malicious attacks and energy consumption is of the utmost importance. It is impossible to prevent malicious attacks without false negatives; thus intrusion detection and detection tolerance are necessary. Models are developed to access the survivability properties of a mobile CPS subject to different scenarios. In [17] security fault is modeled with a Byzantine fault model, and the security failure occurs if one third or more of the nodes are compromised. It is impossible for the system to reach a consensus if it has more than one-third compromised nodes. A mathematical model is developed to evaluate survivability capacity of a mobile CPS subject to energy exhaustion and security failure.

## 3.9 Conclusions

We have discussed the main properties of mobile CPS as a category of CPS that has inherent mobility in the physical system. Limitations in current technologies pose a number of challenges in building efficient and powerful mobile cyber-physical systems. Efforts will be needed to overcome the unique challenges posed in mobile CPSs.

## References

1. B. D. Noble and J. Flinn. 2007. Wireless, self-organizing cyber-physical systems. *NSF Workshop on Cyber-Physical Systems*. Austin, TX.
2. H. Gharavi, K. V. Prasad, and P. Ioannou. 2007. Scanning advanced automobile technology. *Proc. IEEE*, 95, 328–333.

3. C. Thompson, J. White, B. Dougherty, and D. Schmidt. 2009. Optimizing mobile application performance with model-driven engineering. In *Proceedings of the 7th IFIP Workshop on Software Technologies for Future Embedded and Ubiquitous Systems*. Springer-Verlag, Berlin, pp. 36–46.

4. P. Leijdekkers and V. Gay. 2006. Personal heart monitoring and rehabilitation system using smart phones. In *Proceedings of the International Conference on Mobile Business*. IEEE, New Brunswick, NJ, 29 pp.

5. H. Kim and K. Shin. 2008. Efficient discovery of spectrum opportunities with MAC-layer sensing in cognitive radio networks. *IEEE Trans. Mobile Comput.* 7:5, 533–545.

6. K.-L. Yau. 2010. Context awareness and intelligence in cognitive radio networks: Design and applications. Thesis. Victoria University of Wellington, New Zealand.

7. B. Crow, I. Widjaja, L. Kim, and P. Sakai. 1997. IEEE 802.11 wireless local area networks. *IEEE Commun. Mag.* 35:9, 116–126.

8. K. Sairam, N. Gunasekaran, and S. Redd. 2002. Bluetooth in wireless communication. *IEEE Commun. Mag.* 40:6, 90–96.

9. F. S. P. T. Force. 2002. Report of the spectrum efficiency. Technical Report 02-155. Federal Communications Commission, Washington, DC.

10. N. Bolívar, J. L. Marzo, and E. Rodríguez-Colina. 2010. Distributed control using cognitive pilot channels in a centralized cognitive radio network. In *Sixth Advanced International Conference in Telecommunications*, IEEE, New Brunswick, NJ, pp. 30–34.

11. I. Doyle and T. Forde. 2007. The wisdom of crowds: Cognitive ad hoc networks. In *Cognitive Networks: Towards Self-Aware Networks*. John Wiley, New York.

12. C. Stevenson, G. Chouinard, Z. Lei, W. Hu, S. Shellhammer, and W. Caldwell. 2009. IEEE 802.22: The first cognitive radio wireless regional area networks standard. *IEEE Commun. Mag.* 47:1, 130–138.

13. A. D. Domenico, E. C. Strinati, and M. G. D. Benedetto. 2012. A survey on MAC strategies for cognitive radio networks. *IEEE Commun. Surv. Tutorials* 14:1, 21–44.

14. R. Arroyo-Valles, R. Alaiz-Rodriguez, A. Guerrero-Curieses, and J. Cid-Sueiro. 2007. Q-probabilistic routing in wireless sensor networks. In *Proceedings of the 3rd International Conference on Intelligent Sensors, Sensor Networks and Information*, IEEE, New Brunswick, NJ, pp. 1–6.

15. H. Zhang. Overview of wireless cyber-physical systems (WCPS). [online]. Available at http://www.cs.wayne.edu/hzhang/courses/8260/Lectures/.

16. M. Tehranipoor and F. Koushanfar. 2010. A survey of hardware Trojan taxonomy and detection. *IEEE Design Test Comp.* 27:1, 10–25.

17. A. J. Nicholson, I. E. Smith, J. Hughes, and B. D. N. LoKey. 2006. Leveraging the SMS network in decentralized, end-to-end trust establishment. In *4th International Conference on Pervasive Computing*, Dublin, pp. 202–219.

# DESIGN PRINCIPLES

# Chapter 4

# Cyber-Physical System Controls

Tony Huynh, Ahmed Alsadah, and Fei Hu

## Contents

## 4.1 Introduction

Cyber-physical systems (CPSs) define integrated components that include both cyber and physical elements. Controls of such systems require reliable sensor data and robust operation to prevent cyber damage from surfacing in the physical world through actuators. In development of such a system, several factors such as the types of attacks and respective countermeasures are taken into consideration. This chapter will detail several motivations for designing CPS controls and then provide greater insight into two types of attacks generally used. Finally, methods of defense against the two will be described.

While there are several instances of CPS controls being used in different contexts, the general issue with security is that while the physical implementation may be under "fence-and-gate" [1], complete protection is impractical against physical subversion. Furthermore, the control systems may be in operation for critical infrastructure or sensitive equipment that must be protected from damage due to economic or strategic reasons.

## 4.2 Data Center Cooling

One such example of critical infrastructure CPS controls is in data center cooling [2]. In today's world, the need for data centers is quickly growing with the expansion of computational resources integrated with everyday life. Demand for resources such as cloud computing and mass data storage requires the use of large data centers with dedicated hardware to support multiple users. Such dedicated hardware creates multiple layers of complexity in maintaining a proper operational efficiency. The following sections will detail operation of the controls system from a CPS perspective.

Utilizing a cyber-physical approach to model the system yields two interrelated models, the computational network and the thermal network, which model the cyber and physical dynamics of the data center. The cyber aspects of the model represent the computational variables, such as data rate and processing speeds. The physical dynamics cover the heat generated by such activity in addition to the computer room air conditioning (CRAC). One of the main points of interest in maximizing efficiency is the quality of service (QoS) of the represented model. In addition to maintaining effectiveness of the center, the QoS must be taken into account.

The current paradigm in controlling data centers revolves around three general models: server level, group level, and data center–level controls. At the server level, control takes place at each individual server machine. Things such as computing resources in terms of central processing unit (CPU) cycles, networking, and memory are maintained to reduce power consumption and, in turn, heat generation. Previous solutions to the server-level model of control include dynamic voltage and frequency scaling, which throttles certain CPU characteristics under idle conditions to achieve better power efficiency. In addition, server specific fans are included in this category, in which constraints on the thermal generation of a unit are applied to the processing capabilities.

Group-level control applies to the "nodes" that describe a single application using multiple servers. In the case of group-level control, performance tends to involve migration of workloads across virtualized environments. This is accomplished by hosting servers in virtual machines (VMs), which can, in turn, be transferred to different hardware. This allows for processing to be distributed across multiple platforms to balance the workload and lower power consumption across the group. Old solutions involve dynamic prediction of incoming workloads and subsequent distribution based on arrival rates while imposing certain constraints in relation to the thermal management.

Finally, in data center–level control, certain capacities, either in information technology (IT) or computer technology (CT), are shared among all the units, elevating control to cover the entire data center. One of the main characteristics in data center–level control is that several aspects from the previously mentioned levels are integrated into this top level. This includes measures such as server-level optimization and group-level migration. In addition, CRAC control is used while enforcing certain constraints to address thermal distributions across the data center. Current solutions at this level implement the previous level solutions with regard to power consumption as well as heat distributions. This can involve consolidation of server loads into a smaller subset of servers to concentrate energy efficiency. This allows idle servers to shut off during idle times. The problem with this approach lies with the unacceptable effects on QoS.

In development of a model for the data center cooling, a data center–level approach is taken [2]. Network servers are represented as nodes with data arriving for computation and leaving by execution or migration to a different node. A graphical representation can be seen in Figure 4.1 with mathematical notations.

This graphical representation is derived from the following set of equations:

$$a_i(t) = \lambda^w(t)S_i(t) + \sum_{j=1}^{N} \xi_{i,j}(t) \tag{4.1}$$

$$d_i(t) = \eta_i(t) + \sum_{j=1}^{N} \xi_{i,j}(t) \tag{4.2}$$

In addition to these equations, a third equation governs the bounds in which they operate:

$$\upsilon_i(t) = \mu_i(t) + \sum_{j=1}^{N} \delta_{i,j}(t) \tag{4.3}$$

where the desired departure rate for a given node, $\upsilon_i(t)$, is given by the desired execution rate, $\mu_i(t)$, plus the required migration rate $\sum_{j=1}^{N} \delta_{i,j}(t)$. In addition to this, $\eta_i(t)$ is defined at any time as either

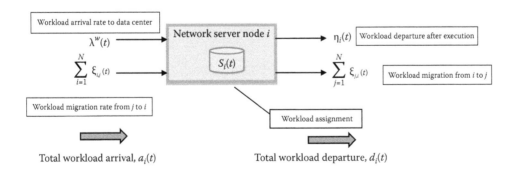

**Figure 4.1 Graphical representation of data center node.**

$\mu_i(t)$ if the total arrival rate $a$ is greater than the total departure rate $d$ or arrival rate $a_i(t)$ otherwise. The thermal network of the data center is represented in a similar fashion with input and output with the following equations:

$$T_{\text{in},i}(t) = \sum_{j=1}^{M} \Psi_{i,j} T_{\text{out},j}(t) \tag{4.4}$$

$$T_{\text{out},i}(t) = -k_i T_{\text{out},i}(t) + k_i T_{\text{in},i}(t) + c_i p_i(t) \tag{4.5}$$

In the above equation, the input temperature for a node is given as the sum of all the nodes' output temperature from $i$ to $M$. Furthermore, the output temperature is given as a linear time-invariant description involving time constant $k$ and power consumption coefficient $c$. Power consumption is proportional to the temperature by the departure rate by execution from a node $\eta$ by a nonnegative coefficient $\alpha$. By solving the above set of equations as a minimization problem for temperature, various results are given for different implementations of the control system strategies.

Simulation of the data center's cooling is completed using three different scenarios: coordinated, uncoordinated, and baseline strategies. The first scenario utilizes data from both the cyber and physical aspects of the data center to achieve optimization. The second scenario implements strategies for both the cyber and physical aspects, but independently from each other. The third scenario implements the baseline strategy as a method of dealing with temperature that is independent optimization. Results for power and average utilization demonstrate the effectiveness of the CPS control in the case of the data center cooling.

## 4.3 Attacks in Control Systems

For a given control system, the types of attacks can be illustrated using a simplified model [1], shown in Figure 4.2.

The first type of attack is not covered as the physical well-being lies outside the scope of control security. The false data injection attack is also known as a deception attack or integrity attack [2],

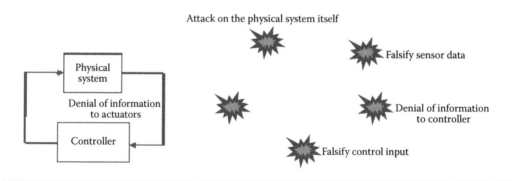

**Figure 4.2 Greatly simplified model of CPS control.**

in which either the sensor data or control input is falsified through injection with the intent of driving the system beyond operational parameters. The denial attack is also known as a denial of service (DoS) attack in which channels are jammed to prevent communication between sensors, actuators, and controllers. For this chapter, the latter two types of attacks are covered in greater detail in the following sections.

## 4.3.1 Denial of Service

DoS attacks can be represented graphically as a switch that opens and closes, effectively shutting out new data from reaching either the controller or the physical system. Two types of mathematical representations can be taken into account for DoS attacks.

### 4.3.1.1 Bernoulli Model

Befekadu uses in his first model [3] an independent Bernoulli process to model a DoS attack against a discrete-time partially observed stochastic system:

$$x_{k+1} = Ax_k + \beta_{k+1}Bu_k + v_{k+1} \tag{4.6}$$

$$y_{k+1} = Cx_k + w_{k+1} \tag{4.7}$$

The first equation represents the state given as a closed-loop system where $x$ is the state, $\beta$ is the DoS sequence {0,1}, $u$ is the control input, and $v$ is a normal distribution to introduce randomness. In the second equation, the observation of the output, $y$, is related to the control state with additional noise or randomness. In addition to this description of the system, a risk control issue is provided as Equation 4.8, where $\theta$ is the risk parameter:

$$J(u) = \theta E\left[e^{\frac{\theta}{2}}\left\{\sum_{k=0}^{T-1}(x_k Mx_k + \beta_{k+1}u_k Nu_{k+} x_T Mx_T)\right\}\right] \tag{4.8}$$

The goal of the control policy is then to optimize itself under the attack. The attack model is represented with Bernoulli probabilistic trials. In the case of success, the switch is opened, preventing flow and thus "1." In the opposite case, a "0" represents failure. Owing to the lengthy nature of the mathematical process of developing the solution to dealing with the attack, only the conclusion is included here. The general solution is developed such that a recursive function (Equation 4.9) provides an information state $\delta$ of the control system based on its own current value in relation to control data, anticipated attack sequence, and observations:

$$\delta_{k+1} = \delta_{k+1}(\delta_k, u_k, y_{k+1}, \beta_{k+1}) \tag{4.9}$$

The information state of Equation 4.9 is based on the information provided by several coupling recursive equations, on the basis of the variance parameters of the normal distribution used for providing randomness/noise to the systems in Equations 4.6 and 4.7, as well as the risk parameter in Equation 4.8. The equations are not shown here for sake of length.

### 4.3.1.2 Markov Model

While the Bernoulli model in the previous section sufficed to model a DoS attack on control systems, it is possible to build a more sophisticated model using Markov hidden variables. This approach allows states to be taken into account, compared to the memoryless Bernoulli model [4]. Given the system in Equation 4.6, the Markov process is

$$Y_k = F_k(Y_{k-1}) + W_k \qquad (4.10)$$

where $F_k$ is a bounded measurable function acting on sensor distribution $Y_{k-1}$, and $W_k$ is a random variable for noise. Befekadu offers the complete mathematical solution where the Markov model is used to derive the optimal risk policy while utilizing cost function (Equation 4.8).

## 4.3.2 Deception or False Data Injection Attack

False data injection attacks tend to be more subtle than their DoS attack counterparts [6] and thus difficult to detect. It is shown by Cardenas et al. [5] that, in some cases, given a linear representation of a CPS of a Tennessee-Eastman Process Control System (TE-PCS) model, it is possible to solve an attack strategy that goes completely undetected by prevention schemas. Mo and Sinopoli [6] formulate the necessary conditions under which an attacker is able to perfectly bypass defensive strategies of a control system scenario defined with a Kalman filter, linear-quadratic-Gaussian (LQG) controller, and a failure detector. The CPS is modeled classically as an linear time-invariant (LTI) system:

$$x_{k+1} = Ax_k + Bu_k + w_k \qquad (4.11)$$

where $x$ is the state variable for a time $k$, $u$ defines the control input, and $w$ describes a certain amount of noise with $N(0,Q)$. An initial state $x_o$ is also given as $N(0,\Sigma)$. The following sections will describe the individual components of the CPS with relation to the LTI system in Equation 4.11.

### 4.3.2.1 Kalman Filter

The Kalman filter accomplishes the task of providing a certain system state estimation $\varkappa$ given in the measurements provided by

$$\varkappa_{k+1} = A\varkappa_k + Bu_k + K[y_{k+1} - C(A\varkappa_k + Bu_k)], \qquad (4.12)$$

$$y_k = Cx_k + v_k, \qquad (4.13)$$

where $K$ is the Kalman filter gain, which varies with time. In Equation 4.13, measurement $y$ is given by the state variable and randomness $v \sim N(0,R)$. $y_{k+1} - C(A\varkappa_k + Bu_k)$ is further simplified to $z_{k+1}$, which defines the residue generated by various processes. The error in state estimation is given by $e_k = x_k - \varkappa_k$.

### 4.3.2.2 *Linear Quadratic Gaussian Controller and Failure Detector*

To ensure system stability, the LQG controller must minimize the function

$$J = \lim_{T \to \infty} \min E \frac{1}{T} \left[ \sum_{k=0}^{T-1} \left( x_k^T W_{x_k} + u_k^T U u_k \right) \right] \tag{4.14}$$

Through this equation and the error of state estimation from the previous section, the stability of the system is assured assuming cov($e$) and $J$ are bounded.

Failure detection is described with a quantified value from a function:

$$g_k = z_k^T \mathbf{COV} z_k, \tag{4.15}$$

where **COV** is the covariance matrix of the residue. Owing to the Gaussian nature of the residue, the value $g$ can be used in comparison to a threshold value such that an alarm is triggered when $g >$ threshold. This further refines the attack sequence to $\beta = P(g_k >$ threshold).

### 4.3.2.3 *Example*

With the above sections on CPS component models, Mo and Sinopoli [6] define a set of theorems describing a perfectly attackable system. Furthermore, an example is given using discrete values to illustrate the function of the model. The example entails the control systems for a vehicle with $x$ velocity and $x$ position. The graphical results of their model in Figure 4.3 best illustrate the results.

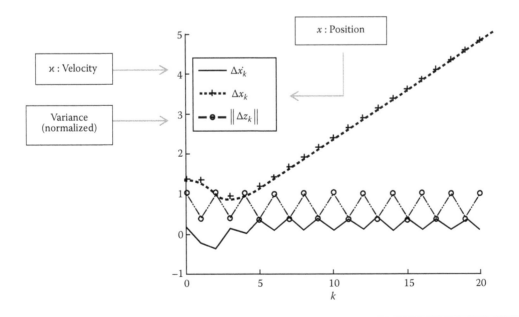

**Figure 4.3** **Application of a false-data injection within bounds of system stability over time.**

Despite having a steady-state velocity and residue (hence avoidance of threshold alarm), the attack is able to alter the path of the vehicle by injection of data into the actuator.

## 4.4 Methods of Defense

There exist several methods to dealing with attacks directed at the control system. Kisner et al. [1] describe several methods toward dealing with secure fail-safe systems, specifically for nuclear applications. Such measures include authentication, redundancy and diversity, secondary systems, noncentralized systems, and recovery. Within the next few sections, some methods of defense will be discussed, such as detection based on cumulative sum and trust anchors, which are based on authentication and secondary system methods.

In this section, an example model, TE-PCS [2], is discussed in relation to both types of attacks mentioned in this chapter. Furthermore, the vulnerabilities in control systems are elaborated on and differences in control security and information security are provided to illustrate the direction undertaken in developing countermeasures for the attacks simulated.

### 4.4.1 Vulnerabilities in Control Systems

Incidents in control systems are typically categorized into three types. The first one is known as *computer-based accidents*, in which errors occur without malicious intent. An example of this is software patches that may cause errors. The next is *nontargeted attacks* where Internet-connected controllers become infected with malicious problems that do not necessarily target the sensor and actuator systems. Finally, there are *targeted attacks*, which are specifically designed with the intention of damaging the CPS. For the scope of this chapter, the latter category of attacks will be focused on. The danger of cyber-attacks is apparent as they are cheaper and can be implemented at a distance (via the Internet) and without as much risk in comparison to a physical attack.

A notable example of a control system attack is the Stuxnet worm, whose goal is to cause damage by altering controllers out of operational parameters. Analysis of this example reveals several vulnerabilities in the present state of control systems. The worm is able to infect embedded systems by first infecting Windows systems. After this, the worm tracks programming/monitoring software for embedded systems and reprograms the controllers. These tasks are accomplished through the use of zero-day exploits, which take advantage of system dates, and rootkits. In addition, trust is achieved between the systems with the use of stolen certificates. Furthermore, evolution of the system is possible through the use of a messaging system, which allows the programmers to adapt to changing conditions.

The approach taken to dealing with the targeted attacks is an anomaly detection mechanism that focuses on the physical system rather than the cyber aspects. It is believed that this approach can result in less false alarms in comparison to more traditional detection methods, especially for certain control systems such as SCAPA.

### 4.4.2 Differences between Control and IT Security

There are several differences between control system and IT security, the main distinction being that control systems influence the physical world through actuators. Another difference is that the method of updates for software tends to be badly suited for control systems. Patching or upgrading a system can take up extended periods of time, as the system must be shut down and

**Table 4.1 List of Security Development Measures**

| Security Measures | Definitions |
|---|---|
| Risk assessment | Takes into account costs associated with tasks. |
| Detection algorithms | Focused on detecting attacks when they occur. |
| Attack response | When detection occurs, appropriate actions are taken to mitigate damage. |

thus be nonoperational. System uptime can be crucial in industrial settings due to economic restraints. Control systems also tend to be real-time oriented, especially in the case of decision critical components.

As mentioned before, an important part of control system security is that the systems have access to and affect the physical world. While IT security is able to protect information, estimation and control algorithms are not taken into account. Therefore there are three main points to be discussed in the development of security: *risk assessment, detection* algorithms, and *attack-response* algorithms (Table 4.1). In the first point, the amount of damage an attack could cause to a control system must be estimated. This allows the more critical components to be protected, for instance, in the power grid and electricity markets. In the second point, attack detection algorithms need to be in place so compromised states can be quickly addressed. For the scope of this chapter, false-data injection attack detection will be used. In the third point, appropriate attack response algorithms need to be used such that components of the system are able to survive attacks without loss of operation. Even in the case of a successful malicious attack, the system must be able to mitigate the damage.

### 4.4.3 Risk Assessment

In order to minimize the impact of a given negative event, risks are evaluated using a metric that calculates average loss by event. The metric is given as $R_u = E[L] \approx \sum_i L_i p_i$, where $R_u$ is the average loss, $L_i$ is the loss given an event $i$, and $p_i$ is the probability of the event. In testing this metric, the experiment involves a sensor network and an attack that compromises sensor $i$, and the corresponding loss $L_i$. Individual sensors in the network are measured as part of a vector $x(t) = \{x_1(t),\ldots, x_p(t)\}$ such that at time $t$ there is a measurement $x_i(t) \in \mathbb{R}$ of sensor $p$ within bounds of $x^{max}$ or $x^{min}$. Furthermore, the controller receives a certain measurement $\sim x_i(t) \in \mathbb{R}$ of sensor $p$. Therefore, under an attack situation the received and actual values may be different. Given an attack duration between times $t_s$ and $t_e$ in vector $T_a$, the value $\sim x_i(t)$ is equal to true value $x_i(t)$ for $t \in T_a$ and attack signal $y_i(t)$ for other times within vector $T_a$. From this model, two types of attacks can be represented: integrity attacks and DoS attacks. In the first type, a given sensor is compromised and arbitrary values are injected. In the latter type, measurements are blocked from the controller, causing a lack of new data. However, in this case, the controller can easily react with the intuitive countermeasure of reusing old data in the presence of the DoS attack.

To test the attacks, the model known as the Tennessee-Eastman Process Control System (TE-PCS) is used. In Figure 4.4, a diagram is shown depicting the process flows with sensors and controllers. Feed 1 is a mixture of two reactants A and C with inert B, and feed 2 is pure reactant A. Within the tank, A and C react with B to form D, accordingly with the formula A + C – (B) –> D. The purge value releases the associated vapors of A, B, and C.

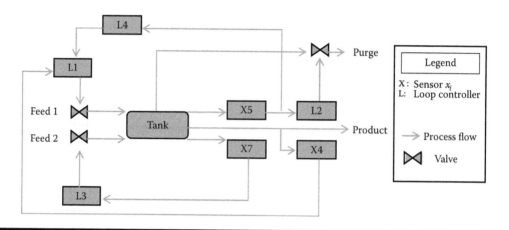

**Figure 4.4   TE-PCS architecture (simplified).**

The goal of the controllers is to regulate several flow rates within the system. First, the flow rate of the product must be maintained at a consistent pace. Also, the operating pressure of the tank must be kept below 3000 kpA because of safety limitations. In addition to this, the pressure should remain as close as possible to the limit without exceeding it. Finally, the operating cost, which is dependent on the loss of reactants A and C through the purge value, needs to be minimized.

The sensor network operates in conjunction with the controller to achieve the above conditions. The last controller L4 controls the pressure by reading data from sensor X5. In addition to this, controller L2 allows further control of pressure through the purge value. Sensor X7 reads the fraction of component A within the tank to allow L3 to increase or decrease the amount. Overall flow rate is determined by reading data from sensor X4 into L1.

On the basis of the safety limitation of the tank, it is assumed that the goal of an attack is to drive the pressure past 3000 kPa, causing damage to the equipment or an explosion. The experiment assumes that at any given time, only a single sensor is compromised by the attacker. Effectiveness of an attack is determined by whether or not the compromised sensor can result in unsafe states or not. In general, testing results showed that min/max attacks were the most effective when the attack signal $y_i(t)$ was equal to either $x^{max}$ or $x^{min}$. As seen in Figures 4.5 and 4.6, an

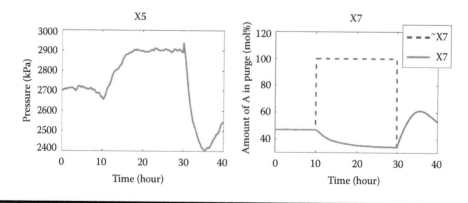

**Figure 4.5   Integrity attack on sensor X7.**

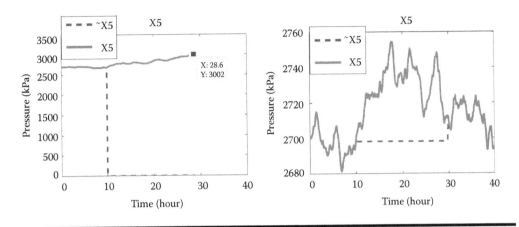

**Figure 4.6   Integrity attack on sensor X5 and DoS attack on sensor X5.**

attack on sensor X7 leads to an increase in pressure, as seen through sensor X5. However, this does not lead to pressure beyond the safety limitations and the system recovers afterward.

An integrity attack on sensor X5, however, led to increases in pressure beyond the safety limit. By enforcing the minimum value of the sensor, the purge valve was disabled to allow the pressure to increase. Given the slow effects (slow dynamics) of the system, it is likely that human operators would be able to notice and minimize damage by such an attack.

Examination of the effects of a DoS attack on the same sensor showed that under the duration of an attack, the pressure never exceeded the safety limitations. In general, DoS attacks were ineffective against the other sensors as well. From the results obtained, the conclusion is that under cost constraints, sensor X5 should be secured under more advanced safety measures. Furthermore, defenses against integrity attacks should be prioritized given their effectiveness over DoS attacks.

### 4.4.4  Attack Detection

Detection of attacks in a control system differs from IT systems in that models can be made of the physical system based on expected reactions to a known input. Given a certain control input sequence, the output sequence can be compared to the expected output to determine if the signal is compromised or not. In solving the problem of detection, two components are needed: a model of the physical system's behavior and a detection algorithm.

Modeling the physical system must represent the behavior of the system, which can take place either through physical representation through the laws of physics or through input and output data by simulation. A linear model is used to model the dynamics: $x(k + 1) = Ax(k) + Bu(k)$, where $x(k)$ represents the state and $u(k)$ represents the control input. $A$ and $B$ represent matrices for physical dependence and control inputs, respectively. From this model, the measurement sequence $y(k) = Cx(k)$ is used, where $C$ is an output matrix.

Two methods of detection are to be used in the simulation: sequential detection and change detection. In optimizing these methods, in the former the goal is to obtain the correct hypothesis within a minimum number of samples, while in the latter the goal is to detect a change at an unknown time. Sequential detection starts with the assumption that observations taken under a time sequence are either under the normal or attack hypothesis. Change detection, on the other hand, starts with the assumption that the observations start in the normal hypothesis before moving into the attack hypothesis.

With sequential detection, if it is assumed that there is a fixed probability of a false alarm and detection, the solution to the problem is a classical sequential probability ratio test (SPRT). The use of SPRT has been known to extend to other problems in security such as worms and port scans. Under SPRT, the following description is made: $S(k+1) = \log \dfrac{p_1(z(k))}{p_2(z(k))} + S(k)$, where $z(k)$ represents an observation generated by the probability distribution $p_i$. The decision is then made, defined as $d_N$ = attack hypothesis if $S(N) \geq \ln \dfrac{1-b}{a}$ or normal hypothesis if $S(N) \leq \ln \dfrac{b}{1-a}$. Variables $a$ and $b$ are the probability of false alarm and missed detection, respectively. $N$ is equal to an infinite set of $n$ such that $S(n)$ is not a false alarm or missed detection.

Change detection can be represented identically to the sequential detection solution using cumulative sum (CUSUM), $S(k+1) = \log \dfrac{p_1(z(k))}{p_2(z(k))} + S(k)$. A simple alteration is made such that $N$ now represents a set of $n$ such that $S(n)$ is greater than or equal to a given threshold. Detection is made by $d_N$ = attack hypothesis if $S_i(k) > \tau_i$ or normal hypothesis otherwise. $\tau_i$ represents a threshold of false alarms. Only small constraints are placed on the observation sequence, taken from the idea of nonparametric statistics, such that assumptions about the probability distribution for an attacker can be avoided.

Further establishing the simulation model, several types of attacks are considered to be used with the detection schemes. For the scope of this chapter, three types of attacks are used: surge, bias, and geometric attacks. Also, each attack is considered to be stealthy where the attacker has complete knowledge of the system parameters such as the linear model matrices $A$, $B$, and $C$. Surge type attacks model an attack whose aim is to inflict maximum damage once access to the system is achieved. Bias attacks describe small modifications made to the system through small disruptions. The last attack describes small modifications initially, before moving to inflict maximum damage once the system is vulnerable.

### 4.4.5 Experiment

In running the experiment, the TE-PCS model is used but replaced with a linear representation. In testing the system threshold, selected values of $\tau$ are tested against stealthy geometric attacks. In Figure 4.7, the states of the sensors for the duration of the attacks are shown. The dotted lines represent attack values, while the solid lines represent true values of the sensors.

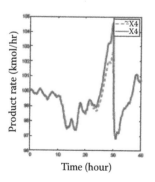 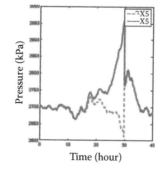 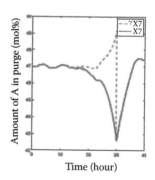

**Figure 4.7  Geometric attacks to sensors 4, 5, and 7.**

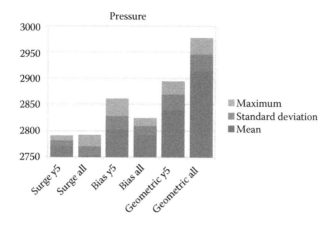

**Figure 4.8    Effects of attacks.**

The final results of the testing are shown in Figure 4.8, which detail the effects of all three types of attacks under two situations. The first situation is described as an attacker who has only compromised the vital sensor X5, and the second situation is the one in which the attacker has the control of all three and subsequently the detection statistic $S(k)$.

As inferred from the graph in Figure 4.8, surge and bias attacks have little effect on the chemical plant. This is due to the slow dynamics of the plant that inhibit the effects and allow the plant to respond and continue operating normally. Alternatively, geometric attacks have much higher effectiveness. Regardless, the detection algorithm still allows the plant to continue operating within safety limits even in the cases of stealthy attacks where the attacker has complete control and is able to solve for the linear model.

## 4.4.6  Attack Response

The proposed architecture for a response mechanism, based on the previous experiments in detection, is detailed in Figure 4.9. The input is fed into both the plant and a linear model that approximates the behavior. In the case of an attack, the anomaly detection module (ADM) can replace values of $\sim X(k)$ with $^{\wedge}X(k)$ if an attack is detected, where $^{\wedge}X(k)$ is the value from the linear model.

With this mechanism in place, attacks on the system are generally unable to push the system past 3000 kPa. This can be seen in Figure 4.10, which compares the systems with and without ADM.

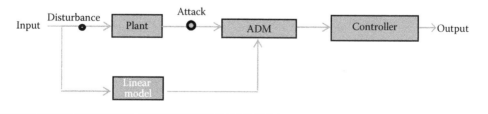

**Figure 4.9    System architecture with ADM.**

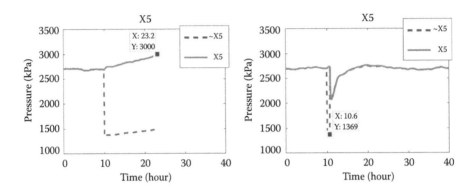

**Figure 4.10  Left: without ADM. Right: with ADM.**

## 4.5  Conclusions

In conclusion, CPSs have become an integral part of many vital areas of both the national and private infrastructure. These systems pose additional risks in that they contain actuators, which can affect their environment, effectively allowing the cyber damage to spill over into the real world. Securing the controls of the systems allows vital operations to remain intact through attacks such as DoS or false data injection attacks. Modeling CPS with LTI systems and attacking with various models that exploit certain vulnerabilities in their parameters allow cost functions relative to risk to be optimized such that both damage and cost are minimized. Given the modeling, security can be greatly enhanced through the use of modules that take advantage of the mathematical models to actively adjust system parameters as needed.

## References

1. R. Kisner et al. 2010. Cybersecurity through real-time distributed control systems. Technical Report ORNL/TM-2010/30. Oak Ridge National Laboratory, Oak Ridge, TN. Available at http://www.ornl.gov/sci/electricdelivery/pdfs/ORNL_Cybersecurity_Through_Real-Time_Distributed_Control_Systems.pdf.
2. L. Parolini et al. 2012. A cyber-physical systems approach to data center modeling and control for energy efficiency. *Proc. IEEE* 100, 251–268.
3. G. Befekadu et al. 2011. Risk sensitive control under a class of denial-of-service attack models. 2011 American Control Conference, San Francisco.
4. G. Befekadu et al. 2011. Risk-sensitive control under a Markov modulated denial-of-service attack model. In *50th IEEE Conference on Decision and Control and European Control Conference (CDC-ECC)*. IEEE, New Brunswick, NJ, 5714–5719.
5. A. Cardenas et al. 2011. Attacks against process control systems: Risk assessment, detection, and response. ASIACCS '11, Hong Kong.
6. Y. Mo and B. Sinopoli. 2011. False data injection attacks in control systems. 50th IEEE Conference on Decision and Control and European Control Conference, Orlando, FL.

## Chapter 5

# Apprenticeship Learning for Cyber-Physical System Intelligence

Kassie McCarley, Joseph Pierson, and Fei Hu

## Contents

## 5.1  Introduction

Mankind's ability to learn and progress their knowledge of a topic has allowed man to consistently achieve things thought to be unobtainable. For decades, machines have been programmed to achieve certain constrained tasks. How revolutionary would it be if machines could actively learn and apply new understandings to better carry out those tasks? Perhaps apprenticeship learning could provide a means to greatly impact the ability of a robot, a typical CPS that has the interactions between cyber algorithms and machine movements.

Apprenticeship learning [1] is the process of utilizing a reward function and a dynamics model to allow a robot to learn from an expert performing a task. A difficulty for apprenticeship learning is defining what doing a task "well" is. For example, what is considered "driving well" to one person is totally different to another. To one person, it may be traveling between two points without being involved in an accident. Another person may see it as getting from point A to point B in the quickest time possible. A third person may describe it as staying within the lane on the road. All of these viewpoints add up to form some of the important aspects of "driving well." Another challenge for artificial intelligence is providing an accurate and complete model in terms of data collection and model construction. To know how to perform a task, it must first be well defined. After the reward function and dynamics model are determined, a control scheme to achieve ideal conduct can then be delivered.

Traditionally, robots have been designated to do specific tasks. These tasks are monitored by controllers such as proportional integral derivative controllers or programmable logic controllers. These systems utilize sensors to verify the operation of the system within a specified range. These control methods limit the applicability of robots and are intended to maintain a more simplified system.

In most controllers, we desire a reward function that will determine a path to enter the needed state to perform our task. However, in many cases the reward function is not easy to specify manually. This is potentially one of the largest obstacles regarding reinforcement learning and optimal control algorithms. Instead of trying to describe doing a task well in a reward function, a demonstration from an expert is much easier to attain. These demonstrations can then be utilized alongside two types of reward functions. The first type compares linearly stated parameters. A discrete feature could be "if a car is in an accident" or "on the road." The second type of reward function is for tasks that must follow a trajectory. Plotting the trajectories manually is very time consuming. The alternative is to use expert demonstrations to find trajectories.

A sufficient dynamics model is also needed for robot control. If these dynamics are not known ahead of time, they must be found by observing the system. Exploration of the system presents a fundamental challenge in obtaining all of the data needed to create a substantial model of all relevant areas to a complete state space model. The $E^3$ algorithm and variations guarantee near optimal performance in time polynomial in the number of states in the system [1]. The algorithm ensures that the attainable portions of the state space model are explored until a sufficient model is in place. The emphasis of exploration is impractical for complex systems and can be very time consuming. Therefore these algorithms will not be used and will be replaced by expert demonstrations to further robot's self-learning. These demonstrations can be paired with a few trials to guarantee strong performance. Theoretically, these control policies will produce performances comparable to that of the expert. Both the expert and the robot performances will be conducted in the same environment and will attempt the same tasks. It is assumed that both performances utilize the same Markov decision process (MDP). Some system identification literature suggests building the dynamics model from a large amount of data. Reinforcement learning shows that massive data collection is unnecessary as an expert demonstration is sufficient. In place of data collection, a detailed dynamics model can apply control algorithms to generate similar performances in different environments from that of the demonstration.

Some applications that will be explored in this chapter involve reward functions for an autonomous helicopter as well as a ground-based automotive system. The car's reward functions depend on the course it is driving through. How close the car will stay to the lines and possibly guard rails is a good indication for the reward function. The combination of reinforced learning and optimal controls provides the best results for the helicopter's flying and stunts. Again, an expert will possibly be observed to demonstrate the applicable abilities of helicopter and remote control (RC) car when there is no specified, close-loop reward function. This procedure of observing the expert's way of piloting the helicopter helps create a reward function scenario. The observation of the expert's handling of the helicopter helps incorporate a control policy system for governing the overall performance and future usage.

In apprenticeship learning, a method of trying to obtain a reward function is to use the inverse reinforcement learning method. This method requires some trials and errors with multiple iterations of the algorithm to find an appropriate reward function. We can model the dynamic system via a first-order Markov model.

## 5.2 Apprenticeship Learning via Inverse Reinforcement Learning

The biggest challenge regarding reinforcement learning and traditional control methods is specifying a reward function. A good reward function needs to define doing a task "well" for all of the various potential states. One solution is to apply the MDP to create a reward function. The MDP is a discrete time stochastic control process in which a step-by-step action $A$ is chosen from multiple available actions in the given state $S$. It then goes on to the next state $S^1$. This new state gives two resulting functions, a reward function $R_a(S,S^1)$ and a state transition function $P_a(S,S^1)$. The probability of moving into the next state is given by the state transition function. A chosen action in the current state determines if the process will move on to the next state. Therefore the path is determined by the actions of a state. However, $S$ and $A$ are conditionally independent of all prior states and actions. Inverse reinforcement learning uses an MDP where an explicit reward function is not given. Instead, we observe the expert to demonstrate the task we desire our device to perform. This expert is considered a linear combination of known features to maximize the reward function. This algorithm should terminate in a minimal amount of iterations as it attempts to achieve a performance similar to that of the expert even though it may never find the expert's reward function.

Some of the difficulty of defining a reward function lies in the need to weight certain aspects of performing a task. This means manual tweaking of the function is often necessary, which creates a large obstacle in the applicability of reinforcement learning. Apprenticeship learning is the process of observing an expert and trying to imitate his or her behavior. This can lead to problems with robotics if the conditions are not exactly the same as the expert demonstration. An example is the constantly changing road conditions and traffic patterns. Therefore attempting to "learn" a reward function provides much greater applicability than simply stating one for a specific circumstance.

An MDP can be considered a tuple $(A, S, \gamma, T, R, D)$ [1], where $A$ is defined as a set of actions; $S$ is similarly a set of states; $\gamma$ is a discount factor where $\gamma \in [0,1)$; $T$ is a set of state transition probabilities dependent on an action $A$ in a state $S$; $T = \{P_{SA}\}$; $R$ is a reward function that must be bounded by absolute value to 1; and $D$ is an initial-state distribution where $s_0$ is determined. MDP/$R$'s, that is, an MDP without a reward function, is simply an MDP without $R$ in the tuple. It is assumed that a vector of feature $\phi : S \rightarrow [0,1)^k$ over many states. A true reward function is shown as $R^*(s) = \omega^{*T}\phi(s)$ given that $\omega^T \in \mathbb{R}^k$. It must also be assumed that $\|\omega^{\wedge *}\|_1 \leq 1$ to guarantee the reward function is bounded by 1. $\Phi$ could be a vector of features denoting different desired parameters used to determine the next step of a trade-off. The value $\omega^{\wedge *}$ is the vector weighting such aforementioned parameters. A policy $\pi$ is used to map the probability distributions of states over those of actions. The policy $\pi$ is defined as

$$E_{s_0 \sim D}[V^{\pi}(s_0)] = E\left[\sum_{t=0}^{\infty} \gamma^t R(s_t)\Big|\pi\right] = E\left[\sum_{t=0}^{\infty} \gamma^t \omega^T \phi(s_t)\Big|\pi\right]$$

$$\omega^T E\left[\sum_{t=0}^{\infty} \gamma^t \phi(s_t)\Big|\pi\right]$$

The state sequence $s_0$, $s_1$,... is formed starting from any state $s_0 \sim D$ while choosing an action according to $\pi$ to get the expectation [1]. The feature expectation $\mu(\pi)$ or expected discounted accumulated feature is defined as

$$\mu(\pi) = E\left[\sum_{t=0}^{\infty} \gamma^t \phi(s_t) \,\middle|\, \pi\right] \in \mathbb{R}^k.$$

This notation allows the policy to be written as $E_{s_0 \sim D}[V^\pi(s_0)] = \omega^T \mu(\pi)$. If the reward $R$ can be expressed as a linear combination of $\phi$, the anticipated sum of discounted rewards is set by the feature expectations of a certain policy $\pi$.

Let $\Pi$ represent a set of stationary policies if two policies $\pi_1$, $\pi_2$ are included in the set. A new policy $\pi_3$ can be created by placing the two policies together. If $\pi_3$ is a probability of two outcomes, a bias $\lambda$ is applied to $\pi_1$ and a probability of $(1 - \lambda)$ is given for $\pi_2$. Therefore $\mu(\pi_3) = \lambda\mu(\pi_1) + (1 - \lambda)\mu(\pi_2)$. This selection of each random step takes place only once at the start of the trajectory. If we want to extend the set of policies up to $\pi_d$ policies, we use a feature expectation vector for

the following convex combination: $\sum_{i=1}^{n} \lambda_i \mu(\pi_i) \left(\lambda_1 \geq 0, \sum_i \lambda_i = 1\right)$. Now all policies between

$\pi_1, \ldots, \pi_d$ are mixed and the probability of $\pi_i$ is found by $\lambda_i$. Assume availability to $\pi_E$, the optimal reward function generated by an expert starting at $s_0 \sim D$. The algorithm utilizes an estimate of the expert's feature expectations $\mu_E = \mu(\pi_E)$. If there are $m$ trajectories created by the expert, the empirical estimate is defined as

$$\mu_E = \frac{1}{m} \sum_{i=1}^{m} \sum_{t=0}^{\infty} \gamma^t \phi\left(s_t^{(i)}\right)$$

The algorithm is assumed to be solved using an MDP/R with a reward function $R = \omega^* \phi$.

The first algorithm is an MDP/R with feature mapping $\phi$, feature expectations for the expert $\mu_E$. Then a policy must be found for a performance similar to the expert's when the reward function is $R^* = \omega^{*T} \phi$. This policy must meet the following criteria: $\|\mu(\pi') - \mu_E\|_2 \leq \varepsilon$. For a policy $\pi'$ with any $\omega \in \mathbb{R}^k (\|\omega\|_1 \leq 1)$,

$$E\left[\sum_{t=0}^{\infty} \gamma^t R(s_t) \,\middle|\, \pi_E\right] - E\left[\sum_{t=0}^{\infty} \gamma^t R(s_t) \,\middle|\, \pi'\right]$$

$$= \left|\omega^T \mu(\pi') - \omega^T \mu_E\right|$$

$$\leq \|\omega\|_2 \|\mu(\pi') - \mu_E\|_2$$

$$\leq 1\varepsilon = \varepsilon$$

$\left|x^T y\right| \leq \|x\|_2 \|y\|_2$ allows the first inequality to appear. Similarly, the second inequality comes from $\|\omega\|_2 \leq \|\omega\|_1 \leq 1$. It simplifies the problem so that we are searching for the policy $\pi'$ that forms feature expectations $\mu(\pi')$ near $\mu_E$. This sets up the algorithm for $\pi'$ as follows:

1. At random, pick a policy $\pi^{(0)}$ and find $\mu^{(0)} = \mu(\pi^{(0)})$ when $i$ is set to 1.
2. Solve for a new attempt at the reward function by computing this convex programming problem:

$$\min_{\lambda, \mu} \|\mu_E - \mu\|_2$$

such that

$$\sum_{j=0}^{i-1} \lambda_j \mu^{(j)} = \mu$$

$$\lambda \geq 0$$

$$\sum_{j=0}^{i-1} \lambda_j = 1$$

Set $t^{(i)} = \|\mu_E - \mu\|_2, \omega^{(i)} = \dfrac{\mu_E - \mu}{\|\mu_E - \mu\|_2}$.

3. Terminate if $t^{(i)} \leq \varepsilon$.
4. Compute the optimal policy $\pi^{(i)}$ for an MDP with reward $R = (\omega^{(i)})^T \phi$.
5. Estimate $\mu^{(i)} = \mu(\pi^{(i)})$.
6. Increment $i$ by 1 and return to step 2.

When the algorithm completes, the attained policy $\pi'$ is a combination of all policies from 0 to $n$. Each policy is weighted by $\lambda_i$ and contains feature counts $\mu$ found within $\in$ of $\mu_E$. At iteration $i$, the algorithm has accumulated the policies for $\pi^{(0)}, \ldots, \pi^{(i-1)}$. The convex optimization problem in step 2 can be thought of as inverse reinforcement learning to attempt to find the expert's reward function. Feature counts near the expert's policy are constrained by the combined policies previously mentioned. When these feature counts are obtained, guessed reward weights $\omega^{(i)}$ are set along $\mu$ and $\mu_E$ [1].

The weight $\omega$ may also be found by solving this convex programming problem:

$$\max_{t, \omega} t$$

such that

$$\omega^T \mu_E \geq \omega^T \mu^{(j)} + t, j = 0, \ldots, i - 1$$

$$\|\omega\|_2 \leq 1$$

This algorithm is attempting to find the reward function $R = (\omega^{(i)})^T \phi$ so that $E_{s_0 \sim D}[V^{\pi_E}(s_0)] \geq E_{s_0 \sim D}\left[V^{\pi^{(j)}}(s_0)\right] + t$. This simply means we want the found reward to be within a margin $t$ to the $i$ policies previously attained.

The support vector machine (SVM) scheme uses an optimization to find a maximum margin hyperplane separating sets of points. To obtain equivalence, we associate $\mu_E$ with a label 1. There is also a label $-1$ consisting of a feature expectation $\{\mu(\pi^{(j)}): j = 0,\ldots, (i-1)$. The desired vector is $\omega^{(i)}$, a unit vector orthogonal to the maximum margin separating hyperplane. Thus, the SVM can solve for $\omega^{(i)}$.

If $t^{(n+1)} \leq \in$ when the algorithm finishes, then the following result occurs due to the optimization formula found in step 2:

$$\forall \omega \text{ with } \|\omega\|_2 \leq 1 \, \exists_i$$

such that

$$\omega^T \mu^{(i)} \geq \omega^T \mu_E - \varepsilon.$$

One policy must be returned from the algorithm. The policy's performance under $R^*$ is at least as good as the expert performance [1]. In other words, we can pick a policy with acceptable performance. As expected from before, the completed algorithm returns policy $\pi'$ with weight $\lambda_i$ and feature counts $\mu$. This policy will achieve performance relatively close to that of the expert. This algorithm does not always recover the reward function correctly. The performance is dependent on matching feature expectations, not uncovering the reward function [1].

To avoid the relatively complex scheme SVM, a simpler approach can be taken. The previous max-margin method can be replaced with a new projection method. In this projection method we can replace step 2 with the following formulae:

Set $\mu' = \mu'^{(i-2)} + \left( \dfrac{(\mu^{(i-1)} - \mu'^{(i-2)})^T (\mu_E - \mu'^{(i-2)})}{(\mu^{(i-1)} - \mu'^{(i-2)})^T (\mu^{(i-1)} - \mu'^{(i-2)})} \right)(\mu^{(i-1)} - \mu'^{(i-2)})$ (By using $\mu'^{(i-2)}$ and $\mu^{(i-1)}$,

the orthogonal projection of $\mu_E$ is computed.)
Set $\mu'^{(i-1)} = \mu'$ if $\mu'$ is found to lie on the convex hull of $\mu'^{(i-2)}$ and $\mu^{(i-1)}$. If not on the hull, set $\mu'^{(i-1)} = \arg\min_{x \in \{\mu'^{(i-2)}, \mu^{(i-1)}\}} \|\mu_E - x\|_2$.
Set $\omega^{(i)} = \mu_E - \mu'^{(i-1)}$.
And set $t^{(i)} = \|\mu_E - \mu'^{(i-1)}\|_2$.

During the first iteration, $\omega^{(1)} = \mu_E - \mu^{(0)}$ and $\mu'^{(0)} = \mu^{(0)}$.

To this point, it has been assumed that the algorithm terminates with $t \leq \in$. If the algorithm does not terminate or does not terminate in a small number of iterations, it loses its applicability. Therefore a Monte Carlo estimate $\mu'_E$ is used from $n$ expert demonstrations as shown in Theorem 5.1.

## Theorem 5.1

Assume an MDP/R with features $\phi \to [0, 1]^k$. One of the aforementioned algorithms may be run using an estimate $\mu'_E$ for $\mu_E$ with $m$ Monte Carlo samples. The algorithm terminates by satisfying the exit condition $t^{(i)} \leq \dfrac{\in}{4}$ at step 3 with a probability greater than $1 - \delta$ after $n$ or fewer iterations and satisfying the condition $\leq \dfrac{48k}{(1-\gamma)^2 \in^2}$. This also returns policy $\pi'$ so the true reward function $R^*(s) = \omega^{*T} \phi(s)$ ($\|\omega^*\|_1 \leq 1$). This gives

$$\left[\sum_{t=0}^{\infty}\gamma^t R^*(s_t)\Big|\pi'\right] \geq E\left[\sum_{t=0}^{\infty}\gamma^t R^*(s_t)\Big|\pi_E\right] - \varepsilon.$$

Then it suffices that

$$m \geq \frac{4k}{(1-\gamma)^2 \in^2}\log\frac{2k}{\delta}.$$

■

An MDP solver is then used when given an MDP/R and a reward function $R^*(s) = \phi^T\omega^{(i)}$ and gives a policy $\pi^{(i)}$ such that

$$\left[\sum_{t=0}^{\infty}\gamma^t R^i(s_t)\Big|\pi^{(i)}\right] \geq \max_{\pi} E\left[\sum_{t=0}^{\infty}\gamma^t R^i(s_t)\Big|\pi\right] - \frac{\varepsilon}{4}$$

If a true reward function $R^*$ does not lie on the span of basis functions $\phi$, this algorithm still has a smooth reduction of performance. If $R^*(s) = \omega^{*T}\phi(s) + \in(s)$ with any residual error $\in(s)$, then the algorithm will perform within the constraint $(O\,\|\in\|\,\infty)$ of the expert [1].

## 5.3 Exploration of Apprenticeship Learning in Reinforcement Learning

Reinforcement learning requires two aspects that are either intricately created per project or already known: (1) the reward function and (2) the dynamics model. The reward function can typically be molded by trial-and-error iterations of mathematical solutions, governed sometimes by an "experts" style. This style may provide reward functions that are unknown to another end user; however, they are developed throughout the course of experimentation and procedures. Examples of experiments given in [1] are the navigation of a car on a highway or parking lot, along with the autonomous piloting of an aircraft. When a structured reward function is developed, it can be used in conjunction with a much needed dynamics model. A dynamics model can also be developed based on observations of an expert's demonstration. Various algorithms are developed and used to create a dynamics model.

A dynamic model for a system can be created through MDP [1]. MDP is a mathematical procedure of decision making based on some known parameters and randomized parameters. The known parameters are controlled by the experimenter, while the randomized parameters can be state space variables from a reward function. A particular algorithm from MDP is used to ensure very detailed and accurate results covering all aspects for "state transition probabilities," known as $E^3$ [1]. This is a polynomial iteration method of finding an accurate representation of the MDP through various inaccuracies called "exploration policy." Afterward the maximum amount of positive rewards can be summed up over time to ensure the reward functions are included. This procedure of maximizing the reward functions is called "exploitation policy." Therefore, to ensure an accurate model, there are multiple attempts of using both the exploration and exploitation policies of $E^3$ to create a useable MDP. Some negatives in using $E^3$ occur in examples of testing an autonomous helicopter, where the exploration policy runs into trouble. When the area to which the helicopter is confined is breached, the helicopter will crash. It is the same when using this principle in regard to a chemical plant, if the limitations of this plant for various reasons are reached, the plant could explode. Therefore, with these instances of limitations when using the exploration policy, the $E^3$ is only mildly used.

To create a more acceptable dynamics model, an expert will demonstrate the basics of the dynamics for a system. The expert's methods and trajectories are recorded so they can later be recreated. Using what we have learned from the expert, a basic dynamic model is created with a reinforcement learning algorithm. Afterward the system has to be tested using the recreated dynamic model, with modifications being required if results are not as expected. Sometimes it is more beneficial to only use expert's demonstrations when the dynamics system becomes too complicated, ultimately reducing time wasted. A model may be used to predict failure if it can work in a simulated scenario but not a real-life scenario. This can help develop a good policy for later use.

More details of the parameters of an MDP for the dynamic model are necessary for understanding the concepts. The specific state space set in MDP can be written as $M = (S, A, T, H, D, R)$. In this tuple, some of the variables are controlled or already known, while one is random or found with some iteration method. The known or controlled variables are $S, A, H, D, R$, while the $T$ variable is more difficult to obtain. $S$ represents state space, while $A$ stands for action space. The $H$ and $R$ are either generated by the expert's demonstration or manually encoded, while the $D$ is also known or solved with the various other parameters. In $M$, we want to approximate $T$ by a closeness of the value $\pi$.

$T$ can be derived by using two different methods. The first uses discrete dynamics model, initializing $S$ and $A$ with values, and determining $T$ based on current $(S)$ and next state value $(S')$. The next method is the linearly parameterized dynamics, where the state–action pair is a continuous value instead of discrete. This is used to create a linear function based on a nonlinear function, which may include some noise.

## 5.4 Gridworld

An interesting application called Gridworld is introduced in [1] to show the efficiency of apprenticeship learning (see Figure 5.1). The experiments were on the parking of a car, possibly in an open environment first then to a very crowded parking area. They also carried out experiments on quadruped locomotion across various types of environments.

To simulate the reward functions and highway driving models, Gridworld uses a graphical user interface that allows various settings and parameters to be adjusted. It was originally used to understand animals and their abilities such as flight. The structure of Gridworld is a square, for example, 128 × 128, for the highway driving simulation. There are four possible movements for the navigation of the car: north, south, east, and west. The mentioned sampling involves taking various movements that use a 30% fail rate to result in a random direction of movement. Within this large 128 × 128 workspace, there are subsections of 16 × 16 grids, called "macrocells." These macrocells make it easier to determine the exact location of reward functions based on the movements generated by the user. Each of these macrocells are given a number from $i = 1$ through 64. The reward is equal to the (randomly generated weight) times the instance of that macrocell. While one method of simulating this driving model used all of the encompassing macrocells, an easier model uses only the macrocells with reward functions. The reason for narrowing the possible iterations using only known reward functions makes the determination of the convergence of the sampling size easier.

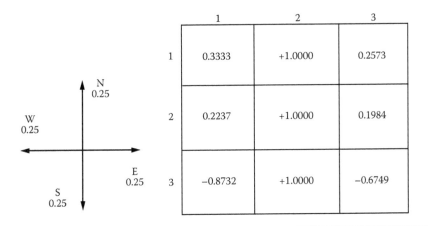

**Figure 5.1   Gridworld diagram. Each direction has a 25% chance of occurring as indicated on the compass on the left. However, different reward function values are assigned to each block so that some moves will be valued more than others. If this diagram was for a car intended to drive in a straight line, the reward functions for going straight are always much higher than those to the sides.**

## 5.5  Car Driving and Parking Application

The car driving simulation model is implemented to compare various driving "styles" [1]. There are various actions that the user can make to the car to avoid traffic, some of which cause the car to drive off road momentarily. Some of the parameters include the distance of a car to another, which lane the car is currently in, and the speed at which the car is traveling. Ultimately, five styles were determined for usage. The first is described as a "nice" style, which is indicated as most appropriate driving, avoiding collisions and staying in the right lane. The second indicated style was called "nasty"; this in a sense is a free-for-all, hit any vehicle style. The third style is similar to nice and is called "right lane nice," where the parameters are tweaked to perform off the road movement to avoid collision. The fourth style is a mixture of the second and third styles called "right lane nasty," where the car can drive off the road on the right but can also cause collisions in the right lane. Finally, the fifth style deviates from the norm and is called "middle lane"; this style stays solely in the middle and can crash into oncoming traffic.

The next phase of the experiments is the parking lot simulation. In understanding the concept of driving in the parking lot, the vehicles' coordinates have to be approximated using the state space idea $<x, \theta, d>$, where $x$ indicates the $x$ or $y$ coordinate, $\theta$ indicates the orientation, and $d$ is the Dirac delta function indicating 0 for forward movement and 1 for reverse. The lanes are given their own coordinates as well, as indicated in $G = <V, E>$, so the $x$ coordinate of the car can be found on the lane by $D(x, G) = \min D(E, x)$, indicating the minimum distance of the car of $x$ in location to the $G$ and $E$ parameters relating to the lanes [1]. The orientation of the car also matters to determine whether the car is going in the right direction in the lane, given by the function $D(x, \theta, G) = \min D(E, x)$. This incorporates the minimum distance of said points of the car with relevance of the lanes, also including the orientation of the car. These formulae can later be used to calculate the weight distribution of various parameters such as the amount of time driving in a certain direction, when lane changes occur, and various other scenarios.

For other experimental purposes, an expert driver brought in his expertise to drive three particular styles for comparative results. These styles, which are similar to those previously described, are "nice," "sloppy," which makes it okay to drive in the other lane at times yet only in the forward direction, and finally "backwards" for a short amount of time [1]. Five iterations were collected for each driving style for learning algorithms. Using various weight formulas, the idea is to try to "match" the simulated models with the experts instead of trying to directly copy it. This matching requires many iterations to emulate the expert's method or style; however, this may ultimately prove more effective than standard apprenticeship learning. In determining the cost function for each demonstration, the following math formulae are used. Originally, just one of the formulas was used, but three are used for a more accurate approximation. The original formula used was the last one, equating $w$ to $\mu - \mu_E$.

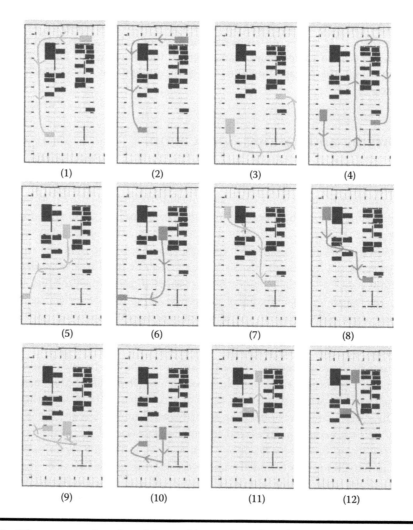

(1)   (2)   (3)   (4)

(5)   (6)   (7)   (8)

(9)   (10)   (11)   (12)

**Figure 5.2  Diagrams are in pairs of the parking styles. The odd diagrams are the expert demonstrations. The even diagrams are the car's attempt to mimic the expert. The first row is "nice," the second is "sloppy," and the third is "reverse."**

$$\min_{w,x} \quad \left\| \omega \right\|_2^2$$

such that

$$\mu = \sum_i x_i \mu^{(i)}; \quad x \geq 0$$

$$\sum_i x_i = 1; \quad \omega \geq 0$$

$$\omega \geq \mu - \mu_E; \quad \omega \in W$$

These formulae limit the weight to a nonnegative value. Figures 5.2 through 5.4 show the learning effects from an expert's demo.

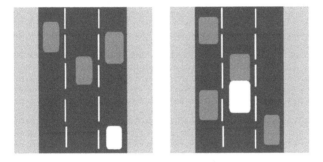

**Figure 5.3** **The left driving simulation shows the white car exhibiting a nice driving style and staying away from traffic and in the left lane. The right simulation shows the white car exhibiting the nasty driving style and crashing into a car in its path while surrounded by other cars.**

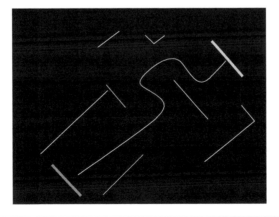

**Figure 5.4** **Car obstacle detection. The car must be able to detect barriers between its starting point, and the end point. There are obstacles that the car must be able to detect and avoid in order to make it to the desired location without a collision. The better the reward functions, the better the performance of the car based on the policies obtained from the expert.**

## 5.6 Helicopter Flight with Parameterization

One application for the reinforced learning model is the helicopter simulation. To approximate a fairly accurate model, a linear system has to be created to understand the principles of helicopter navigation. Some aspects of the helicopter flight cannot be properly simulated because of the non-linear properties such as "sideslip," which is the helicopter's ability to move in a certain direction but not maintain a straight-line path. This nonlinear concept is due to inertia. Another concept in forming a dynamic system for a helicopter's flight pattern is the use of global nonlinear parameterization that allows a general structure for parameterizing a dynamic system such as the helicopter, in search of more detailed and accurate representations of such values, leading to localized nonlinear parameterization, which uses the weight system to determine more accurate ways to portray the helicopter simulations.

There are certain standards and guidelines to go by in developing a helicopter or rotating in-flight vehicle, such as Comprehensive Identification from Frequency Responses (CIFER). To obtain linear models to accurately represent the helicopter, CIFER predicts models with the frequency response of a system. It first finds the test data to characterize a model. CIFER uses data to invert their learning by modeling what they learned, through the above-mentioned apprenticeship learning algorithm (Figure 5.5). More specifically, CIFER uses a MIMO (multiinput/multioutput) nonparametric frequency response system [2].

To further simplify helicopter simulations and create an accurate representation, using certain dynamic system parameters can help eliminate possible headaches. First, the state of the helicopter, $s$, involves 12 total parameters. The first set, $\{x, y, z\}$, indicates the position. The second set $\{\Phi, \theta, w\}$ represents roll, pitch, and yaw, respectively, for orientation of the helicopter. More specifically, roll, pitch, and yaw entail the center of mass of the helicopter to help balance out the weight and center it while at flight. The next set of space parameters are for velocity $\{\dot{x}, \dot{y}, \dot{z}\}$ for each direction of motion. Finally, the last set of parameters used to describe the state space model is for angular velocity $\{d\Phi, d\theta, dw\}$. These parameters are now understood in a four-dimensional space to account for motion. The first two motions $u_1$ and $u_2$ account for left/right and backward/

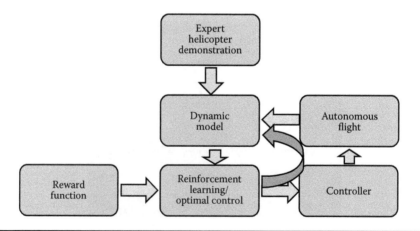

**Figure 5.5 Logic diagram of helicopter apprenticeship learning. The helicopter observes the expert's flight and builds a control scheme based on the reward function and dynamic model. It then uses a loop, in which it attempts a flight, and if it is successful, the model is done; if not, more data must be used to enhance the flight of the helicopter.**

forward motions. The value $u_3$ identifies the yaw or turn radius of the helicopter. The last space variable $u_4$ identifies with the thrust produced by the pitch of the helicopter blades. To help organize and simplify this large list of variables, the 12 can be cut down to eight. The remaining eight parameters are $\{\Phi, \theta, \dot{x}, \dot{y}, \dot{z}, d\Phi, d\theta, dw\}$. Specifying the eight parameters helps understand the location accurately of the helicopter to derive a more understandable approach to its flight pattern.

So how does someone accurately represent a helicopter's ability to sideslip? Using an acceleration prediction model, one can access how the helicopter moves in a more linear fashion even with nonlinear characteristics, such as moving in a direction without the helicopter's alignment following suit. A prediction-based model is the simplest form of helping understand a concept such as sideslip.

Minimizing the possibility of errors in a helicopter's flight trajectory is the key in creating an accurate model. In doing so, simulations may be needed to provide accurate results. The time duration for a simulated model in terms of real-life representations is skewed; therefore a math model has to be created to synch these concepts. Instead of using a short instant of a simulation, a larger time span may be needed for further understanding of errors. The equation used for this is called lagged criterion [1]. Essentially, this represents the summation of time $T$ with horizon time $H$, and calculates the "one step error" of each scenario for simulation and real-time scenario.

Experiments are necessary to ensure the models used for representing a helicopter are fairly accurate. In order to make accurate representations, two model RC helicopters were tested with various dynamic models to compare simulation results. Hardware was also used along with the RC helicopters to track GPS positioning with the Novatel RT2 and determine the orientation for helicopter body coordinates using Microstrain 3DM-GX1. Some of the models used for real-world simulation were the linear one step and the standard linear CIFER as previously mentioned. The duration of the flight for RC XCell Tempest was 800 s, and second run was 540 s. The other helicopter, Bergen Industrial Twin, had two runs for 110 s. From results, the best linear result producing model was that of the linear-lagged, while for a nonlinear acceleration model, the acceleration-lagged model performed the best. These lagged modeling systems seem to most accurately model the helicopter's flight pattern.

The two RC helicopters have to be programmed and used autonomously to fulfill the apprenticeship learning and reinforced learning procedures. The GPSs implemented into the RC helicopters help maintain a level of boundaries for the autonomous flight patterns and maneuvers. An antenna is used to track the orientation of the helicopter, with a small frequency of 10 Hz [1]. Also, two cameras are used to determine the positioning of the helicopter while in flight. A radio transmitter is used to send control signals to the helicopters for the four parameters $u_1$, $u_2$, $u_3$, and $u_4$ previously described. There are serial port connections to a PC to create control signals for the flight patterns and a base station to record the data after the helicopter makes its flight.

## 5.7 Conclusions

In this chapter, for typical CPS robots, many things were explored in understanding robotic controls and learning algorithms. We need to simulate both linear and nonlinear functionality of autonomous-by-nature robotic controls. An expert presents a demonstration in the apprenticeship learning model. This model can be used with reinforced learning to develop necessary reward functions and develop a useful dynamic system for an applicable model. The demonstrator for apprenticeship learning model helps develop a "teacher's policy." The applications presented here include car navigation, car parking, and helicopter flight.

The apprenticeship model requires an expert to properly demonstrate certain maneuvers for simulation purposes. The use of the software Gridworld helped develop policies for car navigation. With the inclusion of reward functions overtime, the accuracy of good driving could be translated. Many driving "styles" are also developed through using Gridworld; these include "nice," "right lane nasty," and "nasty." All have various parameters or weights indicating the distance the car is from the other lane or driving off the road.

The next application developed with apprenticeship learning was car parking. Because car navigation required certain parameterization, so did parking. The main focus was to avoid obstacles based on teacher's policies and reward functions.

Last, the best application to use with apprenticeship learning was helicopter flight. The flight pattern required expert demonstration resulting in teacher's policy, with a good understanding of actual helicopter flight. Parameterization for the helicopter's flight pattern was the big challenge. A GPS was used to locate the helicopter and determine its flight pattern. The used helicopters for demonstration purposes were the Bergen Industrial Twin and XCell Tempest.

# References

1. P. Abbeel. 2008. Apprenticeship learning and reinforcement learning with application to robotic control, PhD diss. Stanford University, CA.
2. University Affiliated Research Center. *Flight Control CIFE*, NASA Ames Research Center. Available at http://uarc.ucsc.edu/flight-control/cifer/index.shtml.

## Chapter 6

# Application of HDP-HMM in Recognition of Dynamic Hand Gestures

Lv Wu, Ting Zhang, and Fei Hu

## Contents

## 6.1 Introduction

Nowadays, stroke is the leading cause of disability among adults in the United States, and more than 4 million people in the United States have suffered from a stroke [1]. Currently, the entire nation spends more than $10 billion per year for visiting poststroke rehabilitation experts such as physical therapists (PTs) [1]. It is important to accurately recognize complex data glove gestures (static) and motions (dynamic). Virtual reality (VR) is a typical cyber-physical system (CPS) because of the interactions of physical objects (humans) with a virtual cyber world. However, most of the existing VR systems can only recognize static gestures very well (such as 26 letters).

Automatic hand rehabilitation training (rehab-training) via VR for poststroke patients could help to significantly reduce the high medical cost as in conventional manual therapy.

More recent studies [2–4] in gesture recognition have focused on the use of hidden Markov model (HMM) and support vector machine (SVM) for pattern recognition. These approaches have produced some good results in terms of recognizing hand gestures [5]. However, they still cannot handle complex hand gestures. In this work, we will show that giving a prior to the HMM could make the results promising. Fox et al. [6] used the improved hierarchical Dirichlet process (HDP)–HMM for tracking maneuvering targets. In addition, she proposed a sticky HDP-HMM [7], which improved self-transition of the state mode effectively. It was shown that the HDP-HMM approach could yield a state-of-the-art diarization. Goldwater et al. [8] proposed a statistical approach to the word segmentation issue based upon the HDP-HMM.

The main focus of this chapter is on the development of a hierarchical Bayesian model for data clustering of unknown gesture movements. Nonparametric models constitute an approach to model selection and adaptation where the size of models is allowed to grow with data size. In this research the unknown cardinality of the number of gestures motivates our nonparametric Bayesian approach, which places a flexible, data-driven HDP prior on the hidden states. The Dirichlet process (DP) simplifies models of gesture being classified. It also allows additional states to be created as new gestures are observed. Its hierarchical structure accounts for temporal correlation of input modes. In our experiment, we explore the use of the HDP as a prior on the unknown number of unobserved input modes in hand gesture applications. When combining with an observation likelihood distribution, we can obtain an HDP mixture model.

## 6.2 Theory Foundations

### 6.2.1 Dirichlet Process

Bayesian nonparametric generalizations of finite mixture models provide an approach for estimating both the number of components in a mixture model and the parameters of the individual mixture components simultaneously from data. Finite mixture models define a density function over data items $x$ of the form $p(x) = \int p(x|\theta)G(\theta)\mathrm{d}(\theta)$, where $G = \sum_{k=1}^{K} \pi_k \delta_{\theta_k}$ is a discrete mixing distribution encapsulating all the parameters of the mixture model and $\delta_{\theta_k}$ is a Dirac distribution (atom) centered at $\theta$. Bayesian nonparametric mixture uses mixing distribution consisting of a countably infinite number of atoms instead of $G = \sum_{k=1}^{\infty} \pi_k \delta_{\theta_k}$, which gives rise to mixture models with an infinite number of components [9]. Being Bayesian, we use a prior over mixing distribution $G$ and the most common prior is DP.

A Dirichlet process DP ($\lambda$, $H$) parameterized by a concentration $\lambda > 0$ and a base distribution $H$ is a prior over distribution $G$. For any finite distribution $A_1,...A_m$ of the parameter space, the induced random vector ($G(A_1),...G(A_m)$) is Dirichlet distributed with parameters ($\lambda H(A_1),...\lambda H(A_m)$). It can be shown that draws from a DP are discrete distributions. The DP also induces a distribution over partitions of integers called Chinese restaurant process (CRP), which directly describes the priors over how data items are clustered under the DP mixture.

The DP mixture model can be depicted as a graphical model (see Figure 6.1) [10]. To generate observations, we choose $\theta_i' \sim G_0$ and $y_i \sim F(\theta_i')$ for an indexed family of distributions $F(\cdot)$. This sampling process is also often described in terms of the indicator random variables $z_i$. Particularly, we have $z_i \sim \beta$ and $y_i \sim F(\theta_{z_i})$. The parameter with which an observation is associated implicitly

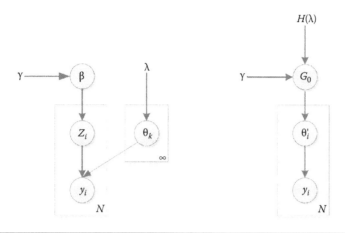

**Figure 6.1    DP mixture models represented in two different ways.**

partitions or clusters the data. When applied to a finite training set, only a finite number of DP mixtures are used to model the data because each mixture part can be associated with multiple data items.

In addition, the CRP representation indicates that the DP provides a prior that makes it more likely to associate an observation with a parameter to which other observations have already been associated. This reinforcement property is essential for inferring finite, compact mixture models. It can be shown under conditions that if the data were generated by a finite mixture, the DP posterior is guaranteed to converge to that finite set of mixture parameters. Inference in the model then automatically recovers both the number of DP mixtures and the parameters of the DP mixtures.

## *6.2.2  Hierarchical Dirichlet Process*

The HDP is applicable to general problems involving related groups of data, each of which can be modeled using a DP, and we begin by describing the HDP at this level of generality, subsequently specializing to the HMM. The HDP model can be depicted graphically in two different ways (Figure 6.2) [10].

To describe the HDP, suppose there are $J$ groups of data and let $\{y_{j1},\ldots y_{jNj}\}$ denote the set of observations in group $j$. Assume that there is a connection of DP mixture models underlying the observations in these groups:

$$G_j = \sum_{t=1}^{\infty} \tilde{\pi}_{jt}\delta_{jt^*}, \tilde{\pi}_j \Big| \alpha \sim GEM(\alpha), j = 1,\ldots,J \tag{6.1}$$

$$\theta'_{ji}\Big|G_j \sim G_j, y_{ji}\Big|\theta'_{ji} \sim F(\theta'_{ji}), j = 1,\ldots,J, i = 1,\ldots,N_j \tag{6.2}$$

We wish to tie the DP mixtures across the different groups such that atoms under the data in group $j$ can be used in group $j'$. The problem is that if $G_0$ is absolutely continuous with respect to

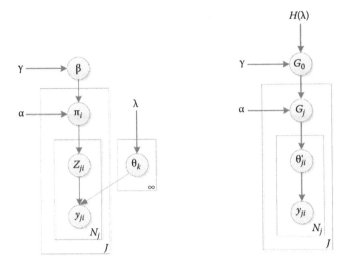

**Figure 6.2 HDP mixture models represented in two different ways.**

the Lebesgue measure, then the atoms in $G_j$ will be distinct from those in $G_j'$ with probability 1. The solution to this problem is to let $G_0$ itself be a draw from a DP:

$$G_0 = \sum_{k=1}^{\infty} \beta_k \delta_{\theta_k}, \ \beta \mid \gamma \sim GEM(\gamma), \tag{6.3}$$

$$\theta_k \mid H, \lambda \sim H(\lambda), \ k = 1, 2,\ldots \tag{6.4}$$

In the hierarchical model, $G_0$ is atomic and random. Let $G_0$ be a base measure for the draw $G_j \sim DP(\alpha, G_0)$, which implies that only these atoms can appear in $G_j$. Thus, atoms can be shared among the collection of random measures $\{G_j\}$.

## 6.2.3 HMMs with HDP

HMMs are widely used to model sequential data and time series data. An HMM is a stochastic Markov chain in which a state sequence $\theta_1, \theta_2,\ldots\theta_\tau$ is drawn according to a Markov chain on a discrete state space $\Theta$ with transition distribution $\pi(\theta_t, \theta_{t+1})$. A corresponding sequence of observations $y_1, y_2,\ldots y_\tau$ is drawn conditionally on the state sequence, where for all the observations, $y_t$ is conditionally independent of other observations given the state $\theta_t$. We let $F_{\theta_t}(y_t)$ denote the distribution of $y_t$ conditioned in the state $\theta_t$; this is referred to as the "emission distributions." Then we can utilize the HDP-HMM to describe our problem.

A classical HMM is used to specify a set of finite mixture distributions, one for each value of the current state $\theta_t$. Given $\theta_t$, the observation $y_{t+1}$ is chosen by first picking a state $\theta_{t+1}$ and then choosing $y_{t+1}$ conditional on that state. Thus, the transition probability $\pi(\theta_t, \theta_{t+1})$ plays the role of a mixing proportion and the emission distribution $F_{\theta_t}$ plays the role of the mixture component. It is natural to consider replacing this finite mixture model by a DP mixture model. However, if these DP mixture models are not tied in some way, then the set of states accessible in a given value of the

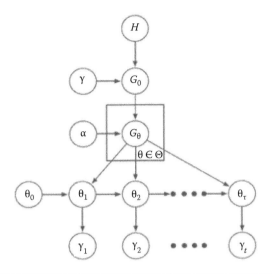

**Figure 6.3   HDP-HMM.**

current state will be disjoint from those accessible for some other value of the current state. Then, we would obtain a branching structure rather than a chain structure. Hence, the solution is to use the HDP to tie the DPs. Figure 6.3 shows the graphical model of HDP-HMM.

More formally, let us consider a collection of random transition kernels, drawn from HDP:

$$G_0 | \gamma, H \sim DP(\gamma, H) \tag{6.5}$$

$$G_\theta | \alpha, G_0 \sim DP(\alpha, G_0) \tag{6.6}$$

where $H$ is the base measure on the probability space $(\Theta, T)$. The random base measure $G_0$ allows the transitions out of each state to share the same set of the next states. We could obtain the following results from the graphical model (see Figure 6.3):

$$\theta_t \big| \theta_{t-1}, G_{\theta_{t-1}} \sim G_{\theta_{t-1}} \tag{6.7}$$

$$y_t \big| \theta_t \sim F_{\theta_t} \tag{6.8}$$

The HDP prior makes states to have similar transition distributions because of $\pi_k \sim DP(\alpha, \beta)$. However, it does not differentiate self-transition from moves between different states. To solve the problem of state redundancy, Fox et al. [11] proposed a sticky HDP-HMM (see Figures 6.4 and 6.5) to solve this problem. The corresponding transition distribution is [12]

$$\beta | \gamma \sim GEM(\gamma) \tag{6.9}$$

$$\pi_j \bigg| \alpha, k, \beta \sim DP\left( \alpha + k, \frac{\alpha\beta + k\delta_j}{\alpha + k} \right) \tag{6.10}$$

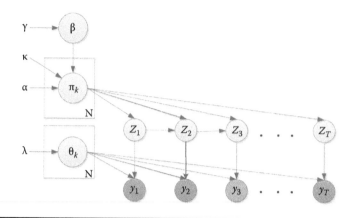

**Figure 6.4   Sticky HDP-HMM.**

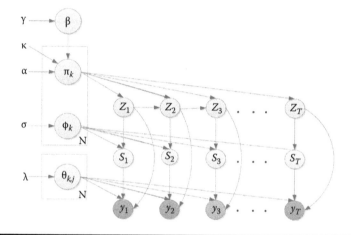

**Figure 6.5   Sticky HDP-HMM with DP emission.**

# 6.3 Experiments

## 6.3.1 Digital Glove

The AcceleGlove [13] is a complete development solution for adding hand motion input or control to a range of applications in the fields of robotics, medical rehabilitation, and telemedicine; and training, gaming, and VR/simulation environments. The AcceleGlove integrates sensors, called accelerometers, within a lightweight, flexible glove. It captures hand and finger positions and motion data. The glove transmits hand motion and orientation data to a PC over a USB cable or uses the optional AcceleGlove Wireless Module over wireless Bluetooth.

The AcceleGlove has six small accelerometers, one for each finger and one for the back of the palm, all integrated into the glove to detect finger and hand positions and motions. Each accelerometer has three sensing elements, corresponding to the $X$, $Y$, and $Z$ signal output from the AcceleGlove. If the hand is horizontal, the $Z$-sensing element is oriented along an axis ("the gravity vector") that is perpendicular to the Earth's surface. $X$ and $Y$ both lay in plane that is perpendicular to the $Z$ axis, offset from each other by 90°.

The AcceleGlove assigns the following coordinates to the system of accelerometers: tilting the AcceleGlove away from level about the $Y$ axis or accelerating the hand in the $X$ direction will result in a change of the accelerometer signal on the $X$ axis (wrist flexion or extension) or moved with acceleration in the $Y$ direction. The $Z$ axis will register a response when the hand is accelerated up and down or is rotated about $X$ axis and/or $Y$ axis.

## 6.3.2 *Using the AcceleGlove Visualizer to Capture and Output Data*

The AcceleGlove Visualizer is a multipurpose application that works with the AcceleGlove. It provides the ability to view the output values of each accelerometer's reading along the $X$, $Y$, and $Z$ axes (see Figure 6.6). The visualizer also can be used to capture and define gesture libraries, recognize gestures, and display gesture recognition probability, data output, and diagnostics.

We use the AcceleGlove Visualizer to capture and export data from the AcceleGlove. In our experiments, we want to recognize the hand gesture of grasping a cup and then putting the cup back to the original place. The glove data are captured as shown in Figure 6.7.

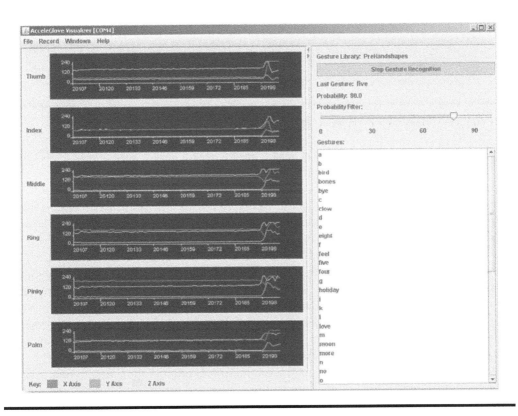

**Figure 6.6 AcceleGlove Visualizer.**

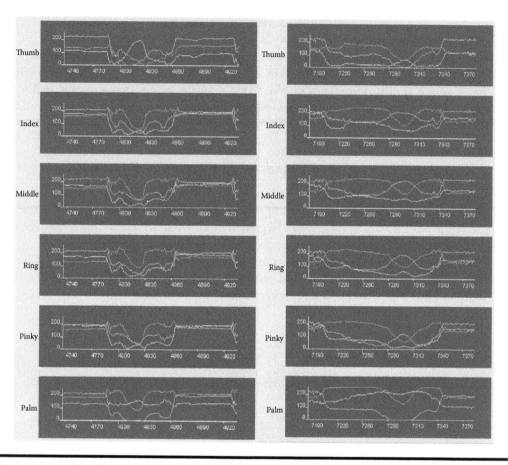

**Figure 6.7   Glove data for grasping a cup.**

## 6.4  Results

We propose a new method for the recognition of complex hand gestures/movements by using an unsupervised Bayesian learning model, that is, HDP-HMM. The construction of an HMM architecture is used to describe the relationship between a complete hand event (such as the process of "grasping cup") and small hand movements. Through an HDP prior that affords a rich set of transition dynamics, the infinite hidden states can robustly adapt to a new observation. Using a truncated approximation to the full Bayesian nonparametric model, we use the blocked Gibbs sampling algorithm, which leverages forward–backward recursions to jointly resample the states and emission assignments for all observations.

At this time, we have explored the performance of the above model in a series of experiments with synthetic data. The real data retrieved from the digital glove are under further experimentation and will be published in other papers.

In addition to the HDP prior, we placed a gamma(1, 0.4) prior on the concentration parameter and a normal inverse-Wishart prior on the space mean and variance parameters, and set the hyperparameters as 0.01 pseudocounts, with 10 degrees of freedom.

Figures 6.8 through 6.11 show the performance of the sticky and original HDP-HMM with Gaussian emissions. In our experiments, we consider the truncation level as five different

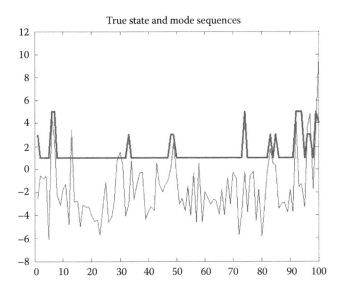

**Figure 6.8**   **Observations with state sequence (bottom curve) and mode sequence (up curve).**

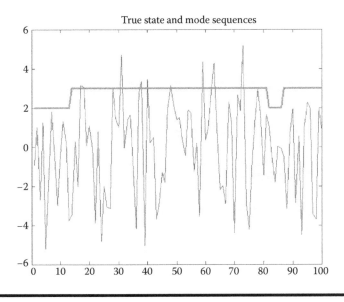

**Figure 6.9**   **Observations with state sequence and mode sequence on sticky model.**

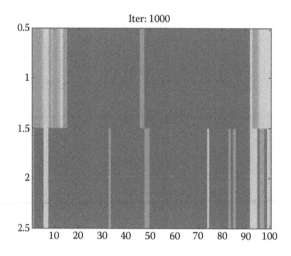

**Figure 6.10    Segmentation of estimated (top with 1000 iterations) and true (bottom) state sequences.**

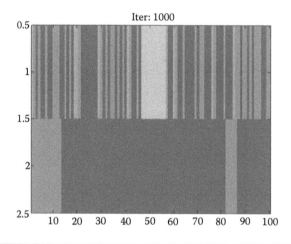

**Figure 6.11    Segmentation of estimated (top with 1000 iterations) and true (bottom) state sequence (for sticky model).**

movements (such as moving forward, moving up, holding up, and moving down). We can see that the sequence in Figure 6.8 is more sensitive than that in Figure 6.9. Any jitter can be captured, which is not suitable to the movement of the hand. At the same time, we could get more hand movement information from Figure 6.8 than from Figure 6.10.

## 6.5  Conclusions

Data clustering is an important activity in many large-scale data analysis problems in engineering and science, reflecting the heterogeneity that is often present when data are collected on a large scale. In this chapter we have shown that clustering problems can be solved within a probabilistic framework via finite mixture models. In recent years we have seen numerous examples of applications of finite mixtures and their dynamical cousins—the HMM in many areas such as bioinformatics, speech recognition, information retrieval, and computer vision. These areas also provide numerous instances of data analysis that involve multiple linked sets of clustering problems, for which clustering methods provide little in the way of leverage [5].

We have described how the Bayesian nonparametric approach could be used in clustering different gestures effectively. We have just sampled part of the data from the digital glove. Nonparametric models constitute an approach to model selection and adaptation, where the size of models is allowed to grow with data size. So the model discussed is flexible and data driven. It means we could still cluster the movements among the fingers, and we will pay more attention to this in the next step. Results on synthetic data clearly demonstrate the practical value of our work.

# References

1. Stroke statistics, published by the University Hospital. See the following site for more details: http://www.theuniversityhospital.com/stroke/stats.htm.
2. A. Akl, C. Feng, and S. Valaee. 2011. A novel accelerometer-based gesture recognition system. *IEEE Trans. Signal Process.* 59:12, 6197–6205.
3. J. Liu and M. Kavakli. 2010. Hand gesture recognition based on segmented singular value decomposition. In *Knowledge-Based and Intelligent Information and Engineering Systems,* Lecture Notes in Computer Science, vol. 6277, R. Setchi, I. Jordanov, R. Howlett, and L. Jain (Eds.). Springer, New York, pp. 214–223.
4. X. Zhang, X. Chen, W. W. Hui, J. Y. Hai, V. Lantz, and K. W. Qiao. 2009. Hand gesture recognition and virtual game control based on 3D Accelerometer and EMG sensors. In *Proceedings of the 13th International Conference on Intelligent User Interfaces Canary Islands*, Spain—January 13–16, 2008, pp. 401–406.
5. J. Yang and Y. Xu. Hidden Markov model for gesture recognition. Technical report. Robotics Institute, Carnegie Mellon University, Pittsburgh, PA.
6. E. Fox, E. B. Sudderth, and A. S. Willsky. 2007. Hierarchical Dirichlet process for tracking maneuvering targets. In *Proceedings of the 10th International Conference on Information Fusion*. IEEE, New Brunswick, NJ.
7. E. B. Fox. 2009. Bayesian nonparametric learning of complex dynamical phenomena. Ph.D. diss. Department of Electrical Engineering and Computer Science, Massachusetts Institute of Technology, Cambridge.
8. S. Goldwater, T. L. Griffiths, and M. Johnson. Contextual dependencies in unsupervised word segmentation. In *Proceedings of the 21st International Conference on Computational Linguistics and 44th Annual Meeting of the Association for Computational Linguistics*. Association for Computational Linguistics Association for Computational Linguistics, Stroudsburg, PA, USA, 2006.
9. P. Orbanz and Y. Whye. 2010. The Bayesian nonparametric models. In *Encyclopedia of Machine Learning*. Springer, New York.
10. Y. W. The, M. I. Jordan, M. J. Beal, and D. M. Blei. 2006. Hierarchical Dirichlet process. *J. Am. Stat. Assoc.* 101:476, 1566–1581.
11. E. B. Fox, E. B. Sudderth, M. I. Jorden, and A. S. Willsky. 2008. A sticky HDP-HMM with state persistence. Paper presented at 28th International Conference on Machine Learning, Helsinki, Finland. sponsored by Microsoft, Yahoo, etc.
12. E. B. Fox, E. B. Sudderth, M. I. Jorden, and A. S. Willsky. 2011. A sticky HDP-HMM with application to speaker diarization. *Ann. Appl. Stat.* 5:2A, 1020–1056.
13. AcceleGlove user guide. AnthroTronix Corporate.

*Chapter 7*

# On Modeling Issues in Cyber-Physical Systems

Michael Johnson, Tony Randolph, and Fei Hu

## Contents

## 7.1  Introduction

Physical processes are made up of a combination of different processes that run in parallel with each other. The job of measuring and controlling these processes by orchestrating actions that influence the processes is a very important task that is performed by an embedded system. Models are a major stepping stone in the development of cyber-physical systems (CPSs), and there are several advantages to working with models. To start, models can be declared to be deterministic. In other words, the output of the model will always be the same for a certain input. Different inputs will have different outputs. Models play an important role in a design process, which include model-based design and model-driven design [1].

Models also show how the design process has evolved, and models help to form the specifications that govern a system. In addition, models allow a design to be tested in a safe environment, which will allow engineers to determine if any design defects exist. This is a great benefit before a prototype of the actual system is developed. To model a CPS, engineers will have to include the models of the physical processes as well as models of the software, computation platforms, and networks.

In addition, feedback loops will have to include any sensors or actuators that will have a direct effect on the system. One example of software that is used for modeling is Ptolemy II, which can model components that use communication ports to communicate with the rest of the system [1]. Ptolemy is an open-source program that allows user access to several different computation models. These models include finite-state machines, discrete time events, process networks, and several others.

CPSs are also being used in wireless networks to assist with data collection, control, computation, and data communication. A major application of a CPS is to recreate an actual physical environment in cyber space and run simulated controls. To be able to represent an environment, several CPS nodes are set up to sample the physical data that are available. However, the number of nodes used can sometimes be inefficient because of cost. As a result, research is being conducted on how to reconstruct a complete system model while using the interpolation method [2]. However, the accuracy of the recreated model will depend on the quality of the data that were available during sampling.

## 7.2 The Concept of Timed Actors

The programming in CPSs requires attention to not only the functional aspects, such as behavior and correctness, but also to the nonfunctional aspects, specifically timing and performance. In [3] it introduces a theory of timed actors that bases its principle on a worst-case design. This theory contrasts with the classical behavioral and functional refinements based on restricting sets of behaviors. The refinement of this theory improves efficiency and reduces complexity by allowing time-deterministic abstractions to be made. It shows how this theory can be used to increase both time and performance of CPSs.

New techniques in the design of sensors, actuators, and computer hardware currently enable the model of *embedded CPSs* (ECPSs). Examples of ECPS can be found in the domains of robotics, health care, transportation, and energy. ECPS interacts very closely with its physical environment by monitoring or even controlling it. This makes ECPS different from a traditional computing system. As mentioned before, the requirements of such a system, referring to ECPS, is not solely functionality but also includes both timing and performance properties. A few examples of these timing and performance properties include throughput and latency.

Two aspects that play key roles in developing these large and complex systems are abstraction and compositionality. Most of the methods that implement these principles exist primarily to deal with the functional properties and pay less attention to timing and performance. The purpose of this work [3] is to provide more insight into the timing and performance aspect of these systems.

Analysis has been performed on high-level models. Refinement and abstraction are used to navigate between high-level models, low-level models, and implementations. This process guarantees that the results obtained from the analysis are preserved during refinement. This chapter defines a general model and a suitable notion of abstraction and refinement that support this process. The model is compositional in the sense that refinement between models consisting of many components can be achieved by refining individual components separately.

The treatment of this theory falls under the category of *interface theories*, which focus on dynamic and concurrent behavior. The interfaces, called *actor interfaces*, are inspired by *actor-oriented* models of computation such as process networks and data flow. Actors generate tokens on their output ports and consume tokens on their input ports. However, because the main goal of this chapter is to focus on timing and performance, the token values are discarded and only the times in which the tokens are produced are kept. Actors are then defined as relations between input and output sequences of discrete events occurring in a given time axis.

The main point of this theory is centered on refinement and is based on the principle *the earlier the better* [3]. For example, if there are two actors, A and B, then A refines B if they have the same input, and A produces no fewer events than B. Also, the events produced by A must be no later than the events produced by B. This notion of refinement contrasts with the typical notion of refinement, which bases its principle on implementing fewer behaviors and being more deterministic. The earlier-is-better refinement principle is interesting because it allows *deterministic abstractions of nondeterministic systems*. Owing to some reasons such as high variability in execution and communication delays, dynamic scheduling, and other effects, it is expensive or impossible to model precisely. Time-deterministic models, on the other hand, suffer less from state explosion problems and are also more suitable for deriving analytic bounds [3].

## 7.3 Event-Based CPS Models

CPSs use a wide range of physical devices as well as computation components, which perform several different duties such as communication, computation, sensing, and actuating. CPSs are considered to be heterogeneous systems of systems because they involve these devices and perform these many different duties. These components used in CPSs are interconnected using both wired and wireless networks. CPS networks are both large scale and well orchestrated. However, several challenges arise when designing CPSs. One of these challenges includes the flexibility of the system [4]. These CPSs must be designed to support a high system level of flexibility, so that the several different components of the system have the ability to dynamically leave or join. Another challenge is the various quality of services (QoS) requirements that these CPSs have to support on every level of the CPS. One example that shows the impact that QoS has on CPSs is a time-related requirement, or deadline, for certain processes such as a control loop. A control loop may be used to indicate that a notable event has occurred in the physical world. Initially, the event has to be sensed and detected by the proper components in the CPS's cyber world. Next, the correct actuation decisions have to be made by the proper components in the CPS, and finally an actuation assignment must be performed by the proper components in the CPS, an actuator, in the physical world. All these duties must be performed within a specific time frame. In reality, the individual timing constraints for each individual subsystem and each individual component vary because of the nondeterministic system delay caused by the several different actions in the CPS, such as sensing, computation, communication, and actuation. When all these systems come together with their own individual timing constraints, the overall timing factor of the CPS becomes a significant verification challenge.

The close interaction that the CPS has to the physical world indicates that the time constraints and other such constraints can be handled by using an event-based approach. An event-based approach uses events in the CPS as units for computation, communication, and control in the system. Several different areas of event-based system design have been studied. Although there have been studies conducted, the approaches used in these studies cannot be directly applied to CPSs. In respect to traditional system designs, there is a consistent view dealing with time and space with respect to a single entity. For this reason, these studies prove to be inaccurate when considering event-based system design for CPSs. CPSs consist of a distributed set of the aforementioned components that operate in their own individual reference frames; therefore a CPS is better characterized by spatiotemporal information. A common frame of reference does not exist because CPSs have a heterogeneous nature. Also, the events of CPSs can be further divided based on different events being on different levels. These events range from lower-level events, such as the actuating and sensing events in the physical world, all the way to higher-level events, such as cyber events that are both machine and human understandable. For the interactions between these different components on these different levels in the system to interact with each other smoothly between the cyber and physical worlds, there must be a unified definition and representation of these events. A certain systematic mechanism must be implemented to compose these CPS events from the higher and lower levels and across the different system boundaries. The event model that results from this composition can both serve as an offline analysis tool and a run time implementation model.

There are two main points made in [4]. The first contribution is a synchronized event structure that has the ability to represent a CPS event instance at the aforementioned different levels. Basically, a CPS event instance has three different components consisting of the event type and the internal and external attributes of the event. When all three of these different components of the event instance are considered together, they explain when and where the event instance took place, when and where the event instance was observed, and who or what the observer was. In addition, the observer, possibly a sensor, can also be defined in the CPS event instance. This allows the observers to dynamically leave and join the CPS at run time. The second contribution is a formal mechanism that is used for defining and composing CPS events from lower-level events. This is done by applying and extending the theory of concept lattice. By using these contributions, a way of accommodating the temporal and spatial constraints in the composition of these events is introduced by using a set of compositional functions.

In conclusion, the fact that CPSs involve so many different processes such as communication, computation, sensing, and actuating makes these systems heterogeneous and widely distributed in the area of physical devices and computational components. The close interactions of these several different processes and subsystems within these CPSs and the physical world make events the major building blocks in the realization of CPSs. Therefore, the design principles of and the several different components in these systems should be approached with a strictly event-based view. The paper [4] introduces a concept, lattice-based event model for CPSs. The CPS event can be represented as three different components, the event type, the events internal attributes, and the events external attributes. These three components together can be used to define the spatiotemporal properties of the event and also can be used to determine the components that observed the event. The event model that results can be used as both an offline analysis tool and as a run time implementation model.

## 7.4 SCADA Model

Today's power grid is a complex system that continuously supplies power to an ever-changing and very nonuniform customer base. In other words, the power grid must supply power to a variety of loads, whose demands will be constantly changing throughout a given day. This constant load change can result in fluctuations in the system voltage and frequency of the power grid. This results in an unstable system. To combat this problem, a system known as SCADA (supervisory control and data acquisition) is used to monitor conditions at high voltage substations. These collected data have been used to determine load forecasts at various parts of the day for different locations [5]. As a result, resources have been made available to maintain system integrity, with extra resources set aside in case of machine failure [5].

However, this system can be considered to be largely an open-loop system without a true feedback path. Large substations provide power to a variety of loads that can include residential, commercial, or industrial loads. As a result, the system continues to have stability issues because true system demands can never be accurately predicted. Furthermore, to increase the overall efficiency of the power grid, more renewable energy sources such as solar and wind farms need to be integrated into the existing system. For this to be successful, there will need to be an increase in the overall intelligence capabilities of the various networks that control the power grid.

First, large-scale sensor networks should be implemented to help gather more accurate data downstream of the substation. These sensors will be incorporated into the already used SCADA system. However, the sensors should not be used to bombard the ones responsible for controlling the flow of power throughout the grid with data. Instead, these sensors would be pulsed at regular intervals to determine the current demands on the power grid. This in a sense will provide real-time data on the demands on the power grid. Using these data, different types of energy sources can also be integrated into the power grid to provide support; examples include solar panels and wind generators. These power systems provide a limited amount of power, while generator outputs at power plants can be increased or more generators can be brought online. Using these real-time data, solar panel or wind farms can be utilized to help with the ever-changing demands. As an example, if the demand for a certain area rises by a few megawatts, a nearby wind farm that is currently capable of meeting this demand can be used. This would result in the use of fewer fossil fuels to increase the output of a generator at a nearby power plant.

Figure 7.1 shows a model of what the system could look like overall. The two generators (G) supply power to the substation. The sensor denoted by the black box represents the current method of sensing the power drawn by the loads downstream of the substation. The new sensors indicated by gray boxes will sense the demands of the loads being fed by the substation. As a result, a more accurate determination of the load on the power grid can be made at any given time. For this system to be correctly implemented, it must first be modeled to correctly simulate the actual physical energy system. Figure 7.1 helps to determine how the sensors will help to monitor the power being drawn from the substation by each of the individual loads.

A CPS will contain the sensors and linear or rotary actuators required to interface with the physical world. Examples of sensors include voltage and current sensors. These sensors are examples of embedded systems, which are programmed to do one specific task. For a CPS to function correctly, any changes in the physical world must be reflected in the digital world in a real-time manner. As a result, the CPS must be able to respond to changes in the system in a real-time manner [6].

Sensors and sensor systems come in a wide variety of shapes and sizes. Individual sensors are associated with a specific portion of the system. One example is an accelerometer that is used to measure the acceleration of an object. In addition, several sensors can be grouped together to create a sensor mote. A sensor mote will act as a hub between several sensors in the field and the central processing unit (CPU). Each sensor mote will contain a small controller or microcontroller unit (MCU) [6]. This device will read the data being supplied by the sensors and will communicate these data back to the CPU. Next, there are actuators that connect the CPS to the physical world. An actuator is a device that can be used by the CPU to imitate a response to a change in the physical system itself.

The changes that occur in a physical system can be classified as two different types of events, temporal and spatial [6]. Temporal events can be further broken down into punctual events or interval events. A punctual event occurs at a specific time, while an interval event occurs at the end of a time interval that has a defined starting and stopping point. A spatial event refers to the actual location of the event, which can be either a point or a field. A point event is classified as an event that occurs at a specific instance or at a specific location. An example can be a point on the $x$ and $y$ plane. A field point is more complex because it involves being a function. A field point represents a time of physical aspect of the physical world.

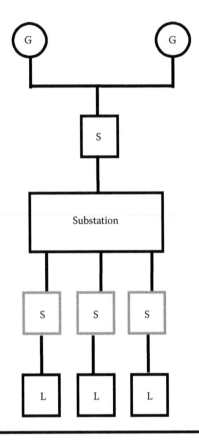

**Figure 7.1   Simple power grid sensing model.**

## 7.5 Medical CPS Interaction Software

For the last 40 years, devices such as pacemakers and implantable cardioverter defibrillators that are used to manage the rhythm of the heart have vastly increased in complexity. These cardiac devices can now have around 80,000 to 100,000 lines of computer code to operate. Between the years of 1990 and 2000, there were safety recalls performed on more than 600,000 pacemaker and implantable cardioverter defibrillator devices. Of the 600,000, around 41% were recalled because of issues with the firmware of the devices. This is a problem that has continued to increase over the years. In 1996, roughly 10% of the recalls on these medical devices were performed because of software-related issues. In June 2006, 21% of recalls were performed because of software-related issues. During the first half of 2010, the US Food and Drug Administration (FDA) issued 23 recalls of defective devices, all of which are categorized as *Class I*, meaning there is a reasonable probability that use of these products will cause serious adverse health consequences or death. At least six of the recalls were caused by software defects [7]. Currently, there are no standards for testing, validating, and verifying the software that these implantable medical devices run. So, taking into account the rapid rate for which these implantable medical devices are failing or becoming a concern for safety because of software-related issues, there is an apparent need for a means of evaluating and testing the safety and functionality of the software that goes into these implantable medical devices.

While the software in an electrical or mechanical system can undergo corrosion and fatigue, or have statistical failure of components, this is not the case for software embedded in a medical device. So, the software failures that take place in a medical device are more commonly from a flaw in the design and development of the system. Also, where the produces such as consumer electronics typically have life spans that are relatively short, the engineering for software that goes into medical devices often have a much longer life span and have a much higher precedence for safety and efficacy. The medical device industry is a regulated industry. Regulatory oversight also governs *how* the device was developed, not just *what* it turned out to be. The belief is that a well-planned, systematic engineering process produces more reliable devices, especially if software is a component of the device [7].

Several industries such as automotive electronics, avionics, and nuclear systems depend on safety and have a strict set of standards for safe software development, evaluation, manufacturing, and postmarket changes. However, the medical device industry has not yet adopted these strict standards for safety. The medical device industry faces several other challenges that are unique. First, currently, these medical devices are being evaluated in an open-loop manner. This open-loop method does not ensure that the device will never malfunction and harm the patient. To remedy this challenge, these medical devices should be subject to closed-loop testing to provide validation based on the context of the patient physiology. Two factors go into the context of the patient physiology; the input from the device's controller and the environment that the device is placed in. Both of these factors must be captured for a proper device evaluation process. Second, there are not many models for the patient and a few clinically relevant simulators for device design. To properly evaluate the safety and efficacy of the device operation, there needs to be high-fidelity model of interaction between the medical device and the patient. Basically, the functional and formal aspects of these models must be integrated so that during testing and validation the patients are evaluated at the same states. Finally, the device should be able to adapt to the specific condition of the patient and its surrounding environment. This would be to ensure that the device can be controlled and optimized to cover a wide range or different patient conditions.

## 7.6 Vehicle CPS Models

One advantage of a CPS is that it is capable of interacting with the system in real time. As a result, a CPS must be able to monitor and react to any constraints that are caused by the ever changing environment. In other words, a well-built CPS will be able to react to any changes in almost real time. CPSs are currently being used, but there are limits to the physical size and complexity of a system that can be achieved.

An example of a CPS that is currently being used is the vehicle-to-vehicle (V2V) system [8]. Any type of vehicle that contains this system will have the ability to monitor its surroundings. As a result, the V2V system will allow a vehicle to determine the presence of objects that are located behind or in front of the vehicle. This can help reduce the risk of an accident because the vehicle can be automatically slowed down to avoid a collision or the driver can receive alerts from a user interface. The driver can then perform the necessary actions to avoid a collision.

In the coming years, as technology begins to advance more and more, V2V systems will begin to grow in complexity and power. According to Steve Goddard, it is not hard to imagine large groups of cars moving together and sharing information on road and weather conditions [8]. In addition, these packs of cars can also share their collected database of information with monitoring stations based alongside major highways. A system of this magnitude is referred to as a spatiotemporal system, which means that all calculations and control will involve both time and space. Furthermore, for this to be a real-time system, any calculations will have to be made fast enough to initiate a fast and accurate response. As a result, there are several fundamental limitations that arise.

As the complexity of CPSs increases, there will have to be changes to the methods of resource allocation, database management, real-time scheduling, and group communication [8]. For instance, in a real-time system, the dynamics of the system are constantly changing and will require fast response times from a CPS. In addition, smart devices can be utilized to help deal with the problems of resource management. For example, certain tasks can be performed by individual processors, which can also receive commands from a higher-order processor that monitors and controls the different scheduling algorithms. As a result, it will be necessary for further research to develop new theories and models to accurately control and monitor any new complex and broad CPSs. Any new models will have to be tested with actual physical systems to determine any problems. Furthermore, this will help developers and engineers to further change and modify models and theories. The end result will be a new technology that has the potential to be a powerful tool available to manufactures and data collection agencies.

## 7.7 Time-Related Modeling Issues in CPS

The passage of time is a critical feature in CPSs. Furthermore, the passage of time is the centermost constraint that distinguishes CPSs from other general systems. Time is critical in predicting, measuring, and controlling the properties of the physical world. This principle revolving around time provides the basic foundations of control theory. Given a (deterministic) physical model, the initial state, the inputs, and the amount of time elapsed, one can compute the current state of the plant. However, for current mainstream programming paradigms, given the source code, the program's initial state, and the amount of time elapsed, we cannot reliably predict future program state [9]. When such a program is integrated into a physical system with physical dynamics, it makes the design of the system hard to accomplish. Presently, engineers have only been able to develop small model-based test designs that are delicate and do not easily handle small changes in the operating conditions and hardware platforms. Furthermore, the differences in the dynamics of the physical plant and the program that is trying to control the physical plant, can lead to several errors including some errors that can lead to catastrophic implications for the system.

Although CPUs have become fast enough to control many physical objects, modern cyber techniques such as CPU instruction scheduling, computer memory hierarchies, memory garbage collection, multithread processing, networking, and reusable component libraries (which do not expose temporal properties on their interfaces) introduce enormous temporal variability [9]. These innovations are based on the assumption that time is irrelevant to correctness and that it is more of a measure of quality. However, CPSs do not rely on fast computing but do require physical actions to be taken at the proper time. For CPSs, time needs to be more of a semantic property rather than a factor for quality.

The integration of computer processes and physical processes is a challenge that has been recognized for some time. This challenge has motivated the emergence of hybrid systems theories. However, progress in hybrid systems theories has been limited because of the combination of ordinary differential equations and automata in relatively simple systems. These models have a uniform notion of time that is inherited from control theory. In these systems, time ($t$) is simultaneously available in all parts of the system. There have been adaptations of these traditional computer science concepts to these distributed control problems, but these adaptations also make the same assumption of time. For example, translation consensus problems from computer science into control systems formulations, shows connections between such consensus problems and a variety of dynamical systems problems (such as synchronization of coupled oscillators, flocking, formation control, and distributed sensor fusion). These formulations, however, break down without the uniform notion of time that governs the dynamics [9]. Such a uniform notion of time cannot be so precisely realized in networked software implementations. An approximate uniform model of time can be used in time triggered networks and time synchronization, but the imperfections that are included in these approximations must be included in the analysis of the dynamics. Without including these imperfections, it is possible that a control system model could be created that violates causality.

The real-time embedded software that is being built today uses programming abstractions that implement little or no temporal semantics. The primary focus of these modern day computer-based systems is multiple computers that interact with each other through a physical network and perform physical processes using sensors and actuators. These systems are commonly referred to as CPSs.

One of the main goals of research into CPSs is to successfully integrate models of the physical world with computational models with the end result being more powerful than the two models in their individual states [10]. To go into more detail, a CPS will integrate digital world models with the physics-based models of real-world systems. As an example, a CPS could integrate a digital device with machinery such as vehicles or robots. Digital embedded systems have helped in the advancement of CPSs by helping to deal with the major problems of allocating resources to manage both the physical and digital aspects of the system. An embedded system can be designed to handle the monitoring of an individual part of the entire system, and this system can report back to the main CPU the information that the programmer believes will be necessary for overall system functionality [10]. One example involves the use of feedback control systems to monitor different aspects of the physical system. In a feedback control system, a comparison between a reference and the system error is made, and this difference is used to adjust the system output to the reference.

A cyber-physical system is made by integrating computational and physical processes. The computational processes include computer and networking systems to monitor and control a physical process [11]. This is accomplished by using feedback control loops. The signals received from the feedback loops will be used to control the computational part of the process. As a result, the controller will cause a dynamic change in the system based on the computations and feedback signals. The process will then repeat. As a result, a designer must have a thorough understanding of the dynamics of software, networks, and the physical process to be controlled.

There are several key problems that emerge when working with CPSs. One of these problems is the time it takes to perform a task. In a general computing task the time to completion is generally considered to be a measure of the system's performance [11]. However, in a CPS the time that is necessary to complete a task could be critical to the overall performance of the system. As a result, it becomes clear that a CPS must nearly resemble a real-time system. Furthermore, a physical process is made up of components that occur at the same time, while software applications will generally run in a sequential and orderly process. To clarify this information, an example of a complex physical process is examined below. A modern aircraft is a complex system of valves, pumps, motors, etc. [11]. These components help control the distribution of fuel to the engines of the plane. In addition, the components listed above aid in moving fuel around the plane to assist with the balancing of the aircraft. Furthermore, the plane's engines can also be cooled by using the fuel to cool the engine oil. This is accomplished by using a heat exchanger. Furthermore, the pressure of each fuel tank must be monitored and adjusted during a plane's ascent and descent to avoid high pressures and possible structural damage [11].

Figure 7.2 is a simplified diagram of an aircraft fuel system. The fuel is transported using pumps. The blocks designated V represent values that will be used to allow fuel to flow to the pumps. The blocks designated P represent pressure sensors that will monitor the pressure of each tank. Finally, different pumps are utilized to pump oil to the engines or to the heat exchanger to cool the engine oil. In addition to monitoring pressure, the temperature of the fuel must also be monitored because the fuel weight and density will change with temperature. Figure 7.3 shows a block diagram of the feedback control system to monitor the above system. It can be seen how the CPS would be used to monitor the fuel tank pressure and temperature, as well as the flow rate and output pressure of the fuel pump.

The physical processes of a CPS can be represented as continuous-time models of dynamics and computations, which result in a hybrid model [11]. There are several software programs that can be used to model dynamic systems. Simulink and LabView Control Design are two examples. In conclusion, a CPS is a type of system that integrates a physical process with the powerful computers to provide monitoring and control. Our example of the aircraft fuel system gives insight into how complex systems can become.

The largest obstacle to creating an effective CPS is system integration. One explanation for this obstacle is the lack of solid scientific theoretical foundation on the subject [12]. System integration is considered to be a monumental challenge because of today's methods, which include making components work together one way or another. However, this method will not work as the complexity of systems continues to grow. Solutions to this problem are very difficult because system integration is the last phase of any project. Up until system integration, the system has

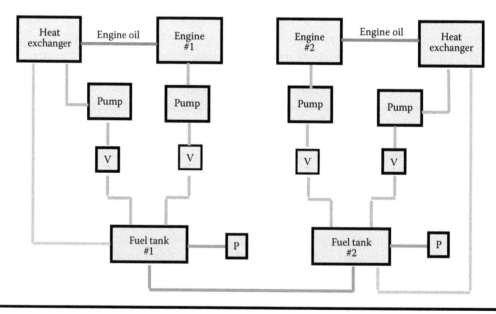

**Figure 7.2  Simple block diagram of aircraft fuel system.**

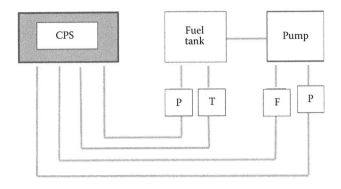

**Figure 7.3   Block diagram of feedback control system.**

been broken up into individual sections, and any or all hidden problems begin to surface during system integration.

A prime example of system integration problems can be found by looking at the automotive industry. As part of the development of a control system, suppliers to the automotive industry develop system components that implement various hardware or software functions. Original equipment manufacturers will take these components and attempt to integrate them. A major problem is that all suppliers have different ideas and methods to create and implement their equipment. As a direct result, the task of system level integration becomes increasingly harder.

# References

1. P. Derler, E. A. Lee, and A. L. Sangiovanni-Vincentelli. 2011. Addressing modeling challenges in CPSs. Technical Report UCB/EECS-2011-17. Electrical Engineering and Computer Sciences, University of California at Berkeley.
2. L. Kong, D. Jiang, and M.-Y. Wu. 2010. Optimizing the spatio-temporal distribution of cyber-physical systems for environment abstraction. In *Proceedings of the 2010 IEEE 30th International Conference on Distributed Computing Systems* (ICDCS '10). IEEE Computer Society, Washington, DC, pp. 179–188.
3. M. Geilen, S. Tripakis, and M. Wiggers. The earlier the better: A theory of timed actor interfaces. Technical Report UCB/EECS-2010-130. Electrical Engineering and Computer Sciences, University of California at Berkeley.
4. Y. Tan, M. C. Vuran, S. Goddard, Y. Yu, M. Song, and S. Ren. 2010. A concept lattice-based event model for cyber-physical systems. In *Proceedings of the 1st ACM/IEEE International Conference on Cyber-Physical Systems* (ICCPS '10). ACM, New York, pp. 50–60.
5. M. D. Ilie, L. Xie, U. A. Khan, and J. M. F. Moura. 2010. Modeling of future cyber-physical energy systems for distributed sensing and control. *IEEE Trans. Syst. Man Cybernetics A* 40(4), 825–838.
6. Y. Tan, M. C. Vuran, and S. Goddard. 2009. Spatio-temporal event model for cyber-physical systems. In *Proceedings of the 2009 29th IEEE International Conference on Distributed Computing Systems Workshops* (ICDCSW '09). IEEE Computer Society, Washington, DC, pp. 44–50.
7. Z. Jian, M. Pajic, and R. Mangharm. 2012. Cyber-physical modeling of implantable cardiac medical devices. *Proc. IEEE* 100(1), 122–137.
8. S. Goddard and J. S. Deogun. 2008. Future mobile CPSs: Spatio-temporal computational environments. Position paper, Department of Computer Science and Engineering, University of Nebraska–Lincoln. Available at http://varma.ece.cmu.edu/cps/Position-Papers/Goddard-2.pdf.

9. J. C. Eidson, E. A. Lee, S. Matic, S. A. Seshia, and J. Zou. 2009. Time-centric models for designing embedded CPSs. Technical Report EECS-2009-135. Electrical Engineering and Computer Sciences, University of California at Berkeley.

10. J. M. Bradley and E. M. Atkins. 2012. Toward continuous state-space regulation of coupled cyber-physical systems. *Proceedings IEEE* 100(1), 60–74.

11. P. Derler, E. A. Lee, and A. S. Vincentelli. 2012. Modeling cyber-physical models. *Proc. IEEE* 100(1), 13–28.

12. J. Stzipanovits, X. Koutsoukos, G. Karsai, N. Kottenstette, P. Antsaklis, V. Gupta, and J. Baras. 2012. Toward a science of cyber-physical system integration. *Proc. IEEE* 100(1), 29–44.

*Chapter 8*

# Cyber-Physical System Modeling on Cognitive Unmanned Aerial Vehicle Communications

Mengcheng Guo, Fei Hu, and Yang-ki Hong

## Contents

## 8.1 Introduction

The unmanned aerial vehicle (UAV) is widely used in military and civilian applications. It can be used for field surveillance through the cameras deployed in the UAV body. However, when the UAVs are deployed in complicated radio signal conditions, it is a challenging issue to maintain the communication performance such as throughput and transmission delay.

This chapter reports part of our work by the UAV video communication testing group at the University of Alabama. We propose to use the cognitive radio network (CRN) with smart radio sensing functions to control UAV communications. The CRN-based UAV ad hoc network is a typical cyber-physical system (CPS) due to its features given as follows: on one hand, the UAVs can use spectrum sensors to detect any unused spectrum (i.e., unoccupied by the licensed primary users [PUs]) in order to deliver their data via those idle spectrum bands; on the other hand, the UAVs access the radio channels by following certain medium access control (MAC) protocols. Their communications can cause interference to the licensed users. In other words, the UAVs can change the radio environment.

It is important to build CPS models to test the UAV communication performance under CRN environments. This is because it will be very costly if we always use dozens of (or even hundreds of) UAVs for real field test. The UAVs can easily crash if not controlled correctly. We first use the Network Simulator (NS2) to simulate the CRN protocols on a computer, and then use the Universal Software Radio Peripheral (USRP) to emulate UAV networks in both the software and hardware level.

## 8.2 Cognitive Radios

Cognitive radio (CR) is a concept that originated from the milestone work of Dr. Joseph Mitola in 1999 [1]. Its kernel idea is to make CR users have the learning capability in order to sense and utilize the available spectrum resources. Meanwhile, it restricts and reduces the interference and collisions to the licensed PUs. By using artificial intelligence or machine learning schemes, a CR user, called a secondary user (SU), can use its past experiences to respond to the time-varying radio conditions. Such past experiences include the understanding of the dead communication zone, radio interferences, spectrum usage patterns, etc. Therefore, an SU can adjust its communication behaviors based on the available spectrum conditions.

There are multiple descriptions for CR implementations. The Federal Communication Commission (FCC) has provided a simple version—FCC-03322. In it they suggest that any kind of radio with the adaptive spectrum sensing ability should be called CR. FCC also points out that the CR should be able to adjust its configurations based on the operating environment, and it should have the functions to sense the surrounding conditions and modify the communication parameters.

As a promising new radio technology, CR can reliably sense the wideband spectrum environment in order to detect the PU's occupied channels and then opportunistically access the available local radio spectrum without causing any interference to the PU's communication. As shown in Figure 8.1, the wireless channels and interferences are changing all the time. Therefore, a CR system is required to have a high flexibility. So far, the general CR applications mostly follow FCC's descriptions. Therefore, CR is also called dynamic spectrum access technology.

As more and more radio applications are rapidly developed nowadays, the spectrum resource in the ISM band is getting busier than ever. On the other hand, the licensed spectrum bands, for example, the TV services, are much less efficiently utilized. Under this situation, the CR concept was proposed to address the issue of spectrum usage efficiency. After then it drew the attention of researchers because of its features such as flexibility and intelligent use of the fixed license spectrum bands.

Some standard organizations have adopted the CR technology and constituted a series of standards to promote the development of CR in different applications. For example, the IEEE 802 group has made the WLAN standard based on the CR technology in the TV service environment. It aims to efficiently utilize the vacant spectrum bands in the TV's VHF/UHF spectrum bands. The IEEE 802.16 WiMAX group is working on the .h version that aims to improve the strategy MAC protocols and fulfill the coexistence of the authorized and unauthorized systems.

The first IEEE standard announced to be based on CR networks (CRNs) is the IEEE 802.22. It is designed for CRN applications under the digital TV public services. Many researchers are proposing new strategies to improve the 802.22 standard.

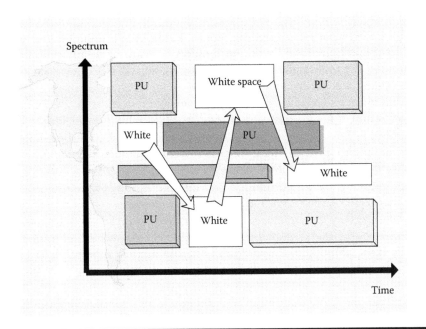

**Figure 8.1   Opportunistic dynamic spectrum access.**

CR is also called smart radio. It has features such as flexibility, intelligence, and configurability through the sensing of the environment and learning via artificial intelligent technologies. The purpose of CR is to adjust the configuration of radio (including transmission power, carrier spectrum, modulation scheme, etc.) to improve the efficiency of the spectrum utilization throughout the entire communications. CR operations include three steps: *spectrum sensing, dynamic spectrum allocation* (DSA), and *spectrum sharing* through the intelligent learning of the system. Figure 8.2 shows the operation principle of CRNs.

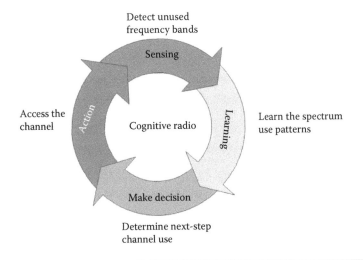

**Figure 8.2   CR strategy.**

CRN is an upswing from CR technology and is now in the front level research of CR applications. Generally, CRN problems include three aspects. The first one is about the spectrum sensing in which the SUs in CRNs need to sense the environment by following the protocol and make sure to highly protect the PUs while in the meantime efficiently using the spectrum in an opportunistic fashion. The second issue is the dynamic spectrum allocation, which is the strategy of how to allocate the spectrum resources sensed in the sensing step without disturbing the PUs' and other SUs' communications. Last but not least, the spectrum sharing and spectrum mobility issue is studied in CRN research to make the transmission go smoothly and back off immediately after the PU's appearance in the spectrum bands. Regarding the spectrum sharing issue, there are some ideas concerning game theory, financial theory, biology model, etc. There are some other problems that need to be improved under the CR environment such as CRN routing protocol design, transportation protocol design, cross-layer design, and CRN security problems.

The ultrawideband (UWB) system is still looking forward to applying CR technology as well. Orthogonal frequency-division multiplexing (OFDM) system meanwhile has features matching the applications of the CRN. Ad hoc emergency and mesh network is proposed to use CRN for improved performance in practice. Most of the applications of CRN are still in the lab experiment level.

How to simulate the CRN efficiently and accurately is an important problem in this area. Below we will introduce the details of the Network Simulator (NS2), a generally used network simulation software, and discuss how to use it to implement the CRN environment.

## 8.3 Network Simulator

NS2 is the abbreviation of the Network Simulator, Version 2. It is the simulation software designed by UC Berkeley, and its maintenance is under the VINT Project group. Now SAMAN and COnsER manage its updates. Meanwhile, the VINT Project designed some tools for the simulation result display, analysis, and transformations. These transformations can convert the topology of the network under NS2 format into the real network topology structure.

NS2 is compiled by C++ and OTcl programming languages and is an objective, event driving network simulator. It implements the TCP, UDP, and other network protocols, data transportation such as the FTP, Telnet, Web, constant bit rate (CBR), and variable bit rate (VBR), routing queuing scenarios such as DropTail, random early detection (RED), and Class-based queuing (CBQ), routing algorithm Dijkstr, etc. NS2 is also used in broadcast and MAC layer protocol simulation.

As shown in Figure 8.3, from the programmer's point of view, NS2 is an objective Tcl script interpreter (OTcl), which has a simulation event scheduler, network module object library, plumbing module, etc. In other words, we can just apply the OTcl script to a program, configure and run simulation networks, initialize the event scheduler, set up the network topology, and announce the time of starting and ending the event scheduler from the network objects and their member functions.

Figure 8.4 shows the basic structure of NS2.

An NS2 user can use OTcl to build the network topology. As a matter of fact, the OTcl is like an objective version of the script programming language Tcl. The Event Scheduler and Network Component are implemented by C++. The network simulation in NS2 is based on the event-driven strategy. For instance, at the timing of 1.2 s, the FTP agent will begin to transfer data, and at 2.0 s this communication will end. All these are controlled by the event scheduler. The network components like the agent (TCP, UDP, etc.) and the traffic generator (FTP, CBR, etc.) are used to attach to protocols and applications in the simulations. We can see from Figure 8.4 that the

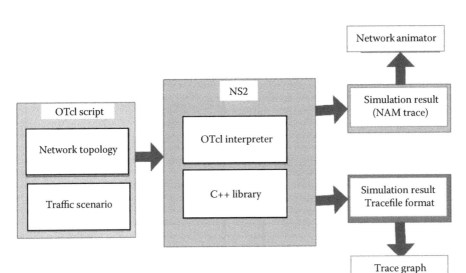

**Figure 8.3  NS2 working procedure.**

bottom of NS2 is the C/C++ part, because the kernel NS2 is based on C++, and users just use OTcl to build and configure the network topology. TclCL is the bridge between the C++ and OTcl. Users will need to connect the C++ and OTcl by using TclCL and a user interface when they want to change the C++ kernel or build their new protocols and network components. By using the simulated object design and the simulator in the OTcl library, the event scheduler and most network components are implemented in C++ and are connected by Tclcl to be usable in the OTcl. We can see that actually NS2 is an object-oriented interpreter (OTcl) with the network component library (C++).

Figure 8.5 simplifies the interactive relationship between C++ and OTcl. The tree structure is the object hierarchy from the object-oriented point of view. There are separated hierarchies in both C++ and OTcl. The dashed lines denote the sharing part of C++ and the OTcl, and the interface part (Tclcl) between them allows them to access each other and get connected in one-to-one correspondence. The object hierarchy defines the inherit relationship of the classes.

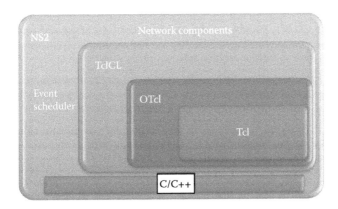

**Figure 8.4  NS2 structure.**

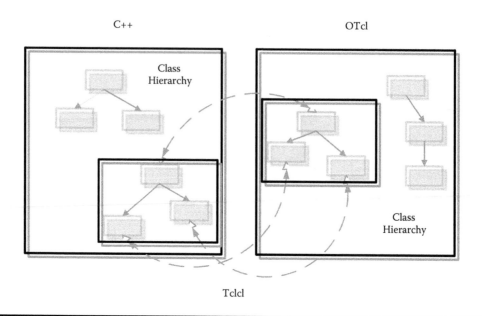

**Figure 8.5  One-on-one correspondence structure.**

Also, these relationships also have the corresponding inherit relationship in both the compiled C++ part and OTcl interpreted part of the sharing area. Why does NS choose two languages? One task is about implementing the components and functions in the lower level by one kind of system programming language. This type of language needs to efficiently control the data bytes and packet header and operate the algorithm with a large data set. For such a task, the running speed is important. On the other hand, the time for changes (run the simulation, find errors, modify the errors, recompile, and multiple time running) is not that important. Because this type of lower-level code will not be changed very often, C++ is suitable for this and it only usually needs to be compiled once. Then it can be used all the time in all simulations. Only when the new function or modification is needed will we use these low-level bases. Otherwise, another compile procedure is not necessary. On the contrary, the OTcl is mainly used for the management of the network environment simulation. By using OTcl scripts we only need to configure the network topology once.

As we can see, the OTcl is suitable to the one-time configuration and setup procedures, and C++ fits to the network components or protocols that would not be changed very frequently. NS2 can assemble the network components with data links and by setting the pointers from one object to another proper object's neighbors. When the users want to create a new network object, a new object from the library is needed. All objects are connected by the links in NS2. This sounds complicated but not when we use the OTcl to configure the network modules.

Other than the network objects there is another important component called the event scheduler. One event in NS2 is just the ID of one packet. This ID is unique for every packet, and it includes schedule time and pointer as well. The pointer is directed to the object that will deal with this packet. In NS2 the event scheduler needs to know the simulation time and introduces different network component objects that follow the schedule time. The simulator will wait until the next component is activated as in schedule. Event scheduler can also be used as a timer. For example, TCP needs a timer to decide the transportation time for the packets.

To enforce the efficiency, NS2 separates the data link implementation from the link control. The network components and event scheduler are compiled by C++ to save the running time. Then the corresponding objects in OTcl and C++ control function and configuration variables are used to bind the objects and variables in C++. In this method, the objects in C++ are controlled by the OTcl. The components that are not needed are not connected to the OTcl.

The usage flowchart of NS2 is shown in Figure 8.6. The Tcl scripts need to be written to complete the network environment simulation. After running the Tcl script we can get a few text files that could be the trace files generated in the simulation procedure. After analyzing the trace files we can get the simulation results.

The simulation results from NS2 can be displayed and analyzed by different software tools, for example, Network Animator (NAM) and Gnuplot. In one case, if we want to observe the packets in the FTP transmission, we need to simulate the environment first. Suppose there are two nodes and we need to run FTP between them. The node, link, agent, and generator are all defined in one OTcl script. Then NS2 will run this script. We can do more analysis by NAM and XGraph software.

The relationship between NS2, TCP/IP, and OSI seven-layer network frame is shown in Table 8.1.

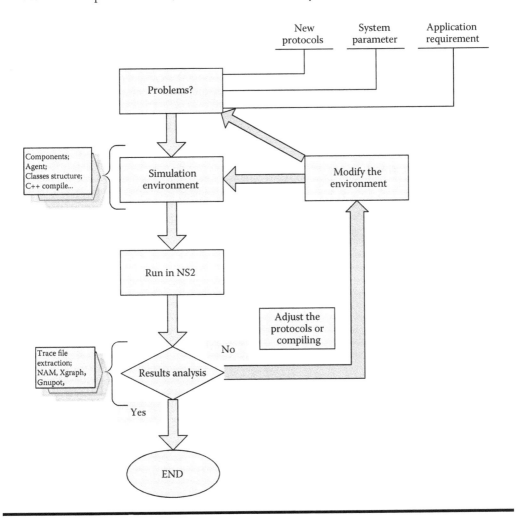

**Figure 8.6   NS2 design process.**

**Table 8.1   NS2, TCP, and OSI Stack**

| TCP Network Framework | NS Network Framework | OSI Network Framework |
|---|---|---|
| Application layer | Application layer | Application layer |
| | | Presentation layer |
| | | Session layer |
| Transport layer (TCP) | Agent | Transport layer |
| Network layer (IP) | | Network layer |
| Physical layer | Node and link | Data-link layer |
| | | Physical layer |

NS2 just uses classes in the Tclcl to make the C++ and OTcl work together. There are six main classes in Tclcl:

1. Tcl class: provide the method from compile part to the interpreter part
2. InstVar class: used to bind the corresponding object member functions
3. TclObject class: the base class of all C++ objects
4. TclClass class: connect the name of the objects in the interpreter with the compile C++ object name
5. TclCommand class: provide the global method for visiting from interpreter to compiler
6. EmbeddedTcl class: convert the OTcl scripts to C++ codes

TclObject classes hierarchy is shown in Figure 8.7.

Here we provide one example of the wireless video transmission extension that will be used for the test of the UAV video transmission system later. For wireless transmission, NS2 only provides two wireless node models: one is CMU's MobileNode and the other is SRnode. The general node structure is shown in Figure 8.8.

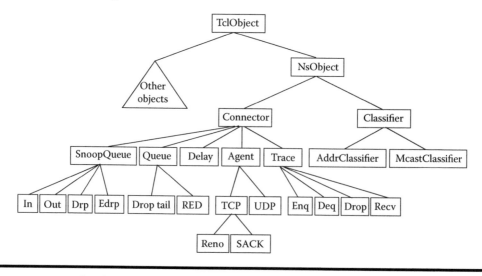

**Figure 8.7    NS2 class hierarchy. (From Shihao, J. et al., *IEEE Trans. Signal Proc.*, 56, 6, June 2008.)**

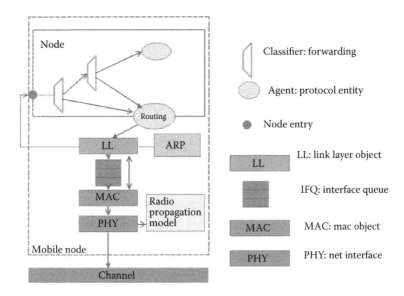

**Figure 8.8   CMU monarch's wireless node (MobileNode) illustration.**

First of all, we need to create the node and configure it in the API. The configure scripts are similar to the following:

```
set val(chan) Channel/WirelessChannel ; #Channel Type
set val(prop) Propagation/TwoRayGround ; #Radio propagation model
set val(netif) Phy/WirelessPhy ; #Network interface type
set val(ant) Antenna/OmniAntenna ; #Antenna model
set val(rp) WCETT ;#Routing Protocol
set val(ifq) Queue/DropTail/PriQueue ;# interface queue type
set val(ifqlen) 50 ;# max packet in ifq
set val(mac) Mac/802_11 ;# MAC type
set val(ll) LL ;# link layer type
set val(nn) 10 ;# number of mobilenodes
set val(ni) 2 ;# number of interfaces
set val(channum) 2 ;# number of channels per radio
set val(cp) ./random.tcl ; # topology traffic file
set val(stop) 50 ;# simulation time
```

For the wireless mobile node we have the following:

1. Link layer: different from wired node and connects to an Address Resolution Protocol (ARP)
2. ARP module: give the address for the destination
3. Interface queue: queue the packets from the routing agent
4. MAC layer: use IEEE 802.11 DCF protocol
5. Tap agents: trace and analyze the simulation procedure by MAC address
6. Network interfaces: the interface for node to visit the channel
7. Radio propagation model: wireless signal transmission models to calculate the power level of the receiver node
8. Antenna: Omni undirected antenna

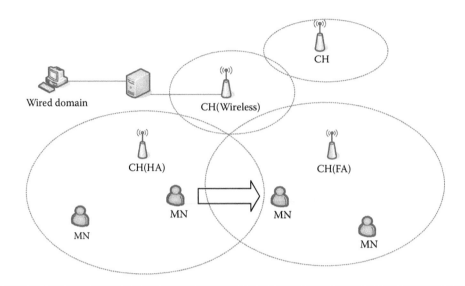

**Figure 8.9   UAV network topology.**

In our application we assume that a UAV is a mobile node in the simulation. We make the UAVs form a cluster-based network to transmit the video back to the wired ground station based on the sensed spectrum information. Figure 8.9 shows our UAV network topology.

The video traffic settings in NS2 are shown below:

```
//  Multimedia Header Structure
struct hdr_mm {
    int ack;      // is it ack packet?
    int seq;      // mm sequence number
    int nbytes;   // bytes for mm pkt
    double time;  // current time
    int scale;    // scale (0-4) associated with data rates

    // Packet header access functions
    static int offset_;
    inline static int& offset() { return offset_; }
    inline static hdr_mm* access(const Packet* p) {
            return (hdr_mm*) p->access (offset_);
    }
};
```

```
//  Multimedia Header Class
static class MultimediaHeaderClass : public PacketHeaderClass {
public:
 MultimediaHeaderClass() : PacketHeaderClass("PacketHeader/Multimedia",
                                             sizeof(hdr_mm)) {
            bind_offset(&hdr_mm::offset_);
    }
} class_mmhdr;
```

## 8.4 Spectrum Information Distribution Theory and Application

CR needs to be very sensitive to the changes of the wireless environment. Based on this kind of information, CR can make a smart decision automatically, and adaptively adjust the sender and receiver side configurations that fit the CR users into the environment properly. CR users need to sense the spectrum bands assigned to the PUs from time to time to see whether the PU has occupied the channels or not. Generally, we can classify the bands into three types:

1. White channel: not used by the PUs at all
2. Gray channel: low energy level/interference, partially used by the PUs
3. Black channel: high energy level/interference; high percentage of the band is used by the PUs

Figure 8.10 shows the typical spectrum sensing technologies.

Next, we will use our research project, which is about the spectrum information compression in CRN. Considering CRN's transmission nature and features, the sensing stage is extremely important and special. In our research, we focus on the spatial–spectral information distribution (SSID) issue, which is about how to exchange the sensing information efficiently to detect the PUs' spectrum use status accurately, while creating less network transmission loads. We use different methods to compress the typical frequency occupancy map and operate the long-term sensing for PU's transmission pattern recognition. We also make the prediction of the PU's behaviors to adjust the SU's adaptive sensing strategy.

In our work, we used the compressive sensing (CS), principle component analysis (PCA), and nonnegative matrix factorization (NMF) algorithms to compress the spectrum information (SI). By using SI compression in CRN, the exchanged information load among SUs is reduced significantly, and the performance of the system is much improved.

Efficient distribution of spectrum information is very important in CRN. Among various spectrum information parameters in CRN (see Table 8.2), the channel sensing information is vital due to the availability of a large number of channels and their frequent variations. The decision about the availability of different channels for CRN users is made based on soft combining, quantized soft combining, or hard combining. In order to reduce the reporting overhead, we have

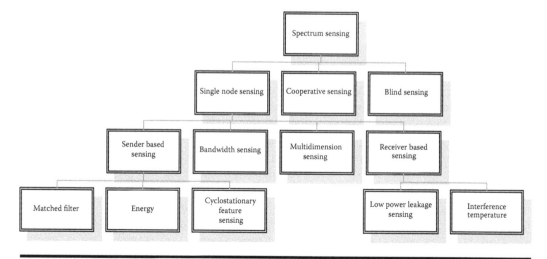

**Figure 8.10  Spectrum sensing technology types.**

**Table 8.2  Parameters for Information Exchange**

| | *Parameter* | *Nominal Value in IEEE 802.22* |
|---|---|---|
| Channel statistics | Center frequency | VHF and UHF TV bands Channels 2 to 69 |
| | Bandwidth | 6, 7, or 8 MHz for TV |
| | RSSI | Minimum value for TV channel is −116 dBm |
| | CINR (or interference temperature) | Depends on interference distribution |
| | Channel holding time | 10 min |
| | Code rate | 1/2, 2/3, 3/4, 5/6 |
| | BER | |
| Node statistics | Modulation schemes | QPSK, 16-QAM, 64-QAM |
| | Transmit power | Max 36 dBm |
| | Available data rate | Peak: 1.5–22.6 Mbps (BER?, TX power?) |
| | Channel sensing time | 30 s |
| | GPS location | |
| | Mobility and direction of speed (Doppler spread) | |
| Application statistics | Goodput, user priority, application data rate | |
| | Packet loss rate | |
| | Delay constraint | H264 video encoding rate |

studied two schemes to compress the channel sensing information for cooperative sensing. These schemes are soft combining with the NMF and wavelet compression.

Next, we will discuss the PU's normal power detection methods with fast Fourier transform (FFT) and discrete wavelet transform (DWT). Also, a novel approach applying the CS in the front end to reduce the computational burden on the CR users will be presented. Following that we will explore PCA and NMF methods to extract and compress the cycle-stationary features. Discussion and comparison of these methods will be stated with the NS2 simulation results as well.

*Normal power sensing (FFT, DWT):* In CRN the first major step is to sense the environment to detect the presentation of the PUs. The goal is to detect the PU fast and accurately in order to implement the cognitive functions. So far there are mainly three categories of detection methods. First, are the power detection methods, which are simple and need little prior knowledge of the PU signals. However, when it is in the color noise environments, the detection accuracy would be low. Second, is the use of a match filter to detect the PU signals. This is an easy and precise method. It needs most of the PU signals at first, such as using the beacon signals. It thus wastes the bandwidth and is not possible sometimes. The last type is the extraction of the cyclostation-ary features of the signals, and learning from that to get the information for our CRN commis-sions. This will need certain knowledge of the PU signals and some computational burden on the learning part. We try to compress the sensing information based on the first and third types of the sensing strategies, since the second type would not need to do much compression work on its quick decision-making.

*FFT:* Normally the original thinking of approach is about using the FFT coefficients of differ-ent frequency bands to get the power level. If it is beyond some threshold we can set the occupancy of channels to be 1, otherwise 0.

*Compressive sensing:* CS is developed in signal processing community and applied for the spare signal compression on the sensing step compared to the traditional Shannon–Nyquist sampling theorem. Other than the Nyquist sampling, which needs more than double sampling rate of the highest frequency component in the signal, the CS uses a much lower sampling rate and randomly collects the information from the whole sparse signal. Then it uses the optimal method to itera-tively recover and reconstruct the original signals with little unimportant data loss. As shown in Figure 8.11, the compressed signal is like a linear combination of the sensed original signal.

How to design the measurement matrix is a basic problem in the CS. After the research of CS the restricted isometry property (RIP) of $\Phi$ as shown in Figure 8.12 is necessary for the measure-ment to collect enough data information for the recovery. We say that a matrix $\Phi$ satisfies the RIP of order $K$ if there exists a constant $\delta(0,1)$ such that

$$(1-\delta)\|x\|_2^2 \le \|\varnothing x\|_2^2 \le (1-\delta)\|x\|_2^2$$

Also, RIP ensures that a variety of practical algorithms can successfully recover any compress-ible signal from noisy measurements.

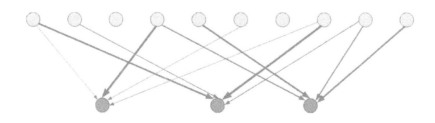

**Figure 8.11   Linear combination of *n*-dimensional signals to produce the *p*-dimensional report.**

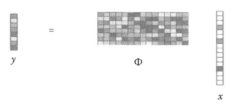

**Figure 8.12   CS measurement matrix.**

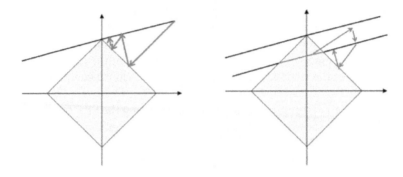

**Figure 8.13** $L_1$-norm algorithm on CS with geometry illustration.

Here, for our CRN system, $y$ is the CS report that has a much lower dimension than the uncompressed result of $n$ channels. Another important issue to be considered in CS is the reconstruction algorithm. In the traditional Nyquist sampling the data amount is much larger than that in the CS case. The recovery is very simple: it just uses the *sinc* interpolation to get the original signals. This is a linear process that requires little computation and has a simple interpretation. However, for the CS, although the data collected are much smaller than the Nyquist sampling case, the recovery algorithm is much more complicated and is a nonlinear optimization process. This type of algorithm (Figure 8.13) usually uses the greedy or $L_1$-norm method to iteratively find the solution that will give much computation load for the system.

There were a few works about applying the CS in the CRN for the channel estimation and spectrum sensing. Two important issues must be considered for the CS scheme. One is the sensing part design and the other is the computational burden for the system.

We have used a different CS scheme to reconstruct the typical frequency-location occupancy map and operate the continuous stream sensing for PU's transmission pattern classification. In this scheme, it is shown in our experiments that the performance is better than those that are conventional, since other than compressing the SSID, we also simplify the reconstruction by the CS processing (CSP) classification. This strategy will markedly reduce the computational load and transmission delay rather than the traditional reconstruction methods.

In our work here, we have used the CS to compress the spectrum by linear combination. Then the fusion center collects them from the group-representative SUs at different locations. After the implementation of the simulation in the NS2 environment, the results give a noticeable improvement of the CRN system performance. Using the compression of the SSID in CRN, the exchanged information load is reduced greatly and the performance of the system is much improved.

### 8.4.1 CRN System Model

We assume that the entire UAV network is organized into different clusters based on their physical proximities. For each cluster there is one fusion center that manages the whole cluster of UAVs. Different clusters may have a different spectrum occupancy situation as shown in Figure 8.14.

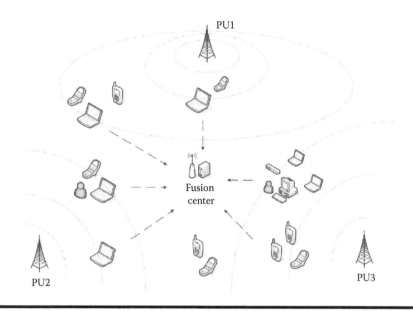

**Figure 8.14 SUs could be grouped into clusters based on different locations.**

There are *n* channels that are sparsely used by the PU(s) (sparse in channel occupancy). Generally, there are two PU traffic models that are often considered. The first one is the On–Off model and the other is the Poisson distribution model. For our system we used the On–Off model with the Gaussian distribution. One demo of the PU is shown in Figure 8.15. To get the sparse feature of the measurement we set the PU utilization efficiency less than 0.05.

We divide the whole cluster into a number, say, *M*\**M* of grids, to build the spatial–spectrum occupancy map (Figure 8.16). We use the safeguard level threshold to make the decision "0" or "1." This binary matrix can be used for our CS scheme (Figure 8.17). For simplicity we pick up one representative SU in every grid to represent the whole group.

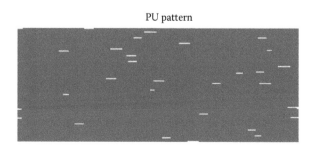

**Figure 8.15 Sparse PU pattern *n* = 100, efficiency is 0.01; Gaussian distribution on–off PU pattern.**

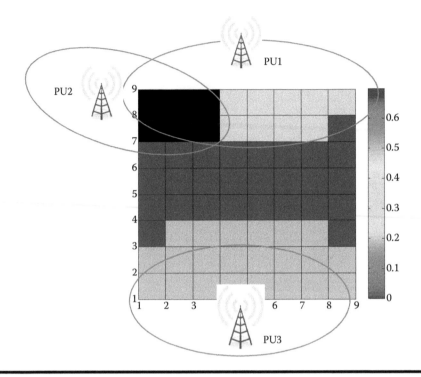

**Figure 8.16   Cluster with three PUs in simulation (the color depth represents channel use percentage).**

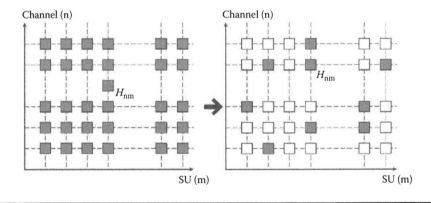

**Figure 8.17   Sparse binary matrix after using threshold.**

## 8.4.2 CSP and Application

There are a few requirements for the design measurement matrix in the CS applications. Through Theorem 8.1 [3], after the $K$th order of the RIP measurement matrix was constructed, we can recover the original signal with certain accuracy as follows:

$$\|\hat{\mathbf{x}} - \mathbf{x}\|_2 \leq C_0 \in + C_1 \frac{\|\mathbf{x} - \mathbf{x}_K\|_1}{\sqrt{K}}, \text{ where } C_0 = 4\frac{\sqrt{1+\delta}}{1-(1+\sqrt{2})\delta} \quad C_1 = 2\frac{1-(1-\sqrt{2})\delta}{1-(1+\sqrt{2})\delta}$$

**Theorem 8.1**

Suppose that $\Phi$ satisfies the RIP of order $2K$ with isometry constant $\delta < \sqrt{2} - 1$. Given measurements of the form $\mathbf{y} = \varnothing\mathbf{x} + \mathbf{e}$, were $\|\mathbf{e}\|_2 \leq \epsilon$, the solution to

$$\hat{\mathbf{x}} = \arg\min \|\mathbf{x}'\| \text{ subject to } \|\varnothing\mathbf{x}' - \mathbf{y}\|_2 \leq \epsilon$$

$\delta$-stable embedding is the additional geometry requirement in the compressed domain for CS. We say that a mapping $\Phi$ is a $\delta$-stable embedding of $(U, V)$ if

$$(1-\delta)\|U - V\|_2^2 \leq \|\varnothing U - \varnothing V\|_2^2 \leq (1+\delta)\|U - V\|_2^2$$

The matrix $\Phi$, satisfying the RIP of order $2K$, is equivalent to being a $\delta$-stable embedding of $(\Sigma_K, \Sigma_K)$ or of $(\Sigma_{2K}, 0)$. There are more random matrix constructions than RIPs in CS, and we assume the matrix will follow the design described below.

Randomize $M \times N$ matrices $\Phi$ by choosing the entries $\Phi_{ij}$ as independent and identically distributed (i.i.d.) random variables. Then we have the following:

Norm-preserving: $E\left(\varnothing_{ij}^2\right) = \dfrac{1}{M}$

Sub-Gaussian: $E(e^{\varnothing_{ij}t}) \leq e^{c^2 t^2 / 2}$

Concentration: $\Pr\left(\left|\|\Phi x\|_2^2 - \|x\|_2^2\right| \geq \delta \|x\|_2^2\right) \leq 2e^{-cM\delta^2}$

Stable embeddings will ensure the stability of our compressive classification, which will control the error below a certain level when we design the measurement matrix as required above. ■

**Lemma 8.1**

Let $U$ and $V$ be sets of points in $R^N$. Fix $\delta$, $\beta \in (0, 1)$. Let $\Phi$ be an $M \times N$ random matrix with i.i.d. entries chosen from a distribution satisfying concentration. If

$$M \geq \frac{\ln(|U\|V|) + \ln\left(\dfrac{2}{\beta}\right)}{c\delta^2}$$

then with probability exceeding $1 - \beta$, is a $\delta$-stable embedding of $(U, V)$. ■

**Lemma 8.2**

Suppose that $\chi$ is a $K$-dimensional subspace of $\mathbb{R}^N$. Fix $\delta$, $\beta \in (0, 1)$. Let $\Phi$ be an $M \times N$ random matrix with i.i.d. entries chosen from a distribution satisfying concentration. If

$$M \geq 2 \frac{K \ln(42/\delta) + \ln(2/\beta)}{c\delta^2}$$

then with probability exceeding $1 - \beta$, $\Phi$ is a $\delta$-stable embedding of $(\chi, \{0\})$.

We use random constructions as our main tool for obtaining stable embeddings. If we fix this probability of error to be an acceptable constant $\beta$, then as we increase $M$, we are able to reduce $\delta$ to be arbitrarily close to 0. We restrict to the deterministic guarantees that hold for a class of signals when $\Phi$ provides a stable embedding of that class.

The signal classification we used here is defined as the signal detection results based on the hypothesis test of distinguishing between $\Phi(s_0 + n)$ and $\Phi(s_1 + n)$.

Hypothesis:

$$\tilde{H}_1 : y = \varnothing(s_i + n)$$

It is the same as

$$\text{Min } t_i = \left\| P_\varnothing TX - P_\varnothing TS_i \right\|_2^2$$

We can equivalently think of the classifier as simply projecting each candidate signal onto the row space and then classifying according to the nearest neighbor in this space.

The classification performance can be evaluated as the distance of

$$d = \min_{i,j} \left\| s_i - s_j \right\|_2$$

It denotes the minimum separation among the $s_i$. For some $i^* \in \{1, 2,..., R\}$, let $y = \Phi(s_{i^*} + n)$, where $n \sim N(0, \sigma^2 I_N)$ is i.i.d. Gaussian noise. Then, with probability at least

$$1 - \left( \frac{R-1}{2} \right) e^{-d^2(1-\delta)M/8\sigma^2 N}$$

the signal can be correctly classified, that is,

$$i^* = \arg \min_{i \in \{1,2,...R\}} t_i$$

Within the $M$-dimensional measurement subspace (as mapped by $P_\Phi^T$), we will have a set of distances between points in $S$ by a factor of approximately $\sqrt{M/N}$.

The SUs in CRN need to detect the PUs quickly and then take the action to protect the PUs' communication. Therefore, the response time is very important. In IEEE 802.22, 30 ms is the maximum response time. Currently, the CS applications in CRN are all about how to improve the reconstruction algorithm. As shown in Figure 8.14, we use CS in each SU and transmit the compressed report to the fusion center. In the fusion center, first we cluster the reports, then we reconstruct the spectrum occupancy map of the whole area.

### 8.4.3 Simulation and Experiment Results

To simulate the UAV-based CPS, we use MATLAB® to simulate the CPS first and evaluate the performance of the PU detection scheme. The system's parameters are listed in Table 8.3. Figure 8.18 is a sample of a spectrum map after we get the binary matrix from the SU's report.

The classifier performance is shown in Figures 8.19 and 8.20. As we can see, our schemes can easily classify different spectrum patterns. Figure 8.20 shows the classification errors with different signal-to-noise ratios (SNRs). Obviously, when the SNR is higher, our scheme has less classification errors.

**Table 8.3    Simulation Parameters for CRN**

| | |
|---|---|
| Sensing duration for each channel | 10.6 µs |
| Number of samples in sensing time | 64 |
| Sensing frequency | 500 Hz |
| Number of observations | 5000 in 10 s |
| Total number of channels | 64 |
| Average percentage of busy channels | 5% (low traffic) |
| Average SNR for busy channels | –1 to 14 dB |
| Frequency | 2–2.4 GHz |
| Channel BW | 6 MHz |

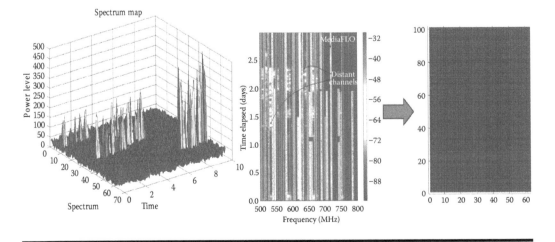

**Figure 8.18    Spectrum occupancy map in one group and binary transformation.**

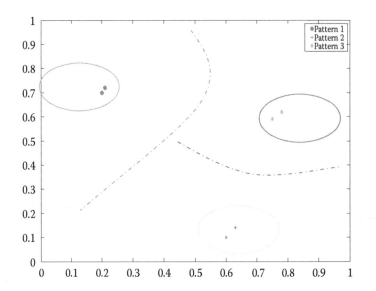

**Figure 8.19 Pattern clustering in compressive domain.**

From Figure 8.21 we can see that the CS scheme will reduce the overload of the whole system noticeably; meanwhile the delay for our CS application is not much worse than the normal sensing.

We have also simulated the system in the NS2 environment as shown in Figure 8.22. Because CS reduces the network load, the overall throughput is stably improved.

After the software simulation in NS2, next we will use USRP to simulate the CRN environment. USRP is used to make the general PC have the function of wideband software radio

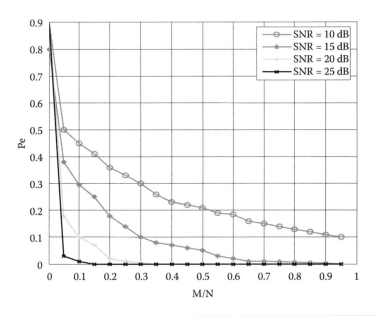

**Figure 8.20 Performance of classifier with different SNR.**

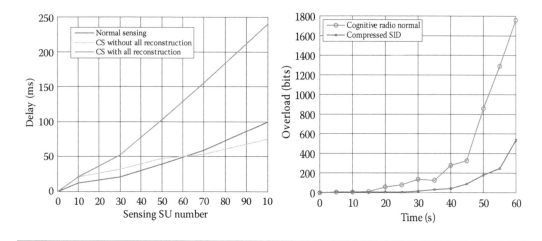

**Figure 8.21** **Overload and delay in CS schemes.**

equipment. It is an open-source platform. USRP supports many software and hardware applications and has a full developing environment where the users can create their own radio for required hardware interface, multithreading, and transplant problems. GNU radio provides the entire general software defined radio (SDR) library including modulation models, error correcting codes, a signal processing module, and a scheduler. It is a flexible system and allows users to use C++ or Python for development. USRP is designed on FPGA with all the general operations in SDR. As shown in Figure 8.23, USRP is composed of a motherboard, some daughterboard, and all kinds of antennas. We are using the USRP to simulate the environment in the lab experiments.

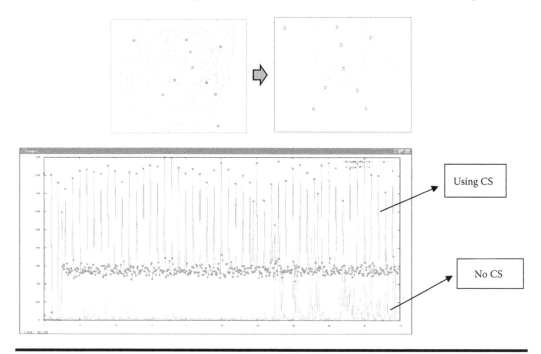

**Figure 8.22** **Throughput in NS2 simulation.**

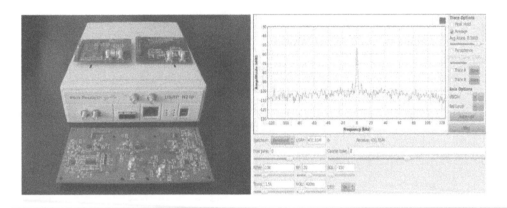

**Figure 8.23  USRP hardware and software.**

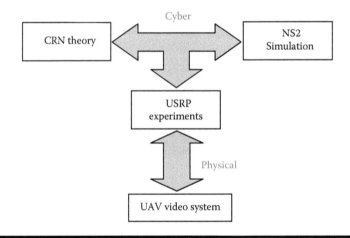

**Figure 8.24  CPS modeling process: in CRN-based UAV communication case.**

## 8.5 Conclusions

In summary, as the flowchart in Figure 8.24 shows, we have followed the CPS modeling procedure from the theory and software simulation (in the cyber world) to the hardware experiments and practice applications, which is the UAV video system (in the physical world). By analysis of the cyber results, our physical system performance can be improved significantly.

## References

1. I. J. Mitola. 1993. Software radios: Survey, critical evaluation and future directions. *IEEE Aerosp. Electron. Syst. Mag.*, 8, 25–36, April.
2. FCC. Spectrum policy task force report. Federal Communications, Tech. Rep. 02-135.
3. M. A. Davenport, P. T. Boufounos, and M. B. Wakin. 2010. Signal processing with compressive measurements. *IEEE J. Select. Topics Signal Process.* 4:2, 445–460, April.
4. E. J. Candès and M. B. Wakin. 2008. An introduction to compressive sampling. *IEEE Signal Process. Mag.* 25:2, March 21–30.

5. I. F. Akyildiz, W.-Y. Lee, M. C. Vuran, and S. Mohanty. 2006. Next generation/dynamic spectrum access/cognitive radio wireless networks: A survey. *Computer Networks: The International Journal of Computer and Telecommunications Networking.* 50:13, 2127–2159, September.

6. W. Yin, Z. Wen, S. Li, J. Meng, and Z. Han. 2011. Dynamic compressive spectrum sensing for cognitive radio networks. March.

7. C. R. Berger, Z. Wang, J. Huang, and S. Zhou. 2010. Application of compressive sensing to sparse channel estimation. *IEEE Commun. Mag.* 48:11, 164–174, November.

8. L. Liu, Z. Han, Z. Wu, and L. Qian. 2011. Collaborative compressive sensing based dynamic spectrum sensing and mobile primary user localization in cognitive radio networks. Global Telecommunications Conference (GLOBECOM 2011), IEEE.

9. S. Ji, Y. Xue, and L. Carin. 2008. Bayesian compressive sensing. *IEEE Trans. Signal Process.* 56:6, 2346–2356, June.

10. Z. Han, Z. Wu, and L. Qian. 2011. Collaborative compressive sensing based dynamic spectrum sensing and mobile primary user localization in cognitive radio networks. Global Telecommunications Conference (GLOBECOM 2011), IEEE.

11. S. Cheng, V. Stankovic, and L. Stankovi. 2009. An efficient spectrum sensing scheme for cognitive radio. *IEEE Signal Process.* 16:6, 501–504, June.

12. M. F. Duarte, S. Sarvotham, D. Baron, M. B. Wakin, and R. G. Baraniuk. 2005. Distributed compressed sensing of jointly sparse signals. In *Proc. 2005 Asilomar Conf.*, Pacific Grove, CA, October–November.

13. B. Ball, A. Brooks, A. Langville. 2007. The nonnegative matrix factorization, A tutorial. NISS, NMF Workshop, Feb. 23–24, Research Triangle Park, NC, USA.

14. J. Pike. 2012. Unmanned Aerial Vehicles (UAVs)/Unmanned Aerial Systems (UASs) sources and resources, Intelligence Resource Program, Federation of American Scientists. http://www.fas.org/irp/program/collect/uav.htm.

15. W.-C. Kao, L.-W. Cheng, C.-Y. Chien, and W.-K. Lin. 2011. Robust brightness measurement and exposure control in real-time video recording. *IEEE Trans. Instrum. Meas.* 60:4, 1206–1216, April.

16. Y. Ebisawa. 1998. Improved video-based eye-gaze detection method. *IEEE Trans. Instrum. Meas.* 47:4, 948–955, August.

17. M. D. Cordea, E. M. Petriu, N. D. Georganas, D. C. Petriu, and T. E. Whalen. 2001. Real-time 2(1/2)-D head pose recovery for model-based video-coding. *IEEE Trans. Instrum. Meas.* 50:4, 1007–1013, August.

18. Y. Wang, R. Fevig, and R. R. Schultz. 2008. Super-resolution mosaicking of UAV surveillance video. *15th IEEE International Conference on Image Processing* (ICIP 2008). October 12–15, pp. 345–348, San Diego, CA.

19. R. Hruska. 2005. Small UAV-acquired, high-resolution, georeferenced still imagery. AUVSI Unmanned Systems North America 2005, Baltimore, MD.

20. R. E. Sward and S. D. Cooper. 2004. Unmanned aerial vehicle camera integration. Technical report, Inst. for Information Technology Applications, Air Force Academy, Colorado Springs, CO.

21. H. Shen, Q. Pan, Y. Cheng, and Y. Yu. 2010. Fast video stabilization algorithm for UAV. *IEEE Intl. Conf. Intelligent Comput. Intelligent Syst. (ICIS),* 4019–4024, 2009.

22. D. L. Johansen. 2006. Video stabilization and target localization using feature tracking with small UAV video. ECE Dept., Brigham Young University, (MS thesis), December.

23. L. Angrisani and A. Napolitano. 2010. Modulation quality measurement in WiMAX systems through a fully digital signal processing approach. *IEEE Trans. Instrum. Meas.* 59:9, 2286–2302, September.

24. K. Sulonen and P. Vainikainen. 2003. Performance of mobile phone antennas including effect of environment using two methods. *IEEE Trans. Instrum. Meas.* 52:6, December.

25. R. X. Gao. 2002. Design of a CDMA-based wireless data transmitter for embedded sensing. *IEEE Trans. Instrum. Meas.* 51:6, 1259–1265, December.

26. B. Wang and K. J. R. Liu. 2011. Advances in cognitive radio networks: A survey. *IEEE J. Select. Topics Signal Process.*, 5:1, 5–21, February.

27. C. Cordeiro, K. Challapali, and D. Birru. 2006. IEEE 802.22: An introduction to the first wireless standard based on cognitive radios. *Journal of Communications*, 1:1, 38–47, April.

28. Y. Wang. 2006. Survey of objective video quality measurements. EMC Corporation Hopkinton, MA 01748, USA. Available at ftp://ftp.cs.wpi.edu/pub/techreports/pdf/06-02.pdf.

*Chapter 9*

# Cyber-Physical System Security

Steven Guy, Erica Boyle, and Fei Hu

## Contents

## 9.1  Introduction

A cyber-physical system (CPS) uses information technology (IT)/data manipulation tasks to interact with a physical area that can be impacted and from which data can be collected (Figure 9.1). CPSs have data-collecting devices that gather important information from a physical region and feed it to embedded processing devices for analysis. The analysis then determines the proper action to be taken on the physical part of the system. The computing region includes the software and algorithms used for analysis as well as the actual computing devices such as a microcontroller and processor. The physical aspect includes the "region of interest" (ROIn), from which data are collected to be used for analysis, and the "region of impact" (ROIm), which the computing side intentionally alters with an effect fitting for the result of the analysis. These systems are becoming more common as people are finding effective cyber-physical solutions to real-world problems. CPSs are often used to perform what is known as "mission-critical" tasks; these are tasks that require a constantly active system that will essentially never fail. For instance, health monitoring systems use embedded chips to analyze data from a physical field (e.g., heart rate and blood sugar). Many diabetic persons use CPSs to monitor their blood sugar and automatically react to extreme situations by injecting substances into the blood stream.

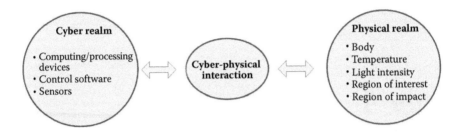

**Figure 9.1  Anatomy of a CPS.**

While CPSs solve many pressing needs, they are also vulnerable to conditions that can cause these mission-critical systems to fail. The three main conditions to be concerned with are safety, security, and sustainability. These three can be abbreviated by "S3."

Safety is indispensable for mission-critical systems. In the past, people have reduced the task of ensuring safety to maintaining effective software, but this does not encompass all safety measures. In order to achieve optimum safety, one should focus on the safety of the interaction between the computing devices and the physical systems. One main safety concern is that of the cyber-physical interactions of separate computing devices interfering with each other. For instance, electromagnetic radiation from one computing device could interfere with the monitoring capabilities of another device. Similarly, the physical effect of a single computing device (e.g., heat produced by a processor) could directly alter the conditions of the physical. Finally, the reverse of this second scenario could happen: physical phenomena resulting from the conditions being monitored could affect the functionality of the cyber devices (e.g., condensation run-off onto a computing device). These kinds of safety issues need to be addressed and prevented in the design of an effective CPS.

The second "S" of S3 is security. Security is defined in [1] as "the ability to ensure that both data and the operational capabilities of the system can only be accessed when authorized." Unauthorized access to any aspect of a CPS poses obvious security threats. Compromised security of a CPS can allow compromised privacy and harmful effects to the physical environment. With more and more industries using CPSs to achieve mission-critical operations, it is imperative that security is always at the forefront of the design.

Finally, energy sustainability is a key concern with CPSs. Sustainability is defined in [1] as "the balance between the power required for computation and the power available from renewable or green sources." There are calls to shift from batteries and generators to environmental energy sources like solar and wind power, but with these efforts come significant challenges regarding how to effectively incorporate them into a system design.

These three topics are the main concerns for CPS design, but in this chapter, we will focus on the issue of security. It is the opinion of most researchers in this area that security must not be approached solely from the cyber side or from the physical side but rather should incorporate elements from both regions that can work together to prevent, detect, and handle any kind of attack by a malicious third party. While this opinion is widespread, the nature of the security system is a topic that covers a wider base of opinions. Many methods of security, though, employ the monitoring system that links the cyber and the physical regions so that security is effectively built into the system, rather than being applied as a patch after completion of the system. Several different methods of providing security to CPSs are covered in this chapter.

There are several different industries affected by CPSs, including body area networking (BAN), smart infrastructure (SI), and data center energy efficiency. BAN applications are mostly in the medical field, involving health monitoring systems. SI systems mainly pertain to smart electric grid infrastructure. Data center energy efficiency involves acknowledging idle and active modes so that nonvital functions are not always on and consuming power.

The information presented in this chapter covers research proposals and summaries on CPS security. Some of the research challenges faced are that the security features can potentially affect the physical ROIn in such a way that corrupts data, as well as the fact that the physical region can be attacked in a way that the computing region cannot monitor, thus causing the computing functionality to not work properly.

## 9.2 Smart Grid Security

Smart grid is a typical CPS. One of the first topics worth discussing is a generalized security method that involves compensating for accidentally observable actions. The purpose of this is to enforce the security properties of information flow in a CPS. If an event occurs within the system that is observable outside the system (thereby posing a potential security risk), then a separate "compensating" action is taken to essentially cover up the offending event. The correcting action can cause the observable event to either resemble a different expected event or simply render its visible effects negligible. This way, even if system changes are viewable by outside parties, the correct information cannot be interpreted to an unauthorized viewer.

Other attempts to enforce information flow include compile-time methods and runtime methods, the latter being more effective. These methods usually involve constant monitoring processes running in the system and prevent execution of a certain process if it is expected to compromise security. The framework discussed here is different in that it can react to a security threat *after* the compromising event has taken place.

A formal summary of the framework can be explained as follows. We want the relevant CPS to adhere to an information flow property (IFP) *P*. However, sometimes there is an event that causes a violation of *P*. When this event, or "violation point," is detected, the system immediately inserts a compensating event or sequence of events into the series of processes currently being executed. Any individual process of this correcting event chain may violate *P* (again), but the chain is executed quickly enough that the violation is not noticeable. At the conclusion of the compensating event chain, an observer would see that *P* is adhered to as if the violating event never occurred. In short, if an event occurs that causes a security violation, a correcting event is immediately injected in order to negate the violating effect and ensure that the end result is the same as if there were no violation to begin with.

One very important application of this security framework is a "smart grid" power transmission infrastructure. Smart grid technology allows for the use of a "flexible AC transmission system" (FACTS) network. FACTS devices can alter particular transmission line characteristics (e.g., impedance). If there is a failure in the electric grid, the FACTS devices can communicate with each other in order to properly alter transmission line characteristics as needed to achieve the necessary power redistribution. FACTS device settings must be kept secret in order to avoid exposure of weaknesses in the network that could lead to physical and control system attacks.

Toward this end of achieving sufficient security in a smart grid, it is important to discuss the correlation between observability and deducibility of CPS information by an outside observer. In [2], there is a lemma that states that "a mix connected network with *n* number of reconfigurable units and *k* number of junctions can be fully deduced with a minimum of $n - k$ observers." This effect is a violation of "nondeducibility security" with respect to IFPs.

We can adopt the concept of mix-connected network (Figure 9.2) to solve the above issue. Suppose two variable resistors are changed in the network, thus causing differences in current. For some of the affected current streams, the current change due to the change in one resistance was opposite to that due to the change in another resistance. This results in no externally noticeable change. These two changes are an example of a compensating couple, which allows changes to be made in the network that are not externally observable. However, there are a finite number of change events that have compensating couples. One resulting strategy from this limitation is to include as many possibilities for compensating changes like this one that have unobservable effects as can be practically implemented.

This goal of applying compensating effects to a system to prevent observable changes to the CPS is much more generalized and therefore very different from other approaches that involve much more targeted and direct approaches to security.

Before discussing these other targeted approaches, we should first go into the ways that the smart grid infrastructure is becoming more advanced. Generation, transmission, distribution, and consumption can all be monitored to make electric services more personalized and convenient. There has been a good deal of research lately involving making the electric grid more "intelligent" by causing generation to adapt dynamically to such factors as consumer demand and varying levels of input from wind, solar, and other alternative sources. Also, researchers want the smart grid to be able to give consumers live information so they know when to use more or less energy. For instance, it would be more economical to perform an energy-intensive task during nonpeak times when price is low. Essentially, we need to incorporate IT systems into the power grid for the purposes of improving price information, meter data, and control commands, among other features (Figure 9.3). The smart grid provides great opportunities, but its complexity and reach introduces significant vulnerabilities to attack. Cyber-physical security is needed to prevent and combat these

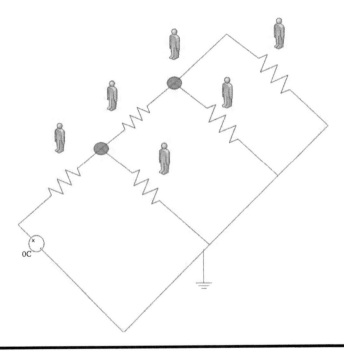

**Figure 9.2 Fully observable mix-connected network.**

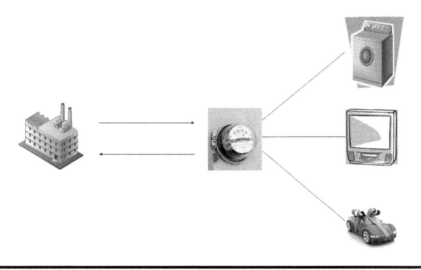

**Figure 9.3   Power plant interacting with appliances through a smart meter.**

potential attacks. We need approaches that can defend against hybrid attacks as well as focused attacks on either the cyber realm or the physical realm.

The first security realm we will discuss is that of the cyber portion of the CPS. There are several requirements in order to achieve effective cybersecurity. First, power consumption information needs to be kept private in order to protect identities as well as personal trends and habits. Additionally, we need to maintain integrity of data, commands, and software [3]. For instance, if price information were compromised, an attacker could change prices to induce a certain effect by smart appliances, which could lead to loss of money for the power company (negative prices) and/ or confusion for customers. Finally, information needs to be accessible at all times, even if there is a denial of service (DoS) attack. Malicious DoS attacks are very common on web servers, and they could certainly be a popular choice by criminals on a smart electric grid.

In order to plan a defense system, we need to anticipate all possible attack models. Intruders would probably identify weak points in the system defense that are vulnerable to attack (e.g., network attack against an ineffective firewall). When some kind of access is obtained, the attacker just needs to alter something within his or her reach to cause potentially disastrous cyber and physical consequences. There are a number of countermeasures to these kinds of attacks that focus on effective defense systems. One method is to have an intelligent key management system. Those with access to change system settings are few and the places from which they can access the system are as secure as possible. Additionally, a secure communication architecture is imperative. All data communication needs to be encrypted and secure in order to prevent intercepted information. This communication system also needs to be protected against DoS and jamming attacks. Finally, system and device security is just as important as the previous two measures. There is always a need in a CPS for well-written software that can prevent injection of a harmful code and detect it when it happens.

All these security measures are very important, but security approaches that only involve cyber solutions will not be effective on a CPS, as an attack on a physical system cannot necessarily be detected or counteracted by cyber systems. Security systems need to also be present in the physical realm of a CPS. The most important requirement is to operate resiliently even during disturbances, attacks, and natural disasters [3]. Reliable operation throughout adverse conditions is essential for a system as integral to our lives as the electric grid. The design of the power system should take into account several specific predicted failure scenarios. Naturally, not every scenario can be predicted, but there are certainly some attack methods that are more likely or more feasible than others, and these should be addressed. Finally, one attack should not cause complete system failure. There should be measures in place to detect and respond to an attack in a way that allows no more than just a localized part of the system to be compromised.

These prevention techniques should be geared toward a probable attack model, as with the cybersecurity techniques. An attack usually involves an intruder altering something about the physical network. For instance, if an attacker were to successfully shut down a generator or change a variable resistance in the network, it could cause important data regarding price and availability of power to be corrupted. The countermeasures against these attacks are contingency analysis and bad data detection. Contingency analysis involves screening and ranking the importance of all monitored conditions in anticipation of an attack, while bad data detection simply refers to the phenomenon that inaccurate information will be identified as such.

Finally, though these cybersecurity and physical security measures can be fairly effective, there also needs to be a presence of a security measure that involves both the cyber and the physical realms, as any given attack may affect both realms while only alerting security from one or the other. A known example of such an attack is the replay attack. This is an attack launched on the cyber side that captures a stream of normal physical data. An attacker would then use these data and "replay" them back to the data analysis in the cyber side, while actually altering certain aspects of the physical system. This way the cybersecurity is fooled, and the physical realm has an unnoticed security breach. In [3] it proposes an algorithm to detect a replay attack. The method attempts to incorporate physical authentication in order to confirm that data received by the processing units are coming from the correct source. In addition to the sensors that are monitoring the physical part of the system, the system has a specially designed estimator, controller, and detector that work together to alert if the presence of a replay attack is detected with high probability.

These CPS security techniques for smart grid infrastructure discussed here are really only the building blocks for a movement that can hopefully produce reliable systems in the future. Smart grid development and security is a very enticing research area, and surely more comprehensive security solutions will emerge in the future.

## 9.3 System Snapshot

There is a need to monitor potentially harmful gases and liquids in order to determine the state of the physical phenomenon, predict the future state, and control the state. In an ideal setting, the CPS would have the functionality to obtain a snapshot of the entire physical setting from all sensors without any kind of delay; however, this is not the case in real-world application. In reality, sensor readings are only available at discrete time instants and only at certain geographical locations. Through research, the Nyquist sampling theorem was developed to formulate the necessary sampling frequency.

CPS designers face a quandary in balancing a high sampling rate in both space and time. A designer can deploy a large number of sensors as a dense network and hence obtain excellent resolution in space (high sampling rate in space). However, if those sensors are in a small geographical area, then no two sensor nodes can broadcast their sensor readings simultaneously (on a wireless channel or a shared bus), so the time required in order to gather all sensor readings becomes large and consequently the sampling rate in time becomes low. A designer can take a diametrically opposite strategy by deploying a small number of sensor nodes and hence obtain a high sampling rate in time but not in space. The paper [4] documents a group of researchers' work in developing an algorithm that allows a CPS to quickly obtain an approximate representation of sensor readings as a function in space in order to achieve a high sampling rate in both space and time. This new algorithm exploits a prioritized medium access control (MAC) protocol to efficiently transmit information of the sensor data.

The main idea for this work is the use of a prioritized MAC protocol—inspired by Dominance/Binary-Countdown protocols. Through these types of protocols, messages or signals are assigned unique priorities, and before nodes try to transmit they undergo a contention resolution phase in which the node trying to transmit the highest priority succeeds. This prioritized MAC protocol will be used for a more efficient distributed computation of aggregated quantities. The research team then used the concept for the prioritized MAC protocol to determine a means for interpolating sensor data with location. The computation first assumes no sensor faults and then later accounts for sensor faults.

## 9.4 Quantitative Modeling of Dependability

Major physical infrastructures such as the electric power grid and water distribution systems are complex and large-scale systems that are required and expected to have high dependability. The principal objective of the proposed research in [5] is to build a qualitative and quantitative understating of dependability in CPSs. The need for control and communication for systems such as the aforementioned creates a CPS with two parallel networks: a cyber network containing intelligent controllers and the communication links among them and a physical network of interconnected components of the infrastructure.

The research team used a physical water distribution system—coupled with the hardware and software that support water allocation—as the model CPS case study. The physical components, for example, pipes, valves, and reservoirs of water distribution networks tied with the hardware and software, provided a very good case study. The primary goal of water distribution networks is to provide a dependable source of potable water to the public. Information such as demand patters, water quality, and water quantity is critical in tracking the efficiency of the system to achieve this goal. Sensors dispersed throughout the physical infrastructure collect the information and feed it through a set of algorithms running within the cyber infrastructure. These algorithms aid in making decisions to hardware controllers in order to manage the quantity and quality of the water.

In order to organize the work, the project was divided into four overlapping tasks:

1. Build a qualitative understanding of interdependencies in CPS.
2. Quantitative representation of these interdependencies.
3. Characterize CPS dependability in terms of attributes of the cyber and physical components.
4. Validate the models using simulations and field data.

The details of their water distribution network case study will be first discussed. As mentioned previously, the research team will be using an urban water distribution network as a basis to delve into a CPS. The water distribution network will be instrumented by sensors that quantify the pressure, volume, and aspects of the chemical composition of water at various locations in the city. The records from these sensors will be relayed to a number of multiplexors for collection and then forwarded to a control center. Valves will use this information in conjunction with software algorithms from the cyber component of the CPS to manage the quality and quantity of water allocated to various areas within the city. The availability of data that captures information such as temperature, precipitation patterns, and usage will facilitate dynamic control of the water distribution network, resulting in an increase in system utilization and improving the reliability of water distribution.

A priority goal is to capture the physical manifestation of failures in cyber control. The purpose for this goal is to define "safe" states where the physical and cyber components are sufficiently operational so that these safe states can be classified or defined as "functional." The Markov Imbedded System model will be used to define these safe states. Once a distinction between safe and "failure" states has been drawn, fault injection will be used to examine failure propagation from the cyber layer to the physical layer. More specifically, faults will be injected in the distributed software algorithm used for control of water quantity and quality, as well as in the supporting of communication and hardware links.

Task 1 of the research team's project plan is accomplished using a simulation of an intelligent cyber-controlled water distribution network. The information gained from this portion of the research will be aggregated into a qualitative agent-based model. Then, the Markov Imbedded System is used to quantify these results in Task 2. The model is refined in Task 3. Finally, the field data and the simulation development in Task 1 are used to validate the models developed for Task 4.

An underlying problem in the study of the dependability in a CPS is quantifying and dividing the interdependencies between physical components and cyber components within the system. This could partly be attributed to the complexity of these systems. Existing modeling techniques for CPS rely on semantics to represent the relationship between the physical and cyber layers.

As part of Task 1, a comprehensive literature review on reliability modeling for CPS was conducted. Furthermore, an extensive study of various simulators for the physical and cyber layers of water distribution networks will be required. One simulator in particular has proven to be an appropriate candidate thus far. For simulating the physical layer, an open-source software package developed by the Environment Protection Agency was found. This simulator includes a wide array of attributes in common water distribution networks including water pressure and presence and amounts of chemical elements such as fluoride or chlorine.

In summary, the purpose of this research proposal [5] is to build both a quantitative and qualitative understanding of dependability in CPSs. The hope of the project is that quantitative modeling of dependability for CPS will ease concerns associated with the conjunction of intelligent cyber control to critical infrastructure systems, thus hopefully paving the way for more widespread applications of CPS.

## 9.5 Conclusions

CPSs are becoming more prevalent in our technologically advanced society. They are used for a wide variety of purposes, ranging from electric grids to health monitoring applications. The continued operation of these systems is imperative to support the wellbeing of consumers across many industries. Therefore, we must always be putting forth renewed efforts to find more innovative and effective methods of ensuring the safety, security, and sustainability of CPSs. This chapter introduces a few selected methods to achieve these ends. CPS research is very lucrative these days, and hopefully CPSs will continue to benefit and improve in the coming years.

# References

1. A. Banerjee, K. K. Venkatasubramanian, T. Mukherjee, and S. K. S. Gupta. 2012. Ensuring safety: Security and sustainability of mission-critical cyber physical system. *Proc. IEEE* 100(1), 283–294.

2. T. T. Gamage, B. M. McMillin, and T. P. Roth. 2010. Enforcing information flow security properties in cyber-physical systems: A generalized framework based on compensation. In *Proceedings of the 2010 IEEE 34th Annual Computer Software and Applications Conference Workshops* (COMPSACW '10). IEEE Computer Society, Washington, DC, pp. 158–163.

3. A. Ashok, A. Hahn, and M. Govindarasu. 2011. A cyber-physical security testbed for smart grid: System architecture and studies. In *Proceedings of the Seventh Annual Workshop on Cyber Security and Information Intelligence Research* (CSIIRW '11), F. T. Sheldon, R. Abercrombie, and A. Krings (Eds.). ACM, New York, Article 20.

4. B. Andersson, N. Pereira, and E. Tovar. 2008. How a cyber-physical system can efficiently obtain a snapshot of physical information even in the presence of sensor faults. *WISES 2008*: 1–10.

5. J. Lin, S. Sedigh, and A. Miller. 2009. A general framework for quantitative modeling of dependability in cyber-physical systems: A proposal for doctoral research. In *Proceedings of the 2009 33rd Annual IEEE International Computer Software and Applications Conference* (COMPSAC '09), vol. 1. IEEE Computer Society, Washington, DC, pp. 668–671.

*Chapter 10*

# Cyber-Physical System Security—Smart Grid Example

Rebecca Landrum, Sarah Pace, and Fei Hu

## Contents

## 10.1 Introduction

As more technology is introduced to the public, the demand for new highly technological devices also escalates, causing more loads on the power system. To keep up with this demand, the smart grid needs to find more ways to provide higher levels of reliability. The US Department of Energy has defined the following seven items as requirements to meet the future demand of the smart grid: resistance to attack, self-healing, motivation of consumers, quality of power, generation and storage accommodation, enabling markets, and optimizing assets. The attacks on September 11, 2001, and other related incidents have made stronger security a higher importance to prevent malicious attacks. This security will need to include cybersecurity to prevent the communication systems from being harmed. Reports have stated that the US power grid has been infiltrated by cyber spies.

Much research has been done to create a more reliable and secure system using a combination of cyber and physical components together to guard the smart grid. This type of system will make it easier to detect and defend against attacks. The combined approach will increase the possibility that the system and its data will remain protected and dependable. The security discussed in this chapter will focus on the following topics: smart tracking firewall, probabilistic dependence graph, anomaly detection, dynamic security, real-time reconfigurable system, and other cyber-physical system (CPS) security issues [1,2].

## 10.2 Smart Tracking Firewall

Smart distribution grids (SDGs) [3] are readily used in applications for power distribution because of the many advantages they display, such as efficiency, resiliency, reliability, and sustainability. In order to facilitate the highly developed traits of SDG, technology must be shaped to support the sophisticated structure. The communication network that fills this role must be adept at sensing information to be interchanged while distributing the appropriate information to specific electric components. Both wired and wireless communications can be chosen either in conjunction or separately in order to effectively carry out these requirements, depending on the circumstance and transmission method.

Some benefits of wireless communications in power distribution networks include component bypassing abilities, lower cost, straightforward design, and versatile networking between neighboring devices. Many technologies exist in the wireless realm, including, but not limited to, wireless local area networks (WLANs), WiFi, cellular networks, and wireless sensor networks (WSNs). The proficiency of these networks alone in an SDG is not desirable because of a variety of performance issues that can result owing to ineffective peer-to-peer communication, inconsideration of varying applications, or rate-distance reliability setbacks. In order to account for the discrepancies of the aforementioned networks alone, wireless mesh networks (WMNs) have been proposed to encompass the technologies in a single integrated network. The WMN is applied to each electric device in an SDG to fulfill the range of demands required such as collecting, sending, and monitoring information to and from electrical equipment.

Security in WMN for an SDG is a concern because of our dependence on electricity supplied by the power distribution networks, and attacks can be made to slow or completely shut down operations. A number of techniques can be implemented to carry out these attacks. Jamming is caused by signals that are of the equivalent frequency to the WMN. Malicious nodes either inside or outside of the WMN have the ability to eavesdrop in order to decode information deceitfully received or overheard by neighboring packets. If a malicious node gains access inside the WMN, protocols such as routing or medium access control (MAC) can be manipulated in a way that packets are misdirected or modified.

Security approaches and techniques are available to ensure attacks will be detected and cleared before some of the dangerous consequences are realized. The physical layer is instrumental in detecting and counteracting attacks made through jamming by way of frequency hopping or direct sequence techniques. When an eavesdropping node is outside of the WMN, encryption measures are the first line of defense, but physical layer security procedures actively being researched act as an effective backup. Malicious nodes inside the WMN that have been authenticated and passed through secure MAC and routing protocols have already penetrated the first stage of security and therefore need to be regarded as exceptionally threatening to the SDG system. For such attacks, a higher level of tracking must be utilized.

Smart tracking firewall is a security measure used to prevent attacks made by malicious nodes that infiltrated the secure WMN. In order to carry out this function, mesh nodes in the network will have the ability to locate and deposit previously detected intruding nodes into either a blacklist or a gray list. Separate lists reduce signaling overhead. When a node is blacklisted by a mesh client, the node cannot communicate with the client by either sending or receiving packets of information. A mesh node has the ability to position a malicious node into their gray list when neighboring nodes send out alerts about a blacklisted node. When a certain number of alerts are sent out or when the intruder node is within the immediate network area, the node is immediately placed in the blacklist, communication is cut off, and a prealarm is sent out to its neighboring nodes. Using this method, zones of defense are quickly and flexibly orchestrated such that malicious nodes have a significantly lower chance of successfully attacking.

As expected, research demonstrates the effectiveness of this scheme. While under attack, simulations utilizing smart tracking firewall have increased packet delivery and throughput as well as lower levels of delay when compared to no detection or individual response methods. Overall, WMNs are effective means of communication among electronic devices in an SDG, and although security attacks are a possibility, when smart tracking firewall is implemented, the risk of danger is significantly decreased.

## 10.3  Probabilistic Dependence Graph

Because power is such a valuable resource, many ways to protect the smart grid have been contemplated and analyzed for reliability purposes. Specifically, fault detection and localization are ways in which dependability can be measured to ensure proper function of power grids. However, fault event diagnosis systems are not readily equipped in today's power grid systems in order to detect widespread fault events due to malicious attacks or naturally occurring physical events. Although this detection is proven to be a rather challenging feat, proven methods do exist and are found to be effective because of increased interest in this area of study.

Some of the challenges include fault recognition and correction into existing power grid systems, the substantial grid size, the inconsistencies when considering systems of this size, and the unavoidable computational and measurement errors when conducting research and experimentation of these structures. Several methods have been proposed to deal with the difficulties such as using sensor networks by embedding them in the system and phasor measurement units (PMUs) by evaluating the phasor information in order to detect and evaluate conditions of transmission lines as they occur. Nevertheless, each of these techniques comes with its own drawbacks, such as the considerable range of variables needed for comparison, which therefore makes them too complex to be incorporated into a real-world scenario sensibly.

To resolve these issues, a probabilistic graphical approach [2] is introduced that utilizes spatially correlated information from PMUs and statistical hypothesis testing. A Gaussian Markov random field (GMRF) is employed to model phasor angles across the buses in a power system in a way that the phasor angles are evaluated as random variables and their dependencies can be studied. The dependence graph illustrates the connections using a Markov random field that is induced by a minimal neighborhood system by inserting an edge between sites that are neighbors. The pairwise Markov property of a GMRF is also exercised such that an MRF is normal with the mean $u$ and variance $J^{-1}$, where $J$ is the information matrix of the MRF, so in this instance $J$ is zero. Also, Gaussian random variables are used to approximate fault diagnosis functions such as flow injection as a result of multiple load requests as well as difference of phasor angles across a nonslack bus.

Several models and hypotheses have been used to further illustrate these concepts. The conditional autoregression (CAR) model is a noteworthy model that explains the conditional distribution where $\mathbf{X}_{-i}$ is an MRF:

$$X_i \Big| \mathbf{X}_{-1} \sim N\left[ u_i + \sum_{j \neq i} r_{ij}(x_j - u_j) \right] \tag{10.1}$$

After research and analysis of the MRF and CAR, an approach was hypothesized such that the null hypothesis is as follows:

$$H_0 : \{\text{there is no change in } r_{ij}, \text{ for all } \{i, j\} \text{ as an element of } E'\}$$

where $E'$ is the edge set.

A challenge faced by these models and hypothesis is the difficulty in correctly estimating the information matrix of the MRF and, in effect, the actual value because of a small sampling number or noise involved. An optimization problem increases the prospect by maximizing the equation $\log |\hat{J}| - \text{tr}(\hat{J}\Sigma)$ while it is also under the circumstance that $\hat{J}_{ij} = 0$, where $(i, j)$ is not an element of $E'$.

In the past, a centralized approach would have been exercised when faced with this problem, but the more sophisticated method of decentralizing is employed in this case. Samples are controlled to the size of the biggest subfield in the decentralized approach as opposed to the entire GMRF as is in the centralized method; therefore the involvement of the computations and measurements is significantly and beneficially reduced. Hypothesis investigation is also decomposed into smaller subproblems, and as a result sites are divided into border sites and inner sites, and the edges are divided into tie-line, border-line, and inner-line edges. Tie-line edges connect subfields, border-line edges connect border sites within the same field, and inner-line edges connect subfields but have at least one of its ends as an inner site. To make the decentralized approach even more reliable, two-scale decomposition is employed to achieve message passing used to accumulate data involving the subclasses of border sites and edges. Just as the two-scale decomposition, a multiscale decomposition can be utilized as necessary for larger-scale systems. Figure 10.1 demonstrates this approach.

Analysis of the techniques used can be summarized by an algorithm that administers the ability to estimate the information matrix as completely as possible given that all phasors at all buses are discernible. It accomplishes this tough task by reconstructing the information matrix by utilizing information from message passing after estimating the subfields. The algorithm is supplied below.

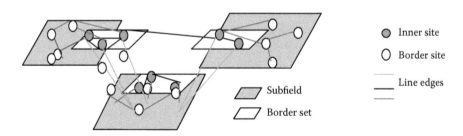

**Figure 10.1   Decentralized estimation with multiscale message passing.**

**Algorithm 10.1: Decentralized *J* Estimation Using Multiscale Message-Passing**

Local estimation: Estimate $\hat{J}_k^l$ for each $X_k^l$ based on local observations.

Down-top message passing: for $l = 1, 2,\ldots,(L - 1)$, the inference centers at $X_k^l$, with $f$ an element of $F(k, l)$, submit $\hat{J}_k^l$ to that of $X_k^{l+1}$.

Top-down reconstruction: For $l = L - 1, L - 2,\ldots,1$, the inference center at $X^L$ reconstructs $\hat{J}^l$ from $\hat{J}^{l+1}$ and $\hat{J}_k^l$, $k = 1, 2,\ldots,K^l$.

Top-down message-passing: The inference center at $X^L$ broadcast $\hat{J}$ to all subfields.

Because many variables exist in a power grid system, a process involving randomly determined sequence of observations where each observation is considered a sample of an element from a probability distribution must be formed. The GMRF model for phasor angles as well as the algorithm successfully performs fault detection and localization.

## 10.4 Anomaly Detection

The importance of security in power systems is fundamental, so a broad assortment of solutions to the cybersecurity at substations has been proposed and considered. It is imperative that anomaly and vulnerability detection due to cyber-attacks be identified and resolved for the protection of substations [1].

First, the type of intrusion methods must be identified such that normal operation of the power system will remain undisturbed. Cyber-attacks can be employed in various ways. Some of them include synchronized attacks on several substations because of the accessibility of the infrastructure at multiple locations. And attacks from a vast array of combinations could remain unnoticed because of intelligence capabilities on the attacker's end. Figure 10.2 illustrates the possible course an attack could take. As shown, intrusions could occur in remote connections through TCP/IP or through DNP/Modbus protocols. If the attacker is successful in gaining contact with C1 or C2, the user interface as well as the substation intelligent electronic devices (IEDs) are vulnerable. The user interface contains a direct contact to overall substation communications and can therefore be used destructively to identify and devastate the exact components used for controlling switches, breakers, and other electrical equipment. When a password for an IED is ascertained, the intruder will have access to important documents such as one-line and data flow diagrams that can then be used to administer commands to circuit breakers that would cause catastrophic events to occur at the substation.

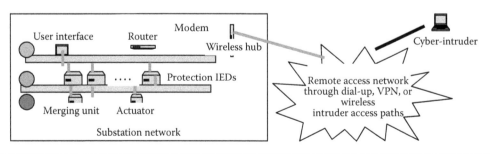

**Figure 10.2   Attack scenario.**

In order to become aware of such intrusions on substation networks made by cyber-attackers, an anomaly detection algorithm that employs benefits such as monitoring in real time, analysis of possible effects of intrusion, and approaches for mitigation is examined [1]. Monitoring the power system in real time enables the algorithm to rapidly and efficiently determine the status of computers and equipment in order to allow a maximum number of connections to be implemented, as well as to authenticate the connection via response times and IP addresses. Such features can detect and track anomalies such as unsuccessful logon attempts in accordance with time and frequency and destructive modifications to files that are vital to the substation wellbeing. These are characteristics of an intrusion being attempted, and if an attack is suspected, an alarm list of possible attackers is created and the device the intruder is attempting to attack is put on lock. Table 10.1 explains the anomaly detection algorithm described. The last column shows where the intruder attempts to change factory settings as well as the location where the event is extracted.

Along with the algorithm, status bits are used to aid in the detection of certain intruder-based irregularities. Below, a row vector is presented to illustrate the weight given to the different bits:

$$\pi_{(I \times k)} = [1 \alpha \pi^{a}_{(T \times L)} \beta \pi^{fs}_{(T \times M)} \delta \pi^{cs}_{(T \times N)} \varepsilon \pi^{o}_{(T \times O)}]$$

where $\pi^a$ is the discovery of intrusion attempts on computers or IEDs, $\pi^{fs}$ is a modification made to a file system, $\pi^{cs}$ is a modification to an important IED setting, and $\pi^o$ is a modification made to switches. The weighting factors are $\alpha$, $\beta$, $\delta$, and $\varepsilon$, and $L$, $M$, $N$, and $O$ are the sizes of the components. $T$ is the number of anomalies in a given amount of time. The resulting matrix is then ranked using the equations below in order to establish whether an intrusion event has taken place at the given substation:

$$\Pi = \frac{\pi}{\|\pi\|2}$$

$$\zeta = \text{rank} \, (\Pi) - 1$$

where $\Pi$ is the normalized vector of $\pi$ and $\zeta$ is the index determined by the rank of $\Pi$ for the substation. If the index $\zeta$ is nonzero, an anomaly event could have occurred, and the substation is then put on the list of possible attacks to be further investigated.

A malicious attack on a substation can be incredibly detrimental, but if an intrusion occurs in critical or multiple substations, the potential outcome can be even more disastrous. In order

**Table 10.1   On the Anomaly Detection Algorithm**

| To/From Control Network or User Interfaces | Attempts of Cyber-Intruders |
|---|---|
| Control | Attempt to connect to IED |
| Setting | Acquire login information |
| Measurement | Successful log in to IED |
| Data log | Gain control of circuit breaker |
| Test/diagnosis | Change setting to factory status |

to detect and diminish such attacks, precautions such as generating an inventory of substations in which deenergizing components would result in complete voltage collapse, ranking suspected intrusion events made to multiple substations in a short timeframe, and designing a system to alert power workers when disturbances occur are accomplished. In addition, an impact factor γ is calculated using the following equation:

$$\gamma = \left( \frac{P_{\text{LOL}}}{P_{\text{Total}}} \right)^{L-1} \tag{10.2}$$

$P_{\text{LOL}}$ is the power flow where loss of load occurs, $P_{\text{Total}}$ is the total power flow, and $L$ is the loading level of the substation. Using this impact factor, the vulnerability of substations is estimated and evaluated so that the least potential damage is realized when an intrusion transpires.

The anomaly detection algorithm presented has been thoroughly tested and analyzed using the well-established Institute of Electrical and Electronics Engineers (IEEE) 118-bus system model in order to establish its credibility and ability to preclude a malicious attack on substations [1]. As expected, the algorithm effectively pinpoints multiple simultaneous logon attempts, impact factors, efforts to manipulate critical files in a detrimental way, and attacks made on multiple and/or critical substations. However, to efficiently implement such a security effort, more research has to be completed in the area of application to already established software and framework currently within the substations [4].

## 10.5 Dynamic Security

Mathematical models called structure-preserving power network models are used at the transmission level to describe dynamic swing equation for the generator rotor dynamics and the algebraic load-flow equation for the power flows through the network buses [5]. This section focuses on the small signal version of the structure-preserving model, which uses the DC power flow equation and linearized swing equation. Most security analysis today is based on static estimation techniques for magnitudes at buses and voltage angles. This is because of the low bandwidth of communication channels for measurements to the control centers, the difficulty in finding and tuning an accurate dynamic model, and dynamic models being more difficult to use. However, with the advancement in recent technology, many of these problems can be solved. This section will provide an integrated modeling framework for dynamic power networks exposed to cyber-physical attacks, show the primary limitations of static and dynamic detection and identification procedures, and offer a detection and identification procedure based on geometrically designed residual filters.

The network studied in this section includes $n$ generators $\{g_1,...,g_n\}$, $n$ generator terminal buses $\{b_1,...,b_n\}$, and $m$ load buses $\{b_{n+1},...,b_{n+m}\}$. The interconnection structure is encoded by a connected admittance-weighted graph. The Laplacian matrix of this graph is $\begin{bmatrix} \mathcal{L}_{gg} & \mathcal{L}_{g1} \\ \mathcal{L}_{1g} & \mathcal{L}_{11} \end{bmatrix} \in \mathbb{R}^{(2n+m)\times(2n+m)}$.

The differential-algebraic model of the power network is provided by the linear continuous-time descriptor system shown below:

$$E\dot{x}(t) = Ax(t) + P(t)$$

In this equation, $x = [\delta^T \omega^T \theta^T]^T \in \mathbb{R}^{2n+m}$ is made up of the frequencies $\omega \in \mathbb{R}^n$, generator rotor angles $\delta \in \mathbb{R}^n$, and bus voltage angles $\theta \in \mathbb{R}^m$. The input $P(T)$ is the acknowledged changes in mechanical input power to the generators or real power demand at the loads. The remaining variables used above are shown below:

$$E = \begin{bmatrix} I & 0 & 0 \\ 0 & M & 0 \\ 0 & 0 & 0 \end{bmatrix}, A = -\begin{bmatrix} 0 & -I & 0 \\ \mathcal{L}_{gg} & D_g & \mathcal{L}_{g1} \\ \mathcal{L}_{1g} & 0 & \mathcal{L}_{11} \end{bmatrix}$$

Let $y(t) = Cx(t)$ represent the $p$-dimensional measurements vector and let $C \in \mathbb{R}^{p \times n}$ be the output matrix. Disturbances that show up in the measurements vector after being integrated through the network dynamics are called state attacks. Disturbances that corrupt the measurements directly are called output attacks. The network dynamics in the existence of a cyber-physical attack can be written as

$$E\dot{x}(t) = Ax(t) + \underbrace{[F \quad 0]}_{B} \underbrace{\begin{bmatrix} f(t) \\ \ell(t) \end{bmatrix}}_{u(t)}$$

(10.3)

$$y(t) = Cx(t) + \underbrace{[0 \quad L]}_{D} \begin{bmatrix} f(t) \\ \ell(t) \end{bmatrix}$$

The inputs $\ell(t)$ and $f(t)$ are the output and state attacks, respectively. Throughout this section, $T \subseteq \mathbb{R}_{\geq 0}$. An attack can be called undetectable if there is an attack set $K$ that has an initial condition $x_1, x_2 \in \mathbb{R}^{2n+m}$, and an attack mode $u_K(t)$ so that, for all $t \in T$, $y(x_1, u_k, t) = y(x_2, 0, t)$. An attack can be called unidentifiable if there is an attack set $R$ and an attack set $K$, with $R \neq K$ and $|R| \leq |K|$, initial conditions $x_K, x_R \in \mathbb{R}^{2n+m}$, and attack modes $u_K(t)$, $u_R(t)$ are so that for all $t \in T$, $y(x_K, u_k, t) = y(x_R, u_R, t)$.

Let $L = \begin{bmatrix} L_\delta^T L_\omega^T L_\theta^T \end{bmatrix}^T$, $F = \begin{bmatrix} F_\delta^T F_\omega^T F_\theta^T \end{bmatrix}^T$, and $C = [C_\delta \ C_\omega \ C_\theta]$. The bus voltage angles $\theta(t)$ can be conveyed via the state attack mode $f(t)$ and the generator rotor angles $\delta(t)$ as

$$\theta(t) = -\mathcal{L}_{11}^{-1} \mathcal{L}_{1g} \delta(t) - \mathcal{L}_{11}^{-1} F_\theta^T f(t).$$

(10.4)

The elimination of the algebraic variables $\theta(t)$ in Equation 10.4 points to the state space system

$$\begin{bmatrix} \dot{\delta} \\ \dot{\omega} \end{bmatrix} = \underbrace{\begin{bmatrix} 0 & I \\ -M^{-1}(\mathcal{L}_{gg} - \mathcal{L}_{g1} - \mathcal{L}_{11}^{-1}\mathcal{L}_{1g}) & -M^{-1}D_g \end{bmatrix}}_{\tilde{A}} \begin{bmatrix} \delta \\ \omega \end{bmatrix} + \underbrace{\begin{bmatrix} F_\delta & 0 \\ M^{-1}F_\omega - M^{-1}\mathcal{L}_{g1}\mathcal{L}_{11}^{-1}F_\theta & 0 \end{bmatrix}}_{\tilde{B}} u,$$

$$y(t) = \underbrace{[C_\delta - C_\theta \mathcal{L}_{1g}\mathcal{L}_{11}^{-1} \quad C_\omega]}_{\tilde{C}} \begin{bmatrix} \delta \\ \omega \end{bmatrix} + \underbrace{\begin{bmatrix} -C_\theta \mathcal{L}_{11}^{-1}F_\theta & L \end{bmatrix}}_{\tilde{D}} u.$$

(10.5)

The reduction of the passive nodes is called the Kron reduction. From these equations the following lemma was deduced.

## Lemma 3.1

(Equivalence of detectability and identifiability under Kron reduction): For the power network descriptor system (Equation 10.3), the attack set $K$ is identifiable if and only if it is identifiable for the associated Kron-reduced system [6].

A static detector is an algorithm that utilizes measurements from the network to check for the attacks at predefined instants of time, without manipulating any relation between measurements taken at different times. Following are two theorems that define how an attack set is undetectable and unidentifiable for a static detector.

## Theorem 3.1 [6]

(Static detectability of cyber-physical attacks): For the power network descriptor system (Equation 10.3) and an attack set $K$, the following two statements are equivalent:

1. The attack set $K$ is undetectable by a static detector.
2. There exists an attack mode $u_K(t)$ such that, for some $\delta(t)$ and $\omega(t)$, at every $t \in \mathbb{N}$,

$$\tilde{C} \begin{bmatrix} \delta(t) \\ \omega(t) \end{bmatrix} + \tilde{D} u_K(t) = 0$$

## Theorem 3.2

(Static identification of cyber-physical attacks): For the power network descriptor system (Equation 10.3) and an attack set $K$, the following two statements are equivalent:

1. The attack set $K$ is unidentifiable by a static detector.
2. There exists an attack set $R$ with $|R| \le |K|$ and $R \ne K$, and attack modes $u_K(t)$, $u_R(t)$, such that, for some $\delta(t)$ and $\omega(t)$, at every $t \in \mathbb{N}$,

$$\tilde{C} \begin{bmatrix} \delta(t) \\ \omega(t) \end{bmatrix} + \tilde{D} u_K(t) + u_R(t) = 0$$

Dynamic detectors check for attacks at all times, $t \in \mathbb{R}_{\ge 0}$. Dynamic detectors are harder to deceive than static detectors, but with this comes more complications. The following theorems define how an attack set is undetectable and unidentifiable for a dynamic detector.

## Theorem 3.3

(Dynamic detectability of cyber-physical attacks): For the power network descriptor system (Equation 10.3) and an attack set $K$, the following two statements are equivalent:

1. The attack set $K$ is undetectable by a dynamic detector.
2. There exists an attack mode $u_K(t)$ such that, for some $\delta(t)$ and $\omega(t)$, at every $t \in \mathbb{R}_{\geq 0}$,

$$\tilde{C}e^{\tilde{A}t}\begin{bmatrix}\delta(0)\\\omega(0)\end{bmatrix} + \tilde{C}\int_0^t e^{\tilde{A}(t-\tau)}\tilde{B}u_K(\tau)d\tau = -\tilde{D}u_K(t),$$

## Theorem 3.4

(Dynamic identifiability of cyber-physical attacks): For the power network descriptor system (Equation 10.3), the following two statements are equivalent:

1. The attack set $K$ is unidentifiable by a dynamic detector.
2. There exists an attack set $R$, with $|R| \leq |K|$ and $R \neq K$, and attack modes $u_K(t)$, $u_R(t)$, such that, for some $\delta(0)$ and $\omega(0)$, at every $t \in \mathbb{R}_{\geq 0}$,

$$\tilde{C}e^{\tilde{A}t}\begin{bmatrix}\delta(0)\\\omega(0)\end{bmatrix} + \tilde{C}\int_0^t e^{\tilde{A}(t-\tau)}\tilde{B}(u_K(\tau)+(u_R(\tau))d\tau = -\tilde{D}(u_K(\tau)+(u_R(\tau)),$$

The following residual filter is presented to answer the attack detection problem.

## Theorem 3.5

(Attack detection filter): Consider the power network descriptor system and the associated Kron reduced system. Assume that the attack set is detectable and that the network initial state $x(0)$ is known. Consider the detection filter

$$\dot{\omega}(t) = (\tilde{A} + G\tilde{C})\omega(t) - Gy(t),$$

$$r(t) = \tilde{C}\omega(t) - y(t),$$

where $\omega(0) = x(0)$, and $G \in \mathbb{R}^{2n \times p}$ is such that $\tilde{A} + G\tilde{C}$ is a Hurwitz matrix. The $r(t) = 0$ at all times $t \in \mathbb{R}_{\geq 0}$ if and only if $u(t) = 0$ at all times $t \in \mathbb{R}_{\geq 0}$ [6].

Next, this section presents a coordinate-free geometric way the main components of this residual filter founded on the notion of condition-invariant subspaces. Let $\tilde{B}_K$, $\tilde{D}_K$ be as classified right after the Kron-reduced model (Equation 10.5) and let $K$ be a $k$-dimensional attack set. Permit $\begin{bmatrix} V_K^T & Q_K^T \end{bmatrix}^T \in \mathbb{R}^{p \times p}$ to be an orthonormal matrix so that

$$V_K = \text{Basis}[\text{Im}(\tilde{D}_K)] \quad Q_K = \text{Basis}[\text{Im}(\tilde{D}_K)^\perp]$$

and let

$$B_Z = \tilde{B}_K (V_K \tilde{D}_K)^\dagger \quad \bar{B}_K = \tilde{B}_K \left( I - D_K D_K^\dagger \right)$$

Identify the subspace $S^* \subseteq \mathbb{R}^{2n}$ to be the smallest $\left( \tilde{A} - \tilde{B}_K (V_K \tilde{D}_K)^\dagger V_K \tilde{C}, \text{Ker}(Q_K \tilde{C}) \right)$-conditioned invariant subspace containing $\text{Im}(\bar{B}_K)$ and allow $J_K$ to be an output injection matrix so that

$$(\tilde{A} - \tilde{B}_K (V_K \tilde{D}_K)^\dagger V_K \tilde{C} + J_K Q_K \tilde{C}) S^* \subseteq S^*$$

Permit $P_K$ to be an orthonormal projection matrix onto the quotient space $\mathbb{R}^{2n} \setminus S^*$, and let

$$A_K = P_K (\tilde{A} - \tilde{B}_K (V_K \tilde{D}_K)^\dagger V_K \tilde{C} + J_K Q_K \tilde{C}) P_K^T .$$

Last, allow $H_K$ and the unique $M_K$ to be so that

$$\text{Ker}(H_K Q \tilde{C}) = S^* + \text{Ker}(Q \tilde{C});$$

$$H_K Q \tilde{C} = M_K P_K .$$

**Theorem 3.6**

(Attack identification filter): Consider the power network descriptor system and the associated Kron-reduced system. Assume that the attack set $K$ is identifiable and that the network initial state is known. Consider the identification filter

$$\dot{\omega}_K(t) = (A_K + G_K M_K) \omega_K(t) + (P_K B_Z V_K - (P_K J_K + G_K H_K) Q) y(t),$$

$$r_K(t) = M_K \omega_k(t) - H_K Q y(t),$$

where $\omega_K(0) = P_K x(0)$ and $G_K \in \mathbb{R}^{2n \times p}$ is such that $A_K + G_K M_K$ is a Hurwitz matrix. The $r_K(t) = 0$ at all times $t \in \mathbb{R}_{\geq 0}$ if and only if $K$ equals the attack set [6]. ■

This section examined the fundamental limitations of dynamic and static attack identification and detection procedures for a power network modeled via a linear time-invariant descriptor system. It has successfully demonstrated that a dynamic detection and identification method utilizes the network dynamics and surpasses the static detection while requiring possibly fewer measurements. By using dynamic residual filters, it was able to explain a possibly correct attack detection and identification procedure.

## 10.6 Real-Time Reconfigurable System

A hardware-in-the-loop reconfigurable system [5], which also incorporates embedded intelligence and robust coordination schemes at local and system levels, can be used to tackle weaknesses of the smart grid. It uses a location-centric hybrid system architecture that distributes processing and coordinates devices that are physically close. This system also assigns most of the intelligence to the lower level of the power grid. A robust algorithm for collaboration is used for neighboring devices to make the local decisions more reliable. A distributed algorithm will also be used to prevent attacks from reaching the system level by recognizing malicious inputs at the lower level. Last, this system uses control-theoretic real-time adaptation strategies to effectively preserve the accessibility of large distributed systems.

Today, many checks on the power grid are run every 5 min and some checks are not even run that often. To create a more real-time operation, the grid needs to guarantee end-to-end real-time responses, carry out distributed processing of security analysis algorithms to contain raw data communication within local areas, and quickly identify and diminish faults. To help increase the responsiveness of the grid, intelligent controllers in power electronics can be used. An example of this would be flexible AC transmission system (FACTS), which can alter the grid into a reconfigurable power system where procedures can be taken in milliseconds to certify reliability and quality of the power grid by controlling the flow of power. Rapid communication across the system has become a major struggle as the number of distributed generation centers has increased because of deregulation. To increase security, smaller collaborative systems would seem to be a better choice. Analysis algorithms should be used to tolerate data faults such as missing data or incomplete data. Portions of the system should be able to isolate itself from the rest of the grid and remain functioning in case of hardware faults such as contingencies and system attacks.

Security for large interconnected systems predominantly concentrates on transient and dynamic stability considerations. Intentional attacks are mostly going to be caused by damaged physical equipment, interruption of the communication network, or other similar problems. Physical interconnection can offer an assessment on the information system in case the communications are compromised, therefore supplying feedback for identifying and isolating an attack on the information flow. However, physical interconnection causes power systems to fail even if the communication is secure; as a result, the physical system and the information system can act together to provide system security. In Table 10.2 we list some examples of attacks that an enemy could instigate on the system. Power system planning is currently considering events with intrinsic equipment failure and possibly simultaneous failures of two or three components. Having multiple lines of defense can be extremely useful. For this system the first line of defense will be to detect and make quick adjustment to potential faults in voltage, current, or frequency. The second line of defense is using robust collaboration between neighboring devices at the local level. Finally, a control-theoretic edition framework at the system level will be the last line of defense.

The durable reconfigurable system presented here will integrate local actuation, sustained by devices with embedded intelligence, and central system-level adaptation, sustained by control-theoretic security solutions. It will use a distributed control scheme where controllers are employed within each facility and are permitted to coordinate among geographically close controllers. Figure 10.3 demonstrates the intention of this location-centric hybrid architecture that supports

**Table 10.2 Attack Examples**

| Attack Locations | Attack Types | Attack Descriptions |
|---|---|---|
| Communication | External | Intercept messages; launch DoS attacks to disable communication between system components. |
| | Internal | Modify or drop transmitted messages; inject false messages; delay message transmission. |
| Component | External | Physical power equipment failure or damage. |
| | Internal | Generate false event; control compromised system components to perform arbitrary tasks, such as coordinated attacks from multiple substations. |

this control scheme. This system further diverges from the existing power grid infrastructure by embedding intelligent controllers in FACTS and distributed generators (DGs), allows collaboration between DGs in the same area and FACTS devices across physically adjacent substations, integrates only the results from a distributed execution of the state estimation at the control center, and uses the control theoretic adaptation scheme. The main goal of the intelligent controllers in DG and FACTS devices is to determine how much real and reactive power to insert into the system when changes are detected.

An interval-based sensor data integration algorithm was developed to accomplish fault resiliency centered on multiple controller coordination even with possibly faulty, incomplete, and

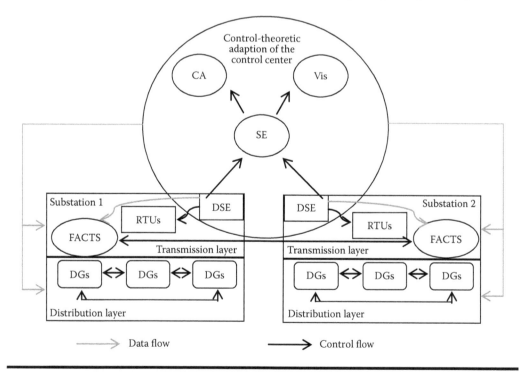

**Figure 10.3   Location-centric hybrid architecture.**

missing data. This integration technique takes an input and produces an interval gathered around the physical readout. A correct device would be one whose interval estimate encompasses the actual value of the parameter being determined; otherwise, it is faulty. The interval measurements can be integrated using a distributed interval integration algorithm where a local device is elected to collect the outputs of the devices. An asynchronous distributed algorithm was also developed to combine the present conventional state estimation and simultaneously introduce distributed processing to take advantage of modern communication capacity improvements and increase in sensors.

To produce analytic assurance on desired dynamic response to quality-of-service (QoS) attacks, we can explain a control-theoretic framework based on advanced control theory such as model predictive control (MPC). An important benefit of control-theoretic framework rests in well-established design methodologies and its capability to provide analytical performance assurance when workload and resource availability fluctuate unpredictably. Utilization control can be used to implement desired utilizations on one or more processors regardless of substantial uncertainties in system workload. On the basis of end-to-end utilization control (EUCON) and decentralized end-to-end utilization control (DEUCON), a feedback-controlled middleware system has been planned to run on QoS-critical security analysis modules.

## 10.7 Other Schemes to Achieve Security in Power System

This section discusses different ways to prevent, ease, and endure cyber-attack on the power grid. A layering technique is presented that identifies the risk for the physical power application and the cyber infrastructure. When the risks are determined, they can then be minimized to lower the risk level. The different elements of the power applications and the supporting infrastructure are discussed individually to determine how each has their own vulnerability and different ways to prevent attack on these applications.

Risk can be determined by running a vulnerability study on a system. This can show what kind of impact a certain attack can bring on the system and how likely such an attack is. A vulnerability study starts with identifying cyber assets [6–12]. Tests are then run to determine how easily these assets can be accessed. When this has been completed, tests should be run to identify the possible impacts of the infrastructure and power system. If the risk from this test is high, then mitigation activities are used to lower the risk to a more acceptable level. This can be done by using a better supporting infrastructure of power applications. There have also been many researchers that have looked into the risk assessment process.

The power system is split into transmission, generation, and distribution. Each has its own control loops that send control messages and measurements to the control center to determine what actions to take. A diagram of a control loop is shown in Figure 10.4. Generation deals with the measurements from the generators. An automatic voltage regulator (AVR) control loop compares current voltage measurements to past voltage measurements. This information keeps the voltage at the desired level. The governor control is used to monitor the frequency. It alters setting on the steam valve according to the speed sensed that accompanies disturbances. Because this control loop is local, attacks are limited. However, attacks can be made through the substation LAN. The automatic generation control (AGC) loop is another frequency loop that adjusts the system frequency to its nominal value. It verifies that each balancing authority fulfills its load condition and the power exchange between tie-lines stays as close as possible to what was scheduled. An attack on AGC would impact the frequency, stability, and economic operation. There is research going on of ways to prevent this.

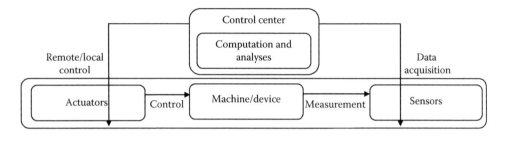

**Figure 10.4  Power system control loop.**

Transmission deals with moving power across the power lines and controlling the switches and reactive power support devices. State estimation is used when measurements from field devices are presumed to be incorrect. The control center uses other measurements to determine an estimation for the measurement that seems to be incorrect. This technique is not designed to be recognized when bad data are put into the system maliciously. Research is being done on ways to tell if data are simply bad or have been purposely changed. Volt-ampere reactive (VAR) compensation controls the reactive power in the system to better the transmission system. One type of attack is denial of cooperative operation, which is an attack where the FACTS devices are jammed leading to loss of important information. Some other attacks are desynchronization and data injection.

Distribution deals with sending power directly to the customers. Load shedding is when load is dropped by the system to avoid the system from failing in times of emergency. This could be attacked by changing the control logic to result in loads being dropped. Advanced metering infrastructure (AMI) uses smart meters at the consumer's location and provides the ability to disable consumer devices when demand spikes.

A secure supporting infrastructure is desired to guarantee that information is correctly saved and transmitted to the proper applications. Secure communication can be accomplished by encryption, authentication, and access control. Using an encrypted virtual private network can protect network traffic. A strong authentication mechanism is necessary to prevent people who do not have clearance to certain information from accessing that information. Access control can prevent people in the company who do not need access to certain functions from accessing those functions. This can help prevent attacks from the inside.

Device security can be accomplished by remote attestation. Security management and awareness can be achieved by digital forensics and security incident and event management. Cybersecurity evaluation is done by first doing a cybersecurity assessment, and then research, test, and evaluate. Intrusion tolerance is accomplished using intrusion detection systems and tolerant architectures.

## 10.8  Discussion

The growing demand for enhancing protection of framework designed for power transmission and distribution systems has resulted in increased enthusiasm surrounding security research and analysis. A multitude of intriguing and promising protection measures have been hypothesized, and some of the most captivating are discussed below. Each experiment has unique qualities that prove to be a possible solution to explicit problems that are fundamental to the security of our smart grid systems such that cyber-physical security, time reconfigurable systems, dynamic security, anomaly detection, probabilistic dependence graph, and smart firewall tracking are incredibly important research topics that should be further considered, experimented on, and eventually integrated into power systems.

In smart distribution grids, wireless mesh networks are exceptionally steadfast and proficient in communicating different aspects of the electrical equipment within each grid system structure. Therefore, the appeal has resulted in a wide use of such systems and, in turn, makes the protection of such systems from malicious attacks an immediate apprehension. In order to prevent, detect, and correct malevolent intrusions, a smart firewall tracking has been proposed. An infiltrated wireless mesh network is monitored such that each mesh node has the ability to detect a possible intrusion and classify it as being in a blacklist or gray list, depending on the severity of the threat, block communication, and then carry out an alert as an admonition for the other surrounding nodes. Research and experimentation was conducted using this method of protection and as expected was successful. Nevertheless, more testing needs to be conducted in order to successfully assimilate this method into a system.

Another scheme that has been adduced is the method of employing a Gaussian Markov random field to quantify phasor angles across a bus in a power system in order to formulate a dependence graph. The dependence graph is subsequently used in order to construct a decentralized method of message passing and classifying sites into tie-line, border-line, and inner-lie edges, which illustrate the effectiveness of the algorithm employed in this approach to detect malicious attacks and prevent harmful effects that can be possible if allowed to be carried out. As expected, intricate computations and extensive research were carried out, and this method of power system security is promising.

Attacks on power systems have the capability to bring to a standstill normal operation of electrical equipment within a system, so the method of preventing such attacks must be proficient in the recognition and reconciliation of intrusions. Anomaly detection monitors systems in real time, analyzes possible attacks, and determines methods of resolution, and as a consequence, logon attempts are monitored, changes in imperative equipment configurations are tracked, and multiple stations are checked for simultaneous attacks. An alarm is energized if a malicious attack or hack has been deemed probable, and impact factors are afterward traced. Much testing has been carried out in this field of study, and though it does appear to be one solution, more should be prepared and completed to determine the best and most likely method of implementation.

Dynamic detection and identification can be extremely useful to prevent attacks. Static detection and identification has been used for several years for many logical purposes. Static detection processes are incapable to identify any attack disturbing the dynamics, and that attacks corrupting the measurements can be designed to be undetectable without difficulty. Using a dynamic algorithm will allow the system to use continuous time instead of predefined instances. However, like the standard static algorithm, dynamic detection and identification still have some limitations. Dynamical filters are used to isolate predefined problems of the network components. The use of dynamic algorithms greatly outperforms its static counterpart.

Switching from the present centralized architecture to the location-centric hybrid system architecture would enable the awareness to fault prevention, mitigation, and detection at different levels with various degrees of association. This can include local actuation, distributed state estimation, and robust local collaboration, supported by devices with embedded intelligence, along with others. The hybrid configuration would provide a resilient reconfiguration capability through coordinated local actions. Having the intelligence near the lower level of the grid gives the resources the ability to make decisions and to respond rapidly to contingencies, and permits a more direct reconfiguration of the physical framework of the power system. Among neighboring devices, a strong collaboration algorithm is used to handle the possibility of incomplete, missing, or faulty data. A distributed algorithm is used to identify damaging inputs and inhibit the attack from spreading past the local level. To offer the preferred dynamic responses to irregular system changes, the system can implement real-time control-theoretic adaptation strategies for analytic reassurance. This will resourcefully preserve the accessibility of large distributed systems.

A dependable smart grid needs a layered protection tactic consisting of a cyber-infrastructure, which restricts adversary access and robust power applications that are able to function properly during an attack. A layer approach would allow protective measures to be implemented in multiple areas of the power system. It is achieved with the control loop at the generation level, the transmission level, and the distribution level. This will prevent many different types of attacks on the smart grid. The progress of a dependable electric grid necessitates a thorough reexamination of the supporting technologies to confirm they correctly achieve the grid's distinctive requirements. This can be accomplished with the use of secure communication, device security, security management and awareness, cybersecurity evaluation, and intrusion tolerance. The combination of these techniques in a layering method should create a system that is exceptionally hard to attack.

## 10.9 Conclusions

The threat of attack on our power distribution and transmission equipment is a constant possibility because of the developing tensions that have arisen not only within our nation but also throughout the world as a whole. Numerous distinct techniques for guaranteeing the security of power systems have been proposed and submitted for testing because the need to protect a structure so vital to contemporary living necessitates it. Studies conducted have presented solutions to the increased security issue such that cyber-physical security, time reconfigurable systems, dynamic security, anomaly detection, probabilistic dependence graph, and smart firewall tracking are all ways in which the detection and correction of security incidents have been proposed to be resolved. Although research and experimentation has been thorough, additional examinations must be made in order to ensure the reliability of the proposed solutions as well as the feasibility of their implementation into existing or new electrical assemblies.

## References

1. C. Ten, J. Hong, and C. Liu. 2011. Anomaly detection for cybersecurity of the substations. *IEEE Trans. Smart Grid* 2(4), 865–873.
2. M. He and J. Zhang. 2010. Fault detection and localization in smart grid: A probabilistic dependence graph approach. *First IEEE International Conference on Communications (SmartGridComm)*. IEEE, New Brunswick, NJ, pp. 43–48.
3. X. Wang and P. Yi. 2011. Security framework for wireless communications in smart distribution grid. *IEEE Trans. Smart Grid* 2(4), 809–818.
4. S. Siddharth, A. Hahn, and M. Govindarasu. 2012. Cyber-physical system security for the electric power grid. *Proc. IEEE* 100(1), 210–224.
5. H. Qi, X. Wang, and L. Tolbert. 2011. A resilient real-time system design for a secure and reconfigurable power grid. *IEEE Trans. Smart Grid* 2(4), 770–781.
6. F. Pasqualetti, F. Dorfler, and F. Bullo. 2011. Cyber-physical attack in power networks: Models, fundamental limitations and monitor design. Technique Report, University of California, Santa Barbara. Available at http://motion.me.ucsb.edu/pdf/2011i-pdb.pdf.

7. O. Kosut, L. Jia, R. Thomas, and L. Tong. 2011. Malicious data attacks on the smart grid. *IEEE Trans. Smart Grid* 2(4), 645–658.
8. D. Wei, Y. Lu, M. Jafari, P. Skare, and K. Rohde. 2011. Protecting smart grid automation systems against cyberattacks. *IEEE Trans. Smart Grid* 2(4), 782–795.
9. Q. Li and G. Cao. 2011. Multicast authentication in the smart grid with one-time signature. *IEEE Trans. Smart Grid* 2(4), 686–696.
10. A. Bartoli, J. Hernandez-Serrano, M. Soriano, M. Dohler, A. Kountouris, and D. Barthel. 2011. Secure lossless aggregation over fading and shadowing channels for smart grid M2M networks. *IEEE Trans. Smart Grid* 2(4), 844–864.
11. L. Xie, Y. Mo, and B. Sinopoli. 2011. Integrity data attacks in power market operations. *IEEE Trans. Smart Grid* 2(4), 659–666.
12. Z. Yang, S. Yu, W. Lou, and C. Liu. 2011. P2: Privacy-preserving communication and precise reward architecture for V2G networks in smart grid. *IEEE Trans. Smart Grid* 2(4), 697–706.

# SENSOR-BASED CYBER-PHYSICAL SYSTEMS

## Chapter 11

# Wireless Sensor and Actuator Networks for Cyber-Physical System Applications

Kassie McCarley, Joseph Pierson, and Fei Hu

## Contents

## 11.1 Introduction

Where would the world be without wireless networking capabilities? The importance that they play in our lives today could not have been imagined 50 years ago. Because of networking and the Internet, we are able to share mass amounts of information across the globe. Sensor networking is just as important to us. Wireless sensor networks (WSNs) allow us to control systems through a cyber-physical world when we take advantage of various pieces of hardware and mix it with software. The steps that it takes to create this technology are not to be taken lightly. There are certain requirements and protocols that must be met. The transport layer plays an important role in sharing data. It involves statistical multiplexing to ensure that data transmission is efficient. The transport protocol is the called the transmission control protocol (TCP). It ensures a reliable way to stream data. Without the TCP, WSNs could not have evolved to where they are today. Also, the quality of the system is an important topic. This is often referred to as the quality of service (QoS). Many factors can alter the state of the QoS such as congestion control, the act of using statistical multiplexing, and the delay jitter possibilities.

## 11.2 Sensor Networks for Cyber-Physical Systems Applications

With the development of the Internet, it has changed drastically compared with 40 years ago. The way information is shared in this new era, however, does contain some missing pieces. The purpose of a cyber-physical system (CPS) is to expand the world's view on how hardware can actually relate to and interact with the physical world we live in. The goal is to have a formatted structure of how mass amounts of information is stored and supplied to the public. A new science must be created for this type of CPS [1]. The following should detail more on the management of the software and the hardware components within the CPS of the sensor networks. The framework described will alleviate the difficulties of integration and management issues.

CPSs are expected to alter the way humans communicate with and view the physical world. The new suggested framework will be able to foresee the integration capabilities of varied applications. For the system to function as specified, there are many system requirements. Early models try to implement the integration of logic with the physical element. The framework will include the following modules: the control module, the application module, the session module, the requisition module, the network module, and the QoS module. The main characteristics of the CPS are specified within the functional requirements. These are the characteristics the system must abide by to accomplish its specified goal. The operation capacity is also identified along with the application environment within this step of requirements. They are considered to be the backbone of the framework. The nonfunctional requirements are also very necessary and important. They relate to the quality of the systems operation. Trust, security, and performance are measured because of these requirements. The difference in the two previously listed requirements is that the functional requirements are set permanently, depending on the behavior of the system. The nonfunctional requirements are geared more toward the quality of the system and can be changed or altered at any time in a way that will benefit the system. Some of the different types of requirements needed for the framework can be seen in Figure 11.1.

The WSN parameters that are relevant to this system include the previously listed requirements. The node hardware designed and used is ultimately what available architecture on the

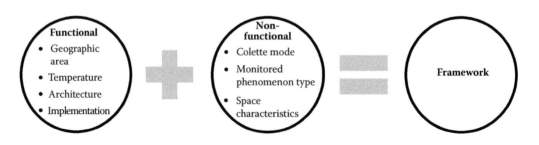

**Figure 11.1 Functional and nonfunctional requirements.**

market is at the time that pertains to the application. It varies with each type of application being created. To determine the nodes, certain measurements are used such as the operation power, the sampling rate and time, the energy restrictions if any, and the possible transmission rate. The topology of each application, along with other nonfunctional requirements, dictates the distribution of the nodes. The ideal topology that would be used is a mesh topology. The sensor network database is a way to store data for the sensor networks. It is a database architecture that is often used to manipulate and sample different features of the network's applications. It will require constant updating due to the heterogeneity of the sensor network. The modeling process will be located within the requirements and often referred to within the knowledge base of the network. The algorithms and other requirements will be used within a knowledge base. The knowledge base can function at the same time as the sensor network database. It will in time provide the necessary mechanisms to merge the sensor networks with the CPSs.

## 11.2.1 Example: Water Monitoring Systems

Another way to use WSNs is to monitor actions that can occur, for example, monitoring water within a pipe and giving feedback. Especially now, reliable water filtration and distribution systems are in high demand. The system consists of the actual water distribution system. When aided with software and hardware intelligence, leaks can be detected and in the case of contaminated water, it can be controlled and localized. Sensors are used to monitor the physical elements and report the data collected to the cyber system. Valves, pipes, and reservoirs are used along with software and hardware components to achieve the satisfaction of operating as a water distribution system that can filter and aid in distributing treated freshwater to a population [2].

To collect the necessary information to maintain healthy drinking water, radio frequency identification (RFID)-based sensors are located within a water pipe infrastructure. These sensors are connected to access points that lead to the Internet. The cyber data server can then utilize the data collected. The algorithms on the cyber data server then operate through hardware controllers on the findings to manage the quality and quantity of water that is produced. In [2], PipeSense was developed. It is an in-pipe water monitoring system that also uses RFID-based sensors. Some things that are measured with the sensors include water pressure, chemical composition, and volume of the water. These measurements can be taken at various places within the pipe. PipeSense uses acoustic signals that are susceptible to noise to detect any leaks within a pipe. As one might predict, this process is very prone to false alarms. The intention is to create a system that monitors a real-time water quality sensing mechanism that will determine the changes occurring within the real-time water distribution network. Their system encompasses a large amount of inexpensive nodes that are able to communicate with each other. This benefits the overall system, making it more efficient and reducing the number of possible false alarms. The CPS is driven by events that take place. The human-centric framework architecture will introduce a human as the means of monitoring the water within the application. Six levels are located within the human-centric CPS framework. These are illustrated in Figure 11.2.

The first tier is the sensing tier. Obviously, it is made up of many sensors that can detect temperatures, pressures, pollution, and, among others, pH levels. They can be classified as either dynamic sensors or static sensors. Dynamic sensors monitor within the pipe through variously placed nodes, and it also identifies the location of the nodes. Static sensors can monitor actuator functionality and temperatures within the pipe. Information collected by either type of sensor is fully analyzed at the server level. The second tier is naturally the processing tier. The information is only partially cleansed when it arrives at this stage. After receiving the real-time information from

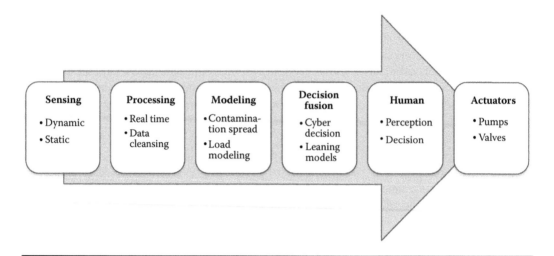

**Figure 11.2　Six levels in a cyber-physical system framework.**

the sensors, it is fully and thoroughly cleansed. It can also be stored to review at a later date. The third tier, the modeling tier, detects the likelihood of a spreading of contaminated elements can be predicted along with different patterns that may occur. The fourth tier is the decision fusion tier. During this stage, the decisions within the cyber systems are made depending on the models that have been taught or previously programmed according to the situations. However, these decisions are not acted on until a later step.

The fifth tier, the human tier, incorporates the input of a human along with the information being decided on. This combination makes up the base for the human-centric CPS framework. A human agent is able to omit the decisions made in the previous stage and override with a new action. This is helpful to prevent false alarms. These instances are recorded by the cyber system and used to alter the training models over time in an attempt to possibly reduce the amount of false alarms. The sixth and last tier is the actuator tier. This is composed of pumps and valves. The valves can either be automatic or manually controlled. They can be used to stop, start, or pause a process. Also, they can be used to alter the pressure or isolate a certain section of a pipe. This may be necessary in the case of a contamination being detected or a leaking pipe. The suggestions to make the changes to a pump or valve are sent to human operators or carried out immediately if operating automatically. The sensing tier then plays the most important step of repeating and analyzing the result of its previous data collected.

In the future, this process can be improved by incorporating a multiinterface data service for the human tier so that the humans can easily make the correct actions occur. Also, these data have the possibility of utilizing a wireless handheld device for processing and decision making. Doing this through Wi-Fi and Bluetooth are being explored in [2]. Ultimately, a CPS framework would positively aid the effort of improving the quality of current water distribution systems.

## 11.2.2　Confidentiality-Aware CPSs in WSNs

When communicating wirelessly, the sense of security is always an important feature. It is a large challenge for CPSs that operate with sensor networks this way. Within [3], a Security-Slack based Heuristic Algorithm (SSHA) is suggested. This method claims to eliminate the slack time for security-critical messages being transferred. Experiments were conducted to prove how their method worked and improved the efficiency of the SSHA.

First, the messages must be classified into different security-sensitive categories. An algorithm will be introduced to increase the level of security necessary, depending on the previously depicted need, when transferring messages. Some examples of security-critical messages can include data used for controls, secret data, and other messages that are subject to security constraints and deadlines. Each message that is determined to be confidentiality sensitive is called a tuple $M_i = (P_i, s_i, l_i, v_i)$. The value $s_i$ is the size of the message being sent. $P_i$ is the period of the message. The value $l_i$ represents the level of confidentiality of the message, and $v_i$ is the impact value of message $M_i$. Each period is assumed to be the same as the deadline.

To determine the level of security necessary for a message, a security profit model is used. In [4,5] the sum of the total security levels, defined as the profit model, is designed to predict the potential level of security-critical messages. This process is not altogether reliable. From [6] the job failure model is used to ensure a more reliable system. The probability of $M_i$ being security critical is found within this model by

$$
P_{\text{risk-free}}(l_i) = 
\begin{cases}
0 & \text{if } l_i < l_i^{\min} \\
\exp\left[-\lambda\left(l_i - l_i^{\min}\right)\right] & \text{if } l_i^{\min} \le l_i \le l_i^{\max} \\
1 & \text{if } l_i^{\max} < l_i
\end{cases}
\tag{11.1}
$$

In the equation, $l_i^{\max}$ and $l_i^{\min}$ are the confidentiality levels that are defined by the message, and $\lambda$ represents the security risk coefficient. The risk-free security profit SP can be defined as

$$
\text{SP}(l_i) = v_i * P_{\text{risk-free}}(l_i)
\tag{11.2}
$$

The SSHA reduces the delay of sending a message that is classified as confidential. When the delay time is altered, it leads to a higher security profit. This is measured by Slack-Security Factor (SSF). The confidentially level is increased while the unit slack time is decreased. Equation 11.3 depicts this process:

$$
\text{SSF}_i = \frac{\text{SP}(l_i + 1) - \text{SP}(l_i)}{s_{i*}[\Phi(l_i + 1) - \Phi(l_i)]}
\tag{11.3}
$$

SSHA, while increasing the security profit, does not disobey the sequential constraints. The main goal of this algorithm method is to find the messages that measure a higher SSF and increase their level of confidentiality as needed.

The simulations executed in [3] confirmed that the security profits were recognized and the confidentiality levels of any security-critical message were raised. They compared the SSHA to the SV-Greedy and SV-RND approaches. Their percentages were not as successful as the SSHA.

## 11.3 Transport Layer

Networked CPSs (NCPSs) are basically systems that include multiple nodes that work together to sense and induce a change to their environment. This is typically performed in hostile conditions [7]. A node can contain devices like sensors and actuators. Each node provides a building block to the entire NCPS. The synchronization of nodes is not always reliable because of the normal nature of the nodes. Within an NCPS, information should be gathered and formed into an action locally at a node; however, an awareness of what else is happening with other nodes may be required. A framework is needed that will integrate system monitoring, balance resources, and adapt to altered environments. In [7] they describe their partially ordered knowledge-sharing model and an application programming interface that will allow contact with the physical world.

The partial knowledge-sharing model is said to be less conceptual than other models that have been developed to connect through end-to-end channels. It is different from previous shared-memory models because it allows a node to have its own view of the knowledge being processed and it can choose whether or not to use the bit of knowledge. This decision is possible owing to the knowledge not having a predetermined destination. An NCPS requires the integration of the physical layer, the network layer, the knowledge dissemination layer, and the application layer. The framework deals with decisions made on each layer. Java was used to create the current application of the framework model featured in [7]. Much of the code featured in their model can be used in different environments. To test this, they used it on an algorithm for optimization and control. The result found was that their model was able to adapt to a variety of operations. If used on a larger scale, it could be applied to computing platforms and applications used on mobile devices that are part of NCPSs.

The Internet was developed as a tool to enable sharing of data on a large scale. When embedded systems were created, the thought was able to control different components of a system with real-time constraints on a relatively small scale. Embedded systems are limited when considering what is being requested by larger networked systems. Now, the new idea is to create a way to control a physical object and its processes at anytime from anywhere on a large scale. The vision calls for a cyber-physical Internet (CPI) [8] to be developed that would encompass and combine the idea of the Internet and embedded systems. This has not previously been considered and if successful will enable large-scale systems to be joined with their physical elements. Before, embedded systems and the Internet were thought of as two totally separate things.

CPI will focus on connecting the physical elements to the computational elements of a system [8]. Already being practiced is radio-frequency identification (RFID). This is considered the most important technology for physical objects within a network. WSNs are the alternative. They focus on the sensing aspect to monitor and control what is going on within the network. The main components of a networked CPS include the following: mobile networks, satellite networks, RFID, the Internet, embedded systems, and WSNs. The combination of CPSs and RFID systems has allowed the use of the Internet to aid in sensor networks. CPSs are mostly scattered networks that perform different tasks related to where they reside. The goal is to connect these CPSs so that they may operate as one larger-scaled network. Many opinions have been voiced when approaching this subject. Some believe that it would be a better option to try and integrate already existing networks into a single CPS. Others think that it cannot be done and the only option is to create a CPS separately while not considering preexisting systems.

To create a large-scale CPS is said to be impractical. In today's society, a technical problem is seen differently depending on the person. Academic researchers care about efficiency, while the consumers are only concerned with the cost and functionality of the actual product. Most of the time when a product is produced and a problem is discovered, a simple patch of the error is preferred over completely recreating the product. This is due to the demand of and cost to the client. It seems that instead of starting from the beginning, most would see it more logical to try and integrate systems to create a large-scale CPS. This complete system is referred to as the CPI. The protocol that will be most important in the CPI will be the Internet protocol (IP).

There are specific requirements that need to be met in order to create the CPI. New protocols must be developed for the CPSs. These protocols must take the elements of the environment into account. This is why an additional protocol layer is required to couple the protocols with their external environment. Instead of having the expected five layers of protocol architecture associated with the Internet, a sixth layer must be added. It is called the cyber-physical layer. This layer will offer a theoretical explanation of properties of cyber-physical data. It will also interact with the other five existing layers. With the physical layer, the interaction can alter properties such as the frequency band of the channel. With the data link layer, it must be adaptive to work with the cyber-physical layer to allow it to make decisions concerning different processes. The network layer will also work with the cyber-physical layer to determine the best routing strategies. The transport layer must encompass protocols that will operate without the need of acknowledgments being returned. The application layer will then combine with the newest layer to deliver standard distributed signal processing algorithms for all possible CPS applications. This will result in decreasing the time it takes to get to market and the cost.

CPSs still face many challenges when considering real-time requirements. This is an important factor for the operating system to support. The operating system TinyOS is among the most advanced solutions to creating an effective hardware and software split; however, it does not support real-time protocols.

As stated previously, the idea of creating the CPI is to connect the physical elements to the computational elements of a system. One protocol being developed for the cyber-physical layer is called the transport protocol for spatiotemporal data. It can monitor the physical behavior of a system within a network. This protocol can send data from nodes in a system to the base station with a constant expected error regardless of its previous error estimation. The new protocol is easy to integrate into existing motes. Researchers in [9] suggest that their protocol may be the first transport layer protocol to include predictive models on motes. They performed tests on MicaZ motes using the LiteOS operating system.

The new transport layer produces a limit on the possible error when estimating the physical actions that occur. It suggests that instead of transporting measurements across a network, perhaps they could be predicted. The new protocol will limit the inaccuracy of the measurements taken of the environment in an unstable network. The limit can be adjusted by the user to their desired percentage. Many sensor network protocols used the end-to-end delivery method to guarantee arrivals. These rely on feedback such as acknowledgment statements to deliver packets. The new idea is to use estimation error as a metric in determining how much information can successfully be delivered. It should be lightweight with a very low consumption of energy. To accommodate various link conditions, the protocol must be able to monitor the quality of the wireless communication channels. It exceeded all of these elements when tested in [9]. The protocol did not require a feedback from the receiver as was required in previous transport protocols. It was also compared to preexisting unreliable protocols and proved to be a better option.

## 11.3.1 Congestion Control through RAPID

For many years, one of the most common networking complications is controlling network congestion. The reason for this is due to the Transport Control Protocol (TCP) using a process that searches for bandwidths slowly. To find the available bandwidth, this process can take very many round-trip times (RTTs). Some take thousands of RTTs. Protocols that are classified as "high speed" also take a long time to determine which spare bandwidth will be used to transmit data. Most of these slower protocols interfere with other traffic on the network. The protocol must first acquire the spare bandwidth; then it must adapt to any changes that are made. The speed of the entire loop is directly proportional to the previous two steps. Responses from the devices in a network cannot reach their destination before a complete RTT is carried out.

Recently, a new way to control congestion has emerged. This new way only requires one to four RTTs to successfully find an available bandwidth compared to a possible thousand RTTs previously. Also, the speed of the transport protocols will increase with the new method. This method is known as RAPID [10]. It stands for RAPID Congestion Control. RAPID depends on a detection system that is delay based. The goal of the RAPID design is to keep a fair timescale when many transfers need to take place at once, to tolerate sending normally low TCP transmissions, and to reduce the amount of congestion while maintaining low buffer occupancy rates.

To achieve the quickness of finding a spare bandwidth, a small probe is sent to a wide variety of contenders. This will result in reducing the chance of overloading a single rate. Also, to reduce the chance of overloading, a packet's average cannot surpass the spare bandwidth that is identified. When there is a packet to be sent, RAPID controls the transmission of data in smaller more reasonable groups. This is known as multiple rate probe streaming. In short, it is called a p-stream. Data are then sent at different intervals until the entire packet is completely received. The sending mechanism is responsible for the rate of the data streaming. When all data are received, the receiver uses all results of the p-stream to identify the spare bandwidth being used. It returns the bandwidth rate in the acknowledgments that are returned to the sender.

When transmitting packets through the p-stream, the system cannot allow for overloads. The average rate at which data are sent must be equal to the spare bandwidth found during the RTT. When these are equal, the occupancy of a link cannot go over its maximum capability. The average rate time also aids the sender in its ability to form to different changes within the bandwidth.

The slow start process in RAPID is the same slow start used in TCPs. It cannot be increased because of the chance that it initially is set to high for a network. The sender acts as it would in any other protocol by sending only as many packets in an RTT as before. The difference happens in the number of RTTs executed. The RAPID slow start ends faster than before because of the multiple probing for available rates. The way a window size is increased is also the same as in protocols before. They are doubled in size once every RTT. Because of the available bandwidth being found quicker, the rate in which it increases window sizes is much faster. Packet losses are similarly found using the same timeout system as TCPs. After they are found, the average rate is reduced by half and then resumes the slow start process.

As mentioned previously, RAPID is not intrusive to sending normally low TCP transmissions. If both types were being sent at the same time and inflict congestion, the RAPID transfer would respond first by decreasing its rate before the TCP transfer would. This is proof that even though both were being sent, the TCP transfer would react normally as if it were being sent as it was before the RAPID congestion control system. The availability of the bandwidth is shared equally.

During experiments, both types of transfers were simulated. RAPID was compared to those techniques used before in other protocols. Most protocols worsened as the speed of the network increased. However, the RAPID protocol acted as it was expected to act. RAPID maintained the loss of packets better than the other protocols, and it was able to find other available bandwidths when the current bandwidth would unexpectedly change. While that was occurring, the amount of packets at the sender refrained from becoming too large. The most difficult challenge in developing RAPID is that it relates to the trends in the packet gaps. This can affect the accuracy of the received times, but most platforms today should still be able to support RAPID speeds. Further research and more experiments of this technique are being performed before RAPID can successfully be implemented.

# 11.4 Quality of System Issues

## 11.4.1 Wireless Sensor/Actuator Networks

The CPSs are a mixture of networking, calculating, and physical dynamics, where embedded devices are networked to monitor, control, and sense the physical world. In the future, wireless sensor/actuator networks will connect a multitude of mobile computing devices and embedded devices. Through these wireless networks, various self-governing subsystems will provide certain services for people. Eventually, the goal is to have WSANs connected to the Internet for global information sharing. WSANs will be what connects the physical world to the cyber system. Sensors will be the tool that gathers information from our actual world, while the actuators will be used to respond to the information being received through certain actions that will be determined by the data that are taken in.

These WSANs are closely related to QoS provisioning. Resource constraints apply to both actuators and sensors. Sensors are normally low cost and offer restricted data processing, battery energy, and memory. The actuators have stronger computation capabilities than sensors along with more energy. Although used together, sensors and actuators are greatly different from each other including their capabilities and what they are used for. Also, comparing different subsystems, hardware, and network technologies used in the WSANs can vary greatly. Another feature of WSANs that is closely related to QoS provisioning is dynamic network topology. Node mobility is very important to many applications. There are times where new sensor and/or actuator nodes may need to be added or removed, and some could even die due to drained battery energy. These factors cause the network topology of WSANs to change dynamically. Last, as the scale of WSANs grows, mixed traffic will become a popular feature.

There are many different applications that will be part of the CPS. Different applications will have different QoS requirements. For instance, a system that regulates the temperature in an office building from the air conditioning unit may allow some packet loss and long delays, but this would not be acceptable for systems that are intended for safety-critical systems.

With supporting QoS in WSANs being an area that is almost entirely unexplored, there are many possibilities to discover new things and there will be many challenges that go along with the discovery of new things. Some areas of interest that are being researched are service-oriented architecture, QOS-aware communication protocols, resource self-management, and QoS-aware power management.

Service-oriented architecture has been around for a while and has been used in many applications including Web service domain. Although it is not new, there are still many aspects that have yet to be explored in CPS. Service-oriented architecture is a style of architecture that has a set of services that allows for complex systems of systems to be built. It allows for cheap configuration of scalable systems that is based on recyclable amenities that are uncovered by these systems. The usefulness of this is great for QoS provisioning in WSANs that are joined into complex CPS. The research that has gone into service-oriented architecture has led to many questions that still need to be answered. For instance, it is still unclear on the number of categories of services to be classified, the quality levels that are applicable and are needed to increase performance from a QoS management point of view, and last, the best way to deal with the variance among actuators and sensors when identifying services. Another area of research that is being looked into is QoS-aware communication protocols.

The performance of the network depends greatly on the core communication protocols. These protocols are one of the most commonly studied features of sensor design. The stack normally consists of five layers that are the physical, data link, network, transport, and application layers, but here more attention is focused on the transport and network layers. For QoS to be supported in WSANs, the design of the communication protocol needs to keep in mind the differences between sensors and actuators that are used in CPS.

It is very important for service differentiation to be supported by the communication protocol when looking at QoS. A CPS may have very different applications with various QoS requirements. Because the current technology that is involved in networking is unable to offer a different QoS for each application, new protocols for WSANs need to be designed. The new design will have to observe the service requirement associated with each application so that a service adapted to each particular application will be provided. Cross-layer design has shown to be very useful in making network performance. It is also a good idea to incorporate cross-layer design into QoS-aware network communication protocol for CPS.

WSANs have serious limits on the amount of computation as well as energy resources. Because of the ever-changing and growing complexity of CPS, dynamic network features, and unpredictable environments, resource management in WSANs becomes challenging. These challenges are taken care of by using self-management technologies; that is, the system can manage resources on its own. This self-management technology will be able to recognize changes and will then adapt resource usage to optimize QoS.

Another important area of research is QoS-aware power management. Being able to conserve power is very important when it comes to WSANs. Battery energy restricts the lifetime of sensor and actuator nodes, so the longer the battery life is, the longer the sensors and actuators can survive. Wireless communication uses way more energy than sensing and actuating. The power used for transmission at the nodes has to be managed so that the consumption of energy is greatly reduced. The transmission power mechanism for nodes associated with actuators can be different from the nodes associated with sensors. QoS can be boosted by manipulating the differences between sensor and actuator capabilities.

## 11.4.2 Delay/Jitter Improvements

Delay/jitter has also been a big issue in QoS support [11]. The difference in time of sending packets to those being received is called a delay. The jitter is the delay difference between neighboring packet arrival events. The jitter can be reduced by holding received packets within a buffer for a certain period of time before they are accessed. That specified amount of time is referred to as the playback delay (PBD). A new approach using the one-way delay (OWD) variation has been taken to improve the possible length of the delay. This new idea of measuring is able to produce a more accurate reading with respect to delay spikes and RTT predictions.

CPSs have been created over time within network control systems (NCSs). This has enabled the network to control and monitor another environment regardless of distance from the system or how they are connected. Both wired and wireless networks have functionality. A sensor network's purpose is to gather data from constrained wireless sensors in an efficient manner. This is often affected when actuators are located within the same NCS. Their interaction produces the choice to make changes to the physical world.

NCSs function in real time and must deal with occurrences of delays, delay jitters, and the loss of packets. The quality of the system is affected greatly by these events. The delays can be sudden and cause a possible prediction of the received time to be incorrect. This is when the previously mentioned PBD is useful. The packets that are received are held for a period of time before they can be accessed. This delay provides some packets, in cases where errors were present, more time to arrive. If it is received after the processing time has passed, it is considered to be a lost packet and a jitter is present where the missing packet should be. Buffers at the physical system send control signals at predictable times in an attempt to soften the jitter. The correct amount of time that the packets are held is very important. If it is not lengthy enough, it is possible to lose more packets. Extending the time, delays the processing time and weakens the quality of the system.

A new way to predict more accurate PBD times is to decompose the variations in the RTTs. There are two components of this. First, is the forward path that is the one-way delay variation of the physical system to the controller. For this review, we will call it $f_{pc}$. The other is the reverse path of the controller to the physical system, or $r_{cp}$. The controller has an advantage in the forward path because it can accurately predict the $f_{pc}$ time. The OWD variation is extracted using the RTTs. Much research was done on forward and reverse OWD variations. The dependency of the forward and reversed variations was also studied. The recently developed method of predicting the PBD value is more advantageous in that it can shrink the delay time window by using known forward RTTs and approximating reverse RTTs. It examines and produces a more acceptable forward RTT than previously used methods.

An NCS includes a physical system that is operated from another location. This can be connected by wire or wirelessly. The signals obtained by the sensors from examining the physical system are sent to a controller. That controller can then generate control signals to send to the actuator. The actuator can change the conditions of the physical system based on the signals it receives. Each signal being sent from the sensors and the controller suffers from delay jitters. A system can produce very unpredictable received times due to the delays. The appropriate PBD determined must be a mix between the packet delay and packet loss. The playback buffer is deployed from the physical system side of the exchange. This is done to smooth the jitters from the transfer, if possible.

Some previous researchers have tried to solve the issue of unreliable network delays by using a monitor that assesses past outputs measurements to predict future controls. All packets were treated the same and assigned the same delay time from the network control. This was not efficient because some packets did not warrant as large of a delay as others. Some focused on the possibility of the way a system was connected being the missed step. They did not even consider the time-varying delays being a factor in the quality of each packet delivery. Some used a PBD to connect end-to-end points within an NCS in hopes to completely eliminate any jitters with buffers. The delay was calculated based on RTT values.

The newest method of PBD prediction is as follows [11]. If a sensor performs a reading $y$ and it is received at the controller after a time $t$, $y$ can either arrive, be lost, or late. If it does arrive at the controller, it is used to control an action $u$. This signal of control can be applied at some unknown time because of delay jitters or be lost. If the signal $u$ arrives, it can be held within the playback buffer at a predictable $t + r$ time. If it does not arrive before $t + r$ expires, the signal is considered lost. The factor of $r$ must be a close approximation of the RTT. If it is too large or too small, the physical system's actions can in no way be predicted. The value $r$ is found at the controller because the controller can then, in turn, produce the needed control signals that can shift the state of the physical system. To find the approximate value based on the RTT and the OWD, the following equation was used:

$$\text{RTT}_{pcp}(i) = \text{RTT}_{pcp}(i-1) + \hat{\text{RTT}}_{pcp}(i) \tag{11.4}$$

$\text{RTT}_{pcp}(i-1)$ represents the RTT of the $(i-1)$th packet that is sent from the physical system to the controller and back. $\hat{\text{RTT}}_{pcp}(i)$ represents the deviation of the previous RTT of the $(i-1)$th packet with respect to $\text{RTT}_{pcp}(i-1)$. Figure 11.3 displays each step where the NCS sends and receives packets from the physical system to the controller. It can be seen at $k_i$ in Figure 11.3 that the delay $r$ is predicted at the controller side because it will use that value to create the $u_i$ value. By observing Figure 11.3 one can also deduce that the controller is able to know the exact value of both $\text{RTT}_{pcp}(i-1)$ and $f_{pc}(i)$. The controller uses these known values to then calculate the only unknown value of $r_i$. This will then produce the value of $r_{cp}(i)$.

In this method, $f_{pc}(i)$ can be predicted from the controller. This can be used by the controller to predict the PBD. This will, in turn, decrease the unpredictability of the PBD time. According to [11], a more specific value for $f_{pc}$ and $r_{cp}$ can be found by using the following equations:

$$\hat{f}_{cp}(i) = f_{cp}(i) - f_{cp}(i-1) = \text{RTT}_{cpc}(i-1) - \text{RTT}_{pcp}(i-1) \tag{11.5}$$

$$\hat{r}_{cp}(i) = r_{cp}(i) - r_{cp}(i-1) = \text{RTT}_{pcp}(i) - \text{RTT}_{cpc}(i-1) \tag{11.6}$$

By using this, the networks do not have to be in synch because only the RTTs of the sender and receiver are used. The value $f_{pc}$ represents the OWD of the $i$th packet and the $r_{cp}$ represents the reverse OWD of the $i$th packet. The value $f_{pc}$ has respect to the clock of the physical system and $r_{cp}$ has respect to the clock of the controller. $\text{RTT}_{pcp}$ represents the RTT of the packet with respect to the clock of the physical system, and $\text{RTT}_{cpc}$ represents the RTT of the packet with respect to the clock of the controller.

By being able to predict the exact value of $f_{pc}$, the controller can monitor the number of delay spikes during the $i$th packets transmission from the physical system to the controller. This will decrease the PBD's amount. Equation 11.7 will produce the estimated PBD $\tau_i$:

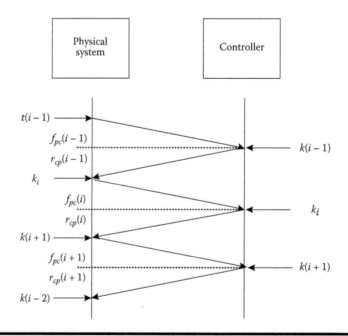

**Figure 11.3  Packet sending/receiving schedule.**

$$\hat{o}_i = \text{RTT}_{pcp}(i-1) + \hat{f}_{pc}(i) + \tilde{r}_{cp}(i) \tag{11.7}$$

The result depends on the approximated value of $\tilde{r}_{cp}(i)$ because $\text{RTT}_{pcp}(i-1)$ and $\hat{f}_{pc}(i)$ are already known values by the controller.

The entire scheme will in turn improve the steps of making predictions of the PBD in an NCS. This is done by not requiring the clocks of each system to be synchronized. The RTT measurements are used from the sender and receiver to predict the OWD time. This process is advantageous because it can produce a more accurate value of the PBD by utilizing the RTT variations, $f_{pc}$ value, and being able to predict the $\tilde{r}_{cp}$ value. The delays are lessened and it contains fewer outliers that can produce further complications.

## 11.5 Discussions

The three main topics covered within this chapter all relate to one another and coexist. To maintain a successful WSN, the transport layer must be designed correctly. To send and receive messages over a sensor network, the transport protocol must be efficient and include a way to securely send a message and control congestion. Even though the hardware that makes up a sensor network may be purchased at a fairly low cost, the transport layer that is implemented must function as if it is a very expensive system. If there is a long transmission time, the quality of the system, commonly referred to as the QoS, is affected. QoS was topic three within this paper. Each must function as expected to ensure a reliable network that is able to transmit data successfully and securely. The different algorithms mentioned in sensor networks touched on the ways a WSN and its contents can be protected.

To create an efficient sensor network, the proper framework must be built. The different tiers mentioned are vital to maintaining a secure and functional network. Without a single step, the entire network could become a complete disaster. The correct transport protocols are required to share information within those sensor networks. If the TPs function as they are normally expected, the QoS of the system will be satisfactory.

## 11.6 Conclusions

In conclusion, we have covered networked CPS and its QoS issues. We have also analyzed the transport layer protocols. Overall, when CPS units are connected into a network, there will be many new issues to be addressed.

## References

1. J. R. B. Garay and S. T. Kofuji. 2010. Architecture for sensor networks in cyber-physical system. Dissertation. University of Sao Paulo, Sao Paulo, Brazil.
2. A. Nasir, B.-H. Soong, and S. Ramachandran. 2010. Framework of WSN based human centric cyber physical in-pipe water monitoring system. Dissertation, Nanyang Technological University, Singapore.
3. W. Jiang, W. Guo, and N. Sang. 2010. Periodic real-time message scheduling for confidentiality-aware cyber-physical system in wireless networks. *Fifth International Conference on Frontier of Computer Science and Technology*. IEEE, New Brunswick, NJ, pp. 355–360.

4. T. Xie and X. Qin. 2006. Scheduling security-critical real-time applications on clusters. *IEEE Trans. Comput.* 55(7), 864–879.
5. T. Xie and X. Qin. 2007. Improving security for periodic tasks in embedded systems through scheduling. *ACM Trans. Embedded Comput. Syst.* 6(3), 1–19.
6. S. Song, K. Hwang, and Y. Kwok. 2006. Risk-resilient heuristics and genetic algorithms for security-assured grid job scheduling. *IEEE Trans. Comput.* 55(6), 703–719.
7. M. Kim, M.-O. Stehr, J. Kim, and S. Ha. 2010. An application framework for loosely coupled networked CPSs. 2010. *IEEE/IFIP 8th International Conference on Embedded and Ubiquitous Computing.* IEEE, New Brunswick, NJ, pp. 144–153.
8. A. Koubaa and B. Andersson. 2009. A vision of cyber-physical internet. Dissertation. Al-Imam Mohamed bin Saud University, Riyadh, Saudi Arabia.
9. H. Ahmadi and T. Abdelzaher. An adaptive-reliability cyber-physical transport protocol for spatio-temporal data. *30th IEEE Real-Time Systems Symposium.* IEEE, New Brunswick, NJ.
10. V. Konda and K. J. Vishnu. 2009. RAPID: Shrinking the congestion-control timescale. Dissertation. University of North Carolina at Chapel Hill.
11. H. Al-Omari, F. Wolff, C. Papachristou, and D. McIntyre. 2009. An improved algorithm to smooth delay jitter in cyber-physical systems. *Eighth International Conference on Embedded Computing.* IEEE, New Brunswick, NJ, pp. 81–86.

# Chapter 12

# Community Sensing

Trenton Bennett, John Grace, and Fei Hu

## Contents

## 12.1 Introduction

The growth in the number of mobile devices and global positioning systems (GPS) over the past decade makes community sensing a reality. Sensors that have been inserted into these devices allow users to gather information. Community sensing will ideally help the community with events such as earthquake detection, environmental issues, traffic issues, and other topics relevant to today's society. By using sensors in mobile phones, GPS systems, and other devices we will be able to gather data to help analyze everyday problems. Abnormalities in data will allow users and the community to know when hazards such as earthquakes pose a threat. Sensors can also be used to gather information on the air quality of communities. Detecting traffic issues will allow users to be notified of where the issue is so they can plan accordingly. Some of these topics cover issues dealing with the health of society. Early earthquake detection would protect the community and allow users to know when to seek shelter from the earthquake ahead of time. Air quality sensing would allow the users to know when they are entering an area of air that poses threats to their health. Motivations for using community sensing include saving lives and protecting society's health. The privacy of the user, however, is a concern when it comes to implementation.

## 12.2 Devices and Programs Involved in Community Sensing

### 12.2.1 Mobile Phones

The acceleration of technology in mobile phones has allowed these devices to use cheap but powerful embedded sensors for different functions. The variety in these functions allows them to be used in personal, group, and community scale sensing applications, which could eventually revolutionize the entire lifestyle of the upcoming generation.

The sensors included in mobile phones (accelerometer, compass, GPS, microphone, camera, etc.) are used across applications for health care, social networking, transportation, and safety/environmental monitoring. The sensors are used to do a variety of functions that help support the user interface. Some examples of these functions are with the compass and gyroscope that help the phone realize its position in the physical world, which could enhance location-based applications. The camera and microphone are likely the most used sensors in the world. Figure 12.1 shows some of the sensors that are used in the iPhone 4S.

One important question concerning sensors is what new sensors are most likely to be introduced for widespread use? By analyzing what sensors could be useful, the phones could be adapted to many new applications.

There have been new companies and corporations helping gain new insight on community sensing in a variety of areas. Main areas of interest are transportation, social networking, environmental monitoring, and wellbeing. We need to consider how to obtain and validate data and also to make sure that user privacy can be protected.

Two sensing paradigms have been presented: (1) participatory sensing where users actively participate and (2) opportunistic sensing where users passively participate. A sensing scale will be used

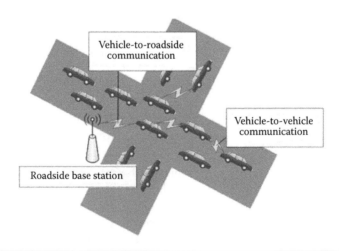

**Figure 12.1   iPhone 4S sensors.**

depending on applications. Personal sensing takes data and gives the information back only to the user. Group sensing usually applies to popular interests in social networking or other groups who could share information to help each other. These data must be taken with a grain of salt unless the group can be implicitly trusted. Community sensing applies to the scenarios where a large number of people participate and the large-scale data can be applied to the whole community. The users participating must be protected to ensure privacy. So far, only limited experience has been gained in building a scaling system. The scaling system may be able to be defined based on how the user is actively giving or using data, which could be seen by the two sensing paradigms. Opportunistic sensing may work better for community sensing because there is such a high volume of users, but still a lot of information could be desired. Participatory sensing could place a high burden on some users to give sensor data, while driving away from other users.

Because mobile phone sensing is still in its early development, there is an issue as to what components should be used on the phone and what should be transported through the cloud to protect privacy. The viewpoint suggested in this chapter is just a beginning to figure out the best architecture for the system. Mobile phones sense raw data from sensors. These data are extracted by applications on the phone or cloud and are still open to research. The next step will be to figure out how to inform, share, and persuade users with the data that have been collected and manipulated by these community sensing applications.

In order for the data to be sensed by the mobile phone, the device needs to be programmed appropriately. This programming will be ever-changing as community sensing becomes more developed. Right now the lack of limits on sensor control does not allow battery consumption to be controlled. Programming could help preserve battery power and allow better accuracy from data. Continuous sensing is an area that could give great insight in many areas such as personal health care. The device in this capability must be able to support multitasking and background processing; however, resource limitations (memory, computation, storage, etc.) will always affect a phone when processing data. For continuous sensing to work in the future, sensing should not affect the applications that the phone needs to complete.

Another problem with sensing is the unpredictability of how a phone will be used. Some phones may be able to share sensors to help capture the data, but researchers are currently looking into reducing the noise in collection through supersampling.

When the data have been collected, they must be interpreted to produce results. This can be accomplished with data mining and statistical tools. Most of these data are interpreted for people, and supervised learning techniques are primarily being used to build algorithms. This technique seems to work well for a small-scale application, but most likely it will not be able to be transferred to a large-scale sensing application. The most successful way to infer everyday human activities to this point has been using multimodal sensor streams. This method uses the sensors in a device to slowly infer complex human behaviors. Another problem with interpretation is how to scale different activities to different people. The key to designing this scalability will be to generalize information that can be used by large-scale communities. The concerns for scalability at this point have been channeled to finding a model that is adaptive. As the model may be constantly changing, the computational needs will significantly increase as the interpretation increases. For this technology to be developed, researchers must build and share models, while then finding large-scale projects to apply techniques and algorithms to evaluate progress.

Now the information needs to be given back to the users so that they can be informed, which could allow for different goals to be achieved. One important consideration is how the information is presented and therefore how easily accessible the data will be. Some users might like personalized sensing that could reflect the preferences and behavior of users. These behaviors could be accumulated, and then your phone could perhaps identify items that are not made using sweat-shop labor. Users could also be persuaded with data to make lifestyle changes. There is research into how people in a community can be used as an iconic influence on that user population/community. These people could be able to influence great change in communities in a variety of lifestyle areas. While all of these demonstrate ways to get the data back to the user, ultimately, no one will participate unless his/her privacy is always ensured. People will always be aware of how these data are being captured and processed, especially so no one can identify the data to a single person. Community sensing always stands the chance of having information leak and revealing personal information. Current work is being done so that information that is shared cannot be reverse engineered to grab potentially damaging information from someone. There are many questions about how to protect users from other users nearby, but very few solutions exist to protect the privacy of users especially in large-scale community sensing projects.

Overall, there has been much progress using mobile phones in community sensing applications, but there are still many obstacles before the concerns can be overcome. The main technical barriers are the privacy sensitivity issue, using the resources optimally to collect less noisy data, and giving the best feedback to users. Eventually, sensing through mobile phones will give useful data in a variety of functions to improve life in every aspect [1].

### 12.2.2 PRISM

Smartphones are one of the main devices considered for community sensing. This is due to the amount of smartphones used in society. Development of the applications needs to be relatively easy in order to make these useful. PRISM or a platform for remote sensing using smartphones [2] was created to test the functionality of applications using remote sensing. During this research, three applications were tested using PRISM: Citizen Journalist, Party Thermometer, and Road Bump Monitor [3]. Sensing would occur through a user's mobile phone solely by running software designed to operate on the mobile device. Three goals are associated with PRISM: generality, security, and scalability. These goals mean that many applications can be used, the user's privacy is not lost when using the application, and many phones can be used and do not affect the infrastructure.

Users begin by installing PRISM onto their cellular device. It is only used in public areas to protect the user's identity. A pull method can be used to gather information from devices. A push method can be used for a predetermined number of devices, which allows for a fast response, efficiency, and scalability. Two challenges arise when using a push method. One is a generic and flexible application programmer interface (API) that needs to be designed so that applications can be effectively pushed to a desired set of mobile phones [3]. The second challenge is to track the amount of phones used without consuming too much energy. When using the PRISM system, registration and updates are supposed to be made by the user to run efficiently. Registering the product allows the server to know when applications can run on this device.

Privacy risks are very prevalent using PRISM. With access to sensors such as the microphone and the GPS of a smartphone, the user's privacy is at risk. The microphone would allow for recordings to occur; while the GPS would allow for the location of the user to be used. Sensor access control is used to regulate the type of data that are gathered. There are three situations: no sensors, location only, and all sensors. No sensors allow for the user to be protected fully. Sensors will not be activated unless the user activates the sensors themselves. Location only tracks the GPS locations. All sensors provide the ability of being accessed.

PRISM also uses metering. Energy metering pertains to the battery life. Battery life is a concern when using PRISM. We may use a simple model where the energy consumed is estimated as a linear function of the amount of time that a particular device is active and the amount of data read and/or written [3]. Through this the battery life will be tracked and will not deplete it. Bandwidth metering is also a factor for cost and privacy.

When implementing PRISM, 15 smartphones were used. All of them have a microphone, a camera, and a GPS. All code used in the implementation was written in C#. PRISM includes an infrastructural and a mobile phone component. All calls are blocked by the mobile phone unless they are noted otherwise. Three applications were tested on PRISM. The first application is called Citizen Journalist. This sends a request to users in the area and asks them to respond. The responses could be composed of many different types of forms such as pictures. Requests are sent out to phones in the area. The 3G phone allows for more signals to be deployed than the 2G phone. This application was done on 10 users. They were asked how heavy traffic was at a certain intersection. A ringtone signifies that the application has launched when the user comes in the area of interest. The user can choose to activate the application, ignore it, or cancel it and give a reason for doing so. Table 12.1 shows the statistics from the Citizen Journalist experiment [3]. One of the volunteers who did this experiment was greatly pleased and asked for a new type of application. That response shows the need for community sensing.

The second application tested was called Party Thermometer. This was used to detect how and where the party was, as well as other information pertaining to parties. Microphones, along with location, are used to detect when a user is in the location of the party. First, the microphone detects music and the party atmosphere. Location is a check to see if the user is actually at the party or if the user is just in the area. Figure 12.2 illustrates the general idea. The sinusoidal waves created by the music are interpreted from the microphone as music. When the microphone comes into certain proximity of the music, indicated by the red circle, the user's location is then verified by the GPS location to see if the user is actually at the party, which is represented with the letter A in this illustration. This application was tested to make sure that only the users that were in the area had the application deployed onto their phones.

The third application was called Road Bump Monitoring. This application does not involve the user. It uses the sensors in the devices and the GPS to gather the data. This experiment was carried out in a neighborhood. To verify the accuracy, the actual locations of the speed bumps

**Table 12.1 Statistics from Citizen Journalist**

| Item | Count |
|---|---|
| Deployed | 417 |
| Launched | 274 |
| Total responses | 235 |
| Response time in seconds (avg, max) | 46, 149 |
| Photo responses | 141 |
| Total canceled | 38 |
| Canceled (too far away) | 9 |
| Normalized deployed distance (avg, max) | 71%, 443% |
| Normalized launched distance (avg, max) | 83%, 100% |

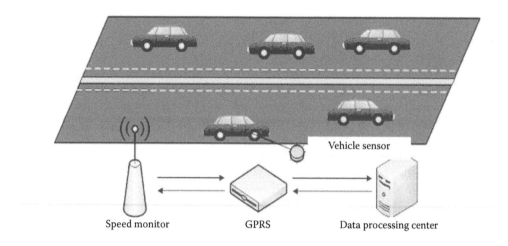

Speed monitor          GPRS          Data processing center

**Figure 12.2   Party Thermometer process.**

were predetermined. It was found that nine bumps were discovered by the application, six of them were actual bumps.

PRISM was created in order to make community sensing through the smartphone simpler. The privacy of the user is a concern along with the battery life of the user's device. PRISM addresses the challenges described in this section in order for community sensing to work more efficiently.

### 12.2.3  Device Control

One main area that community sensing can be developed is through the use of privately held sensors. These sensors are everywhere today and are contained in personal cell phones and GPS devices. One main reason these sensors are not used in collecting any data, however, is because of privacy concerns. Sending data could be intrusive to the original purpose of the device, and battery and network resources may also be an issue. Another factor in the design of community sensing is what the "cost" will be to get the information needed for an application. At first, the primary focus is on selecting the best observations to get the information needed. However, these observations must be balanced with the cost needed to make these observations. Also, the sensing may have to distinguish between high versus low cost sensors.

For privately held sensors to be used a main focus must be "optimal sensor polling policies" (TCS 1). The approach taken is getting sensed data from a variety of sensors populated in an area and computing information with inferences about how available and useful the information can be. The main focus will be on using private sensors for traffic monitoring between characterizing traffic and forecasting traffic ahead.

Main contributions from this research [4] are determining how to value information to make different models have near optimal sensing policies with maximal utility that satisfies the constraints placed by the private individual and devices. There is a case study showing how traffic monitoring data can be evaluated. These models are proven through advanced algorithms and mathematics. The models introduced will then be shown in the case study of traffic monitoring. The goal for this case study was finding the normalization of road speeds in a certain section of a road.

In order for community sensing to work best, the best subset of sensors needs to be used. To find this subset, a model needs to be created to get the most optimal information possible. This model must balance the availability of sensors, which will always be uncertain. The first model, that is, the phenomenon model, reduces the variance between locations. It predicts values and different locations shown in Equations 12.1 and 12.2. The case study shows how a Gaussian process finds the normalized road speeds over a certain section of road. The Gaussian process is then fit to the traffic data and then a kernel is taken from those data. The data from the case study for phenomenon modeling are discussed and shown through a series of charts and graphs.

Equation 12.1 shows the model to predict variance at each location.

$$\text{Var } X_s \,|X_A = x_A = E[X_s - E[X_s|X_A = x_A]]^2 \,|X_A = x_A|$$ (12.1)

Equation 12.2 shows the quantifying value of sensor locations.

$$\text{Var}(X_s) - \text{Var}(X_s|X_A = x_A)$$ (12.2)

Demand modeling deals with taking values where they are needed most and then getting the highest reduction in variance at these locations shown below in Equation 12.3. For the case study, each road segment must have its demand defined as the number of cars moving over the section of road. That demand is then modeled with a Poisson random variable. When modeled and normalized, the demand model can show the road section observations that can efficiently improve the demand-weighted prediction accuracy. The demand modeling data are discussed and shown through a series of models and graphs.

$$R(A) = \sum_{s \in V} E\left(D_s(\text{Var}(X_s) - \text{Var}(X_s|X_A))\right)$$ (12.3)

With modeling availability and privacy we will not be able to take samples at a precise location at all times, so a main focus in developing community sensing will be integrating models of sensor uncertainty. Here, the model assumes a set of possible locations sensing might take place. By distinguishing the possible locations from the measurements taken, the final objective is to maximize the variance reduction (see Equation 12.4). To help achieve privacy and protect from inference attacks, selection noise can be introduced through two methods: spatial obfuscation and sparse querying. In spatial obfuscation a certain road section is determined and then certain sensors are determined by a probability distribution. The sensor reveals the information needed for observations but not identifying information. In the case study, a cell phone company could keep the identifying information so that there is no leak of information. Spatial obfuscation is better for a smaller set of observations. Sparse querying is for the scenario where a certain pool of users volunteers to have observations taken from sensors. The case study has a certain pool of drivers from Microsoft participating. The users are not continuously monitored but instead are sent sparse and random queries. The application would have a time-dependent availability distribution that determines which cars should be queried. The data for spatial obfuscation and sparse querying are shown and discussed through models and graphs.

Equation 12.4 shows the expected demand-weighted variance reduction.

$$F(\beta) = E_{A|\beta}[R(A)] = \sum_A PA|\beta R(A)$$ (12.4)

For each model the data are taken and applied to the appropriate algorithms and theorems throughout the chapter. From the data, certain sensors are selected to get data that can then be turned into a set of road speed sensor readings. The predicted values for each model are then compared to the values taken from the sensors. The error between those two values is called the demand-weighted expected RMS error (DRMS) and is shown in the following:

$$\text{DRMS} = \sqrt{\frac{1}{n} \sum_s \lambda'_s r_s^2} \qquad (12.5)$$

The experiment was then done without considering noise and therefore only done on the phenomenon and demand models. The predicted data are then compared to the test errors and is evaluated as the accuracy of the model. As you can see in Figure 12.3, the predicted error and the test error match very well so the models are in very good shape. The next experiment shows the difference between spatial obfuscation versus purely random selection and sparse querying. These data portray the traffic results using the two different techniques over rush hour around Seattle. The map below was taken from Google Maps to try and show the main loop from which the data were taken. The thicker the lines in Figure 12.3 means the higher the intensity of the traffic (therefore more data that were accumulated throughout this test). The main loop goes from Lynwood in the north to Kent in the south.

Besides traffic monitoring, there are many other different applications community sensing could be applied to. Some include fitness and recreation with community sensors in place instead of using private sensors. Other sensors could be used for consumers to get information about different aspects of businesses. Businesses could use community sensors to figure out what customers are thinking about their product and how different franchises are doing throughout a day. Traffic sensing could give drivers an alert to accidents ahead on the road they are traveling, or an opportunity to see what the traffic looks like on the route home. Overall, this model shows that community sensing could make an impact in traffic sensing. With an optimized selection of sensors used, there could be a minimized amount of queries taken to get a specified level of accuracy. This accuracy would give great improvements to citizens of these areas employing traffic sensing.

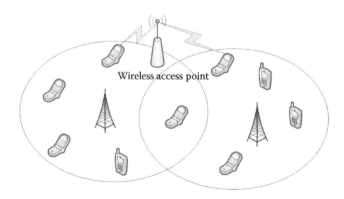

Wireless access point

**Figure 12.3   Areas testing occurred in rush hour (intensity based on thickness).**

## 12.3 Wireless Community Networks

The main future of the Internet is how network infrastructures will work throughout different communities and provide information through mobile devices that will communicate as wireless community networks (WCNs) [5]. The community applications that will be used with these devices will be able to help residents, local authorities, and places of public interest to get information that will help them the most. This section will focus on the application scenarios that WCNs will use and identify the challenges that will most likely come in security and privacy of these scenarios.

There are different networks that can provide the structure for WCNs. These two networks are wireless mesh network (WMN) that allows the different wireless network devices to communicate, and the wireless sensor network (WSN) that uses sensors to measure and compile data from information gathered in an environment to turn on appropriate actions. See Figure 12.4. The WSN should use the WMN to provide information to the WCN that will allow the users of mobile devices to gain a vast amount of information that could run in different applications. Sensing applications are already used in WCNs for a wide variety of measurements and purposes. The hope is for these sensors to one day improve the overall condition of a variety of everyday activities for people in society by allowing them to easily access the data gathered.

The three application sensing scenarios for WCN explained in the paper [5] are personal, designated, and community sensing. The differences in these scenarios come from the user's role and how the user can access the data gathered in each scenario. In the first scenario, personal sensing is used. Personal sensing is when the user has individual access to a household's WSN. The WSN will provide information that is useful only to the household and should not be seen by anyone else. The second scenario is designated sensing where the information gathered will be accessible to a few users in the community specifically given access. In-house WSN nodes could gather information useful to local authorities and other agencies to alert them to action. An outside operator or authority rather than individual living in the house most likely would control these nodes. If someone inside a house were controlling a node, it could lead to a lot of privacy issues. The third scenario is community sensing where the public will have access to certain WSN nodes. These nodes would provide functions useful to the entire community where they are located. Community sensing could be broadened to the gathering of information through mobile devices

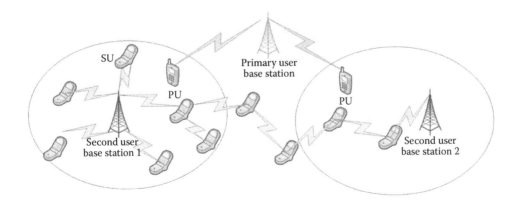

**Figure 12.4   Structure of WCN with user interaction.**

that are shared when in close proximity of a certain location. The diagram below shows how these three different sensing applications with user interaction might work with WCNs.

The main issue with WCN models is providing the users with the amount of security and privacy needed for adequate protection. Security and privacy may have to be added for the roles of different users to make sure users do not overstep their bounds and attack WCNs from outside or inside the network. Control mechanisms can be used to give the user privacy control in their network, but preserving that access can be difficult. Also, it will be difficult to guarantee privacy to certain nodes while allowing partial access to other nodes. Encryption methods could also be used for the privacy and designated sensing networks. However, those methods would have to deal with different authorities and keys. Privacy for community sensing is different as the issue is not keeping the data from someone but making it anonymous. The main problem with this is a user could perhaps trace back the node being used to share the information to trace the user. Differential privacy could provide interesting solutions for a variety of community sensing applications [6].

## 12.4 Areas Community Sensing Could Be Applied

### 12.4.1 Environmental Applications

A project [7] was created to involve the community in taking measurements and using these measurements to compare values. This project was aimed to allow everyday people, with no expertise in the subject, to be able to work on these devices. These devices, when created, were used on street-sweeping vehicles. Ideally, researchers wanted to implement this technology onto mobile phones. Allowing these devices to capture, process, and disseminate sensor data allowed the community to produce credible data. This environmental information would benefit not only everyday people but also government agencies and other organizations.

GPS was incorporated so that while data were taken the location was also tracked. Electronics were isolated from the environment through a split board design [7]. This design allowed air to reach the sensors, while the other side encased the electronics to protect them from weather and other debris. Testing occurred on street-sweeping vehicles in San Francisco. Air quality values were taken as these street-sweeping trucks performed their daily functions. Figure 12.5 shows the sensors that would be attached to the street-sweeping vehicles. It also shows the location being updated through the use of GPS as values are being taken with the sensors. These values were then sent to a database where eventually the researchers created a Web site that allowed users to view the data.

**Figure 12.5   Street-sweeping implementation.**

Another environmental project was created to involve the community in taking measurements to bring awareness to air pollution and the effects of air quality on communities. Through this project, people will be allowed to take their own measurements of air that they encounter on a daily basis. This action will allow for more locations to be evaluated. These devices will be integrated into mobile phones so users will be able to use them relatively easily.

These sensor readings along with GPS space-time stamps are sent to a database server using a GPRS radio [7]. The researchers constructed three parts to their demonstration: a handheld component, a Web portal, and a mobile phone. The handheld component and Web portal will allow for the data to be viewed in real time. The Web portal will also allow viewers to determine the data from what time frame they would like to view. The mobile phone will implement an application that allows users to view the air quality results around them. These air quality results will be uploaded to the mobile phone via Bluetooth and viewers will then be able to see the pollutants that surround them.

## 12.4.2 Air Transportation

Mobile autonomous and semiautonomous systems are being used today that are performing projects over a vast range of applications. As technology progresses, these systems have progressed and now the developers of these systems are trying to push the systems past the evolved sensory, mobility, and intellectual capabilities of humans [3]. These systems are allowing humans to quit performing the main tasks of flying or driving vehicles. And now they are performing calculations for other tasks needed during navigation. Safety applications have become more reliable by putting embedded cyber-physical systems (CPSs) within the vehicle. These embedded systems can provide us with process controls and other functions, but there are still difficulties translating human intent into priorities and actions to be planned and carried out in a certain order.

One main problem for these CPSs is coordinating simultaneous activities and sharing knowledge at the correct time without the system becoming overloaded. This problem is exacerbated when the system is run in harsh environments and even worse when performing in an extraterrestrial environment where the system cannot work with real-time Earth-based signals because there will obviously be delays and bandwidth issues in the communication. The paper [3] talks about many different tasks and anomalies they are investigating with different systems and the main system talked about is the air transportation network. The limitations, research challenges, and possible innovations for the air transportation network are explained further.

## 12.4.3 Earthquake Detection

Community sensing has recently been used to detect events and accidents in the community. These occurrences are often being tracked with mobile devices. This makes researchers optimistic that tracking earthquakes is also possible, but challenges arise when trying to implement this. The approach to implementation is to allow the sensors to learn the normal data as well as anomaly detection. The goal of this system is to detect seismic motion using accelerometers in smartphones and other consumer devices [8]. This early detection will allow the users of the devices to be warned prior to the earthquake occurring. Through these early warnings, lives may possibly be saved, because the users will potentially get to a safe place.

Some problems arise with creating the approach to earthquake detection. These problems include the sensor types, locations, and ambient noise. With many sensor types, the characteristics cannot all be collected at the same time. This collection has to be controlled with algorithms, and the data also have to be sent out, outside of the earthquake zone, to a reliable network. Decentralized detection is the method used to detect earthquakes. To control the amount of messages, messages are sent using binary sensing. This means when an event occurs or $M_{s,t} = 1$ and when an event does not occur a 0 is sent. Along with sending messages when an event occurs, the amount of false negatives and positives is also tracked. False negatives mean the earthquake was not detected and lives could not be fully protected. False positives mean warnings are being given out to the community when earthquakes are not actually occurring (see Table 12.2).

Online decentralized anomaly detection was implemented with deferring the treatment of spatial correlation to future work such as a rock concert [8]. $P[X_{s,t} | Et = 0]$ represents when an event has not occurred, while $P[X_{s,t} | Et = 1]$ represents when an event has occurred. On the basis of the observations and the great amounts of data collected at one point in time, the detection of an event should be rather easy. A threshold $T_s$ is created in order to decide if an event occurs. This allows the decision to not only be made on the likelihood, $L_0(x)$, of the event occurrence, but the event would only occur if it was under the given threshold:

$$L_0(x) \leq \tau_s$$

The less probable $x$ is under normal data, the larger the likelihood ratio gets in favor of the anomaly. The latter is the assumption on which most anomaly detection approaches are implicitly based [8].

$$L_0(x) \leq L_0(x') \Leftrightarrow \frac{P[x | E_t = 1]}{P[x | E_t = 0]} \geq \frac{P[x' | E_t = 1]}{P[x' | E_t = 0]}$$

Online density estimation is used to decide when the normal data are distributed. The threshold also has to be estimated. This can be done by using a uniformly consistent density estimator, which will allow the false positive probability to be the following:

$$P[L_0(X_{s,t}) \leq \tau_s] \text{ is } |p_0 - p_0| \leq 2\varepsilon$$

If the estimate $L_0(x)$ above converges uniformly, then the percentiles converge as well. The false positive rate will then only change by $\varepsilon$ [8]. In order to fully perform tests, the true positive rate also needs to be found. Receiver operator characteristic (ROC) curves monitor the relationship between the false positive rates and the true positive rates. These curves can help detect whether the anomaly method is working correctly or not. A conservative curve allows for an operating point $p_0$ to be chosen to maximize a lower bound on the true detection performance [8].

**Table 12.2  Classification of Events**

| Scenario | Meaning | $M_{s,t}$ |
|---|---|---|
| False negative | Earthquake undetected | 0 |
| False positive | False alarm | 1 |
| True positive | Earthquake detected | 1 |

A community seismic network is also being created to allow for early warning signals when an earthquake has been detected. It will also allow users to view where the areas most affected by the earthquake were. With the cloud computing system, locations all over the world can have access to it. Two sensors are being used: one as USBs in computers and sensors within an Android phone. The USB products would have to be purchased by the users and would require an extra fee for the users. With the sensors used in the phone this would allow owners of the phone to not purchase any other products. The phones were used to perform tests to see if they would be able to detect earthquakes. It was found from the data that the phones would be able to detect large earthquakes; this was found due to their difference in acceleration values. Testing is being done by the research team currently. There are hopes that the system can be used by the overall community.

Messages are sent to the Cloud Fusion Center (CFC). Heartbeats are sent so that the CFC can tell which phones are running the application. A user interface displays the recorded waveforms [8]. Power usage was also a factor considered with implantation. The application did not need to drain the battery too quickly. Users who charged their phone on a daily basis would not be affected by the power usage. The Google App Engine was used because of many factors. The ease of scalability was one reason it was chosen. Another reason was data security. To risk not losing data to the earthquake that we want to track, the Google App Engine was chosen. The last reason was its ease of maintenance.

There were, however, two drawbacks to the Google App Engine. One drawback was the loading requests. Approximately 10% of the requests processing time were increased. Another drawback was design implications. The data store function cannot be used for short-term reasons. These are stored in the memory cache. It can be faster. However, owing to memory limitations, some data might be lost to make room for incoming data.

The experimentation process included the research group. They carried their phones around, during daily activities, to gather the normal data. Data were gathered from previous earthquakes to determine observations that occur during an event. The USBs were tested to work more efficiently than the phones. The baseline approach worked for the USBs but not the phones, while all the other approaches achieved 100% for the true positive rate.

In conclusion, the study of detecting rare events was researched in [8]. Through the use of normal data and anomaly detection, these rare events can be detected. Ideally, the network will be open to the community, but only initial testing occurred within the research group. Cell phones and USB devices will be used when implemented into the community to gather data necessary to track and detect earthquakes that occur. This is important for the wellbeing of society and our community.

## 12.5 Field Experiments

After an earthquake devastated Haiti on January 12, 2010, there was a need to catalog the recovery effort throughout the various communities that were affected. Seven communities in particular were chosen to have data analyzed before and after the earthquake and then throughout recovery. This study is researching the effects of remote sensing on areas affected by natural disasters and trying to create a scale so that recovery results can be compared throughout the various communities. The main focus was physical damage and disruption in these communities.

The team used the data retrieved from remote sensing with various other tools and interviews with Haitian officials to try and catalog the disruption of the communities due to the earthquake. Three communities were inside the capital city of Port-au-Prince, and four were outside the capital.

The team was deployed throughout the Haitian communities from May 6 to May 16 to collect data on the damage and recovery of each community. In each individual community the ratio for the damage buildings went from 2% to 21%. The data collected from remote sensing were verified in the field, and the disruption information was taken with the interviews conducted while out in the field.

The scientists in Haiti took over 1000 km² of optical aerial imagery and assessed the damage with this imagery. Owing to the damage throughout the area, only buildings that were identified as having heavy damage or completely collapsed were classified as damaged and equate to Level 4 or 5 in the EMS-1998 rating. There were a total of 1670 buildings that were determined to have a Level 4 or 5 rating and then were given a recovery score based on the information as shown in Table 12.3 [9]. Interviews to gain information on disruption were taken with various community representatives and government officials. Disruption is based over 11 different sections in the community: drinking water, energy/fuel/utilities, sanitation, education, health care, shelter, food, and food availability. Each section was ranked prior to the earthquake, immediately after, one month after, and four months after the earthquake. The scale for disruption is shown as in Table 12.4 [9].

**Table 12.3   Recovery Scale Given to Damaged Buildings throughout Haitian Communities**

| Recovery Score | Description |
| --- | --- |
| 1 | Structure unchanged since the earthquake |
| 2 | Structure intentionally demolished, but not cleared |
| 3 | <50% Rubble removed |
| 4 | >50% Rubble removed |
| 5 | Structure under construction |
| 6 | Structure rebuilt on same footprint |
| 7 | Structure rebuilt on different footprint |

**Table 12.4   Disruption Scale for 11 Sections in Community**

| Recovery Score | Description |
| --- | --- |
| 1 | No availability |
| 2 | Minimal availability |
| 3 | Poor availability |
| 4 | Moderate availability |
| 5 | Good availability |
| 6 | Almost full availability |
| 7 | Full availability |

Each community had the total number of buildings cataloged along with the percentage of damaged buildings. Remote sensing shows that the change in the damage throughout the communities does not have a pattern whatsoever. Recovery varies with place, but typically the communities inside the capital (Port-au-Prince) have seen a faster recovery than those communities outside of the capital.

The data gathered from the field interviews indicated that the disruption rates before the earthquake were very poor with average ratings. The overall community rating was given a rating of "moderate availability" of 3.8/7. The officials said that after the earthquake there was a substantial increase in the disruption for service provision.

Interviews to gain information on disruption were taken with community representatives, National Government Officials (NGO), United Nations (UN) Clusters, and utility agencies. Disruption is based over 11 different sections in the community: drinking water, energy/fuel/utilities, sanitation, education, health care, shelter, food, and food availability. Each section was ranked prior to the earthquake, immediately after, one month after, and four months after the earthquake. The data gathered from the field interviews indicated that the disruption rates before the earthquake were very poor with average ratings. The overall community rating was given a rating of "moderate availability" of 3.8/7. The officials said that after the earthquake there was a substantial increase in the disruption for service provision.

## 12.6 Comparisons and Discussions

One of the main issues in community sensing is that there have not been enough programs to see what is successful and what is not. Because community sensing is just starting out, there are a lot of ideas and opinions as to the direction it should go. We have seen a few of the ways people are looking to use data from sensors in cell phones, GPS, and other devices with traffic, environmental issues, and detection of natural disasters. Some people believe that sensing could eventually predict routines and the habits of people, so that it could be applied to nutrition and other aspects of daily life. Others see it providing businesses with information about the products they sell so they can begin to understand why and when consumers buy certain products.

A main concern with the data collection was how to protect the user's privacy and security from other users or certain people trying to misuse the data through hacking. While community sensing is designed to help everyone involved, there are always people who will try to use the data for harm or find out information on someone. The algorithms involved to protect the user cannot be too similar, however, so that if a hacker does figure out how to extract information, he/she cannot do it on a wide variety of programs. There is a lot of work into programming these security algorithms so that they can evolve over time and not allow any information to be leaked. With the algorithms that will be used to create these applications another concern is how to test and create the brains of each application. The developers of these programs may not have the money and other resources to go down one path of testing and realize that another path would be much more appropriate [10].

We believe that the future of community sensing runs hand in hand with the production and innovation of mobile phones. Mobile phones continue to push the envelope with their capabilities as new applications are being made every day to enhance the usefulness of the device. These mobile phones may also begin to create new sensors that will allow sensing to be done in a variety of areas. Obviously, because these sensors have not been created yet, we are not sure what their function might be, but we believe an emphasis will be put into protecting the environment and the human body. As better applications are made for community sensing that does not put too much of a burden on the user (time, energy usage of battery), the amount of people who will participate in these applications will skyrocket. One main area that community sensing could be applied are college campuses. Campuses provide a flurry of activity with a group of users who love participating in new programs that could potentially help themselves. If a developer or researcher would use the abundant resources on a college campus, they also could show their applications to students who potentially could take that application across the country after graduation if it has not spread already. College students also might come up with new thoughts as to what the applications could potentially do.

In some areas of community sensing the purpose was to allow the applications to do all the work after the algorithms can be worked out correctly. The data collection for these processes could become quite burdensome and overload the system that is collecting all of the data. In areas such as transportation, the data collection is paramount and if some data do get lost, the consequences could be disastrous. Whoever creates these applications must have plenty of resources available to take in data and the appropriate steps to know when it is time to discard data after their usefulness has been used.

One area that community sensing is just being applied is to help communities know how disaster relief is making progress. In the earlier example where the group went into Haiti after the earthquake, significant data could be taken in order to show what places needed more help. Because Tuscaloosa, Alabama, was ravaged by a tornado last April, we have seen the need for programs like the one described to show what areas need help. The people of Alabama and the United States gave so much in the immediate time period after the tornado, but after a while the donations stopped coming in even though the need was still there. We believe that community sensing programs could continue to show people the need in places struck by natural disaster so that aid may continue to come to those in need of help.

## 12.7 Conclusions

In conclusion, community sensing still needs a lot of progress to be made in order for its full capability to be fulfilled. There are so many areas that community sensing can be applied in everyday life. It can be used to make users aware of many things. Many of these include environmental issues, some being life threatening. Early earthquake detection could possibly help save lives. Air quality tests would also allow users to know of areas with potential health hazards. Other community sensing uses include issues that users are concerned with every day. Allowing users to know when traffic and accidents are in the area will allow for them to plan ahead before reaching this area of congestion. Another everyday use would be indication of party activity and indication of speed bumps. Full implementation will allow users to be notified of many common issues that arise in one's life. As mobile phones and other devices contain more and more sensors, applications and programs can be designed specifically for these new sensors, helping save lives and keeping communities safer and more in tune with nature.

## References

1. P. M. Aoki, R. J. Honicky, A. Mainwaring, C. Myers, E. Paulos, S. Subramanian, and A. Woodruff. 2008. Common sense: Mobile environmental sensing platforms to support community action and citizen science (demonstration). In *Adjunct Proceedings, 10th International Conference on Ubiquitous Computing (Ubicomp '08)*, pp. 59–60.

2. D. Tathagata, P. Mohan, V. N. Padmanabhan, R. Ramjee, and A. Sharma. 2010. PRISM: Platform for remote sensing using smartphones. In *Proceedings of the 8th International Conference on Mobile Systems, Applications, and Services (MobiSys '10)*. ACM, New York, pp. 63–76.

3. E. M. Atkins. 2006. Cyber-physical aerospace: Challenges and future directions in transportation and exploration systems. In *Proceedings of the 2006 NSF Workshop on Cyber-Physical Systems*, Austin, TX.

4. P. Dutta, P. M. Aoki, N. Kumar, A. Mainwaring, C. Myers, W. Willett, and A. Woodruff. 2009. Common sense: Participatory urban sensing using a network of handheld air quality monitors. In *Proceedings of the 7th ACM Conference on Embedded Networked Sensor Systems (SenSys '09)*. ACM, New York, pp. 349–350.

5. D. Christin. 2010. Security and privacy objectives for sensing applications in wireless community networks. Proceedings of 19th International Conference on Computer Communications and Networks, Zürich, Switzerland.

6. A. Krause, E. Horvitz, A. Kansal, and F. Zhao. 2008. Toward community sensing. In *Proceedings of the 7th International Conference on Information Processing in Sensor Networks (IPSN '08)*, IEEE Computer Society, Washington, DC, pp. 481–492.

7. P. M. Aoki, R. J. Honicky, A. Mainwaring, C. Myers, E. Paulos, S. Subramanian, and A. Woodruff. 2009. A vehicle for research: Using street sweepers to explore the landscape of environmental community action. In *Proceedings of the SIGCHI Conference on Human Factors in Computing Systems (CHI '09)*. ACM, New York, pp. 375–384.

8. J. Bevington. 2010. Uncovering community disruption using remote sensing: An assessment of early recovery in post-earthquake Haiti, Tech. Rep., Disaster Research Center, University of Delaware, Newark. Available at http://dspace.udel.edu:8080/dspace/bitstream/handle/19716/5850/Misc%2069.pdf?sequence=1.

9. M. Faulkner. 2011. The next big one: Detecting earthquakes and other rare events from community-based sensors. 10th International Conference on Information Processing in Sensor Networks (IPSN), Chicago, IL.

10. N. D. Lane, E. Miluzzo, H. Lu, D. Peebles, T. Choudhury, and A. T. Campbell. 2010. A survey of mobile phone sensing, *IEEE Commun. Mag.* 48:9, 140–150.

*Chapter 13*

# Wireless Embedded/ Implanted Microsystems: Architecture and Security

Derek Chandler, Jonathan Pittman, Jaber Abu-Qahouq, and Fei Hu

## Contents

## 13.1 Introduction

Wireless embedded microsystems (WEMs) have been popularly used in sensing and control of physical objects. For example, in a structural health monitoring application, a fiber-optic strain gauge sensor (a type of WEM) could be embedded into the elastic layer of a bridge to detect bridge fracture. In a medical application, the implanted medical devices or biochips such as pacemakers can be implanted into the human body for organ control. In a machine maintenance application, tiny sensors may be embedded into a machine to wirelessly report the equipment's operation and corrosion status.

Here, we call the object into which a WEM is embedded into the inside of its host. Typically, WEM applications have three important operations as follows:

(1) Wireless power charge: Because a WEM is embedded or implanted into the inside of its host, its battery is not easy to replace. Thus, a WEM is typically equipped with a rechargeable battery. Today, wireless power charge has become a popular, convenient way to recharge the battery. Its principle is to utilize the resonance frequency coupling to transfer the electromagnetic energy from a charger to the WEM's inductor in the battery.

(2) Anytime access: An external WEM reader often needs to access the WEM at any time. First, because a WEM can sense the host's status such as the bridge's cracks (via a gauge sensor) or the heart rhythm (via a pacemaker), we need to read out the sensing data at any time, including some emergency access scenarios such as reading a pacemaker in the ambulance. Second, a WEM can also use its actor(s) to change the features of the host. For example, in a medical emergency, we may speed up the pacemaker to accelerate the heartbeats. In summary, the WEM access includes read/write (R/W) actions at any time (including emergency access cases).

(3) Logging: A WEM should be able to log its access (R/W) operations in a brief format for future references. For example, a log could be like, "Dr. Smith speeds up the pacemaker to 80 triggers per minute at 15:10:20 on 12/30/2012."

Implantable medical devices (IMDs) are a special kind of WEM, defined as man-made medical devices that have one of three functions: replacing a missing biological structure, supporting a damaged biological structure, or enhancing an existing biological structure [1]. Thus, one can see that IMDs have a wide variety of potential uses. From prosthetic limbs to artificial pacemakers, IMDs work in concert with the human body to achieve many biological tasks.

There are two basic types of IMDs. The first type is biomedical material tissue. Unlike the tissue used in transplanted biomedical tissue IMDs, the tissue used in biomedical material tissue IMDs is not taken from a person or animal. Instead, it is created in a laboratory using either synthetic or biological components [2]. For example, it is currently possible to grow a human jawbone in a laboratory using stem cells. The second type of IMD is active implant electronics. This type of IMD includes some type of electronic circuit. Examples include the pacemaker and retinal implants [2]. Active implants will be the main focus of this survey chapter.

Active implants typically perform their tasks by doing one or both of the following: applying a therapy to an organ (usually by way of electrical signals) or monitoring certain parameters in a patient's body [2]. An example of a device that both applies therapy and monitors body parameters is an automatic pacemaker. This type of device monitors a patient's heartbeat. If the pacemaker senses that too long of a time period has passed between consecutive heartbeats, it applies a low voltage pulse to the ventricle of the heart. This action causes the heart to beat. Thus, this type of device both monitors a patient's body and applies a therapy. Pacemakers will be examined in more detail later in this chapter.

It is often necessary for a doctor to be able to communicate with an active implant without actually making contact with the device. This requires wireless communication, which can be difficult to implement. The implementation of secure and reliable wireless communication is one of the most difficult problems to overcome in IMD design [3]. Medical Implant Communication

Service (MICS) is a communication standard that uses frequencies between 402 and 405 MHz to communicate with electronic implants [3].

One of the greatest challenges in IMD wireless communication is information security [4]. The four components of information security are confidentiality, integrity, availability, and accountability [2]. Security prevents an unauthorized party from determining if a patient has any IMDs, and it also prevents unauthorized retrieval of data from said IMDs. There are many reasons that patients would not wish that their IMD information be displayed or broadcasted. Therefore IMD information should be disclosed only to authorized entities and the patient. Of course, secure wireless communication can be difficult to achieve while maintaining a high data throughput [3]. While communication protocols like Wi-Fi Internet are attractive for their high throughput, it would not at all be secure to transfer IMD information using this protocol. Every Wi-Fi device has a particular IP address, which could be used to determine the patient's location and other private data. For this reason, security measures have been developed specifically for IMDs. These security measures will be discussed in more detail later.

The remainder of this chapter will explore the technical aspects of IMDs in more detail. First, we will introduce two types of IMDs: the pacemaker and the neurostimulator. We will discuss the applications and overall design for each of these IMDs. Then we will discuss several issues related to wireless communication with IMDs. After that, we will discuss a circuit design that implements wireless communication with an IMD. Finally, we will look briefly at the future of IMD technology.

## 13.2 IMD Examples

Although there are many specific types of IMDs, here we only introduce two that are typical: the pacemaker and the neurostimulator. Both are examples of active implants, and both provide good examples of the modern challenges facing IMDs. Specifically, both the pacemaker and the neurostimulator involve challenges regarding wireless communication, power, safety, and reliability. Thus, these two types of IMDs help to provide a good survey of the entire IMD field, even if they do not represent every IMD challenge.

### 13.2.1 Pacemaker

A pacemaker is an implantable device that uses electrical impulses from electrodes to communicate with heart muscles. The pacemaker is placed in the chest region, near the heart, to help control abnormal beats of the heart. The electrical devices inside the pacemaker are used to cause the heart to beat at a normal rate or to control arrhythmias. Arrhythmia means that the heart may beat with an irregular rhythm: too fast or too slow. When the heart has a heartbeat that is too fast, it is called tachycardia, and when the heartbeat is too slow, it is called bradycardia [5]. When arrhythmias occur, there may not be enough blood pumped to the body from the heart. Arrhythmias can cause tiredness, passing out, shortness of breath, or even vital damage to the body's organs and death.

The pacemaker was one of the first IMDs ever created. External pacemakers were originally designed as simple devices. They were designed and used to control the rhythm of the heart without any direct side effects. Whenever the battery dies in a pacemaker, they just simply need to be recharged and recalibrated for the intended user.

A pacemaker is usually inserted into the patient through simple surgery procedures using either general or local anesthetic. Typically, the patient is administered a drug dosage for relaxation before the surgery, and an antibiotic is given as well to prevent infection. An incision is usually made just beneath the collar bone on the left shoulder to create a small pocket for the pacemaker to be inserted into the patient's body. The leads are pushed into the heart through a large vein in the body by using a fluoroscope to monitor the progress of the insert. The number of leads used usually depends on what type of pacemaker is being installed. For example, the right ventricular lead should be positioned on the interventricular septum away from the tip of the right ventricle to enhance the strength of the heart. Usually, the surgery for installing a pacemaker lasts around an hour to an hour and a half [6]. The pacemaker is typically placed underneath the left shoulder below the collar bone.

A patient should proceed with caution after pacemaker surgery. The patient should monitor and tend to the wound as it heals [6]. There is always a follow-up session in which the doctor checks the pacemaker's status using a type of program that communicates wirelessly with the pacemaker and allows the system's use [6]. The programmer also determines different settings of the pacemaker such as the pacing voltage output. The patient should have the pacemaker tested or analyzed once every 1 or 2 years to make sure that the placement of the right ventricular lead has not caused the left ventricle to become weakened or damaged [6].

Once a pacemaker loses battery power, the battery must be replaced through a surgical procedure [6]. Fortunately, the battery replacement procedure is simpler than the original insertion of the pacemaker because it does not require leads to be implanted. In most battery replacement procedures, an incision is made to remove the existing pacemaker; the leads are detached from the original pacemaker. They are then attached to the new device, and the new device is then inserted into the body [6]. Ideally, the new device works just as the original device had.

There are several possible safety concerns with any pacemaker. One of these concerns is pacemaker-mediated tachycardia (PMT) [6]. PMT can occur with dual-chambered pacemakers and is a form of reentrant tachycardia. In pacemaker-mediated tachycardia, the artificial pacemaker forms the atrium to ventricle limb and the atrioventricular node forms the ventricle to the atrium of the circuit [6]. Treatment of this condition is usually simple and requires reprogramming of the pacemaker.

According to a statement made by the Heart Rhythm Society, it is legal and honorable to deactivate implanted cardiac devices [6]. Lawyers use the example of a feeding tube to be similar to pacemakers. A patient, or one with legal authority to make decisions for a patient, has the right to discontinue or refuse treatment (including treatment that keeps the patient alive). A physician may refuse to turn off a pacemaker, but he or she is required to refer the patient to another qualified physician who would [6]. In some cases, patients who have hopeless conditions (strokes, dementia, heart attack, etc.) prefer not to progress through the often painful and stressful last stages of their lives with supportive devices such as pacemakers.

There are also some privacy and security concerns with wireless pacemakers. Unauthorized individuals (third parties) could possibly read the patients records contained inside the pacemaker [4]. A team of researchers has demonstrated this weakness [6]. This weakness could allow the third party to potentially reprogram the pacemaker and even could potentially cause a great deal of harm to the patient (up to and including death). The research team only conducted an experiment dealing with short-range wireless security. However, as technology evolves, it is possible that private records will be able to be accessed from longer distances [6]. This calls for better security and some type of patient alert system in accessible medical implants such as pacemakers.

### 13.2.2 Neurostimulator

A neurostimulator is a device designed to help patients manage intense pain. It works by delivering electrical impulses to the spinal cord. These impulses prevent pain signals from reaching the brain. Instead of experiencing pain, patients with a neurostimulator experience a tingling sensation, much like one experiences when a limb goes numb. Obviously, this is not ideal for a patient (ideally, a patient would simply experience normal feeling), but it is preferable to intense pain. Neurostimulators are most effective when used to treat pain in the back, legs, or arms.

Neurostimulators make use of a battery-type device, called a pulse generator, to deliver electrical impulses to a lead that is placed near nerves along the spinal cord. The exact placement of the lead varies based on the area of the body that is experiencing pain. The lead is always implanted under the patient's skin [7]. The pulse generator, however, is not always implanted. In fact, the placement of the pulse generator is the main difference between the two main types of pulse generators: implantable pulse generators (IPGs) or radiofrequency generators (RFGs) [8].

In IPG systems, which are also known as internal systems, the pulse generator is placed under the skin, usually in the buttock or abdomen. The generator contains a battery, and in many older systems, surgical methods are required to replace the battery. In newer neurostimulators, however, the battery is recharged by placing a wireless charging device over the area where the generator is implanted. Many times, patients are kept awake while IPG systems are implanted. This allows the patient to provide feedback about the device's effectiveness so that the surgeon will know how to program the device. Some newer IPGs can be wirelessly reprogrammed after the surgery is complete. Many patients prefer IPGs to RF neurostimulators because the system is never visible [4]. Also, IPGs can be used while showering and swimming because they are fully covered by skin [9]. The major disadvantage to using IPGs is that they do require a more complicated surgical procedure than RF neurostimulators. Of course, every surgery procedure entails risk.

In RF systems, which are also called external systems, the pulse generator is not implanted into the patient's body (the spinal leads, of course, are implanted). RF systems' function uses a small external device called a transmitter. A small receiver device is implanted in the patient's body. The transmitter is connected to an antenna, which must be placed on the patient's body over the receiver. Pulses are transmitted wirelessly through the patient's skin from the transmitter to the receiver. The battery for an RF neurostimulator is located in the transmitter, which allows for easy, nonsurgical battery replacement or recharging [8]. Unfortunately, though, these systems cannot be worn while the patient is in water (bathing, swimming, etc.) since the transmitter and antenna are not implanted [9].

As we have seen, there are advantages and disadvantages to using IPG and RF neurostimulators. Table 13.1 summarizes the information for each type of neurostimulator. It allows for an easy comparison of the two types.

Neurostimulators are generally considered safe and have been used successfully for over 30 years [9]. There are, however, some common safety concerns associated with neurostimulators. First, neurostimulators do require a surgical procedure, and any surgical procedure does involve risks (infection, bleeding, adverse reaction to anesthesia, etc.). For this reason, many doctors will not suggest a neurostimulator unless all nonsurgical methods have failed. Second, all neurostimulators carry the risk of failure. This risk is especially dangerous if an implanted component fails. In this case, surgical procedure is required to perform maintenance. Also, if the device fails while the patient is in a high-risk situation like driving, the resulting pain could be enough to cause further injury (lose control of a vehicle, etc.). Finally, if the device is implanted improperly, it can cause permanent damage to the spinal cord over time. Of course, many patients find the benefits provided by neurostimulators to be worth these potential risks.

**Table 13.1 Neurostimulator Comparison**

|  | *IPG Systems* | *RF Systems* |
|---|---|---|
| Implanted components | Pulse generator, spinal leads. | Receiver, spinal leads. |
| External components | None. | Transmitter, antenna. |
| Battery replacement/recharge | Surgical procedure required for older systems. Newer systems make use of wireless recharge. | Battery located in transmitter. No surgery required for replacement or recharge. |
| Wireless communication | Only used in newer systems for reprogramming. | Pulses sent wirelessly from transmitter to receiver. |
| Maintenance | Often require surgical procedure. | Rarely require surgical procedure. |

Thus, we have seen that neurostimulators are active implants (i.e., they provide a therapy to the body), they involve issues regarding battery recharging, and they involve issues regarding wireless communication. The specific circuit design for some of the components used in neurostimulators will be discussed in later sections of this chapter. Also, issues regarding wireless power and wireless communication will also be discussed.

## 13.3 Wireless Power Transfer and Communication

Many IMDs make use of wireless technology to transmit power and/or data. The benefits of wireless communication with IMDs are numerous. The wireless transmission of power to IMDs reduces the need for intrusive battery replacement surgeries. Also, wireless transmission of data allows for IMDs to be monitored without resorting to surgical means. This section will investigate the two main technologies used to communicate wirelessly with IMDs: magnetic coupling and the MICS. Also, this section will investigate several issues related to wireless security.

### 13.3.1 Magnetic Coupling

Traditionally, IMDs have used magnetic coupling to achieve wireless power and data transfer. These types of IMD systems consist of two parts: an implantable stimulator and an external device that drives the stimulator. The external device transmits power and data over a magnetically coupled link, and the implantable device makes use of the power and data to provide therapy to the patient [10]. These systems are very similar to the RF neurostimulators that were discussed in the preceding section (Figure 13.1).

In a wireless charge circuit, modulation and demodulation occur on both the external and internal devices. Of course, this is necessary because they both must be able to send and receive messages using whatever protocol is established. The internal device must have the ability to receive and regulate power from the external device; a rectifier and regulator are often used to accomplish this task. Also, note that the internal device makes use of a controller to drive the patient's therapy.

While magnetic coupling is a well-known and well-tested method for transmitting signals wirelessly, it does have several major drawbacks. First, it requires the use of an external device that must be

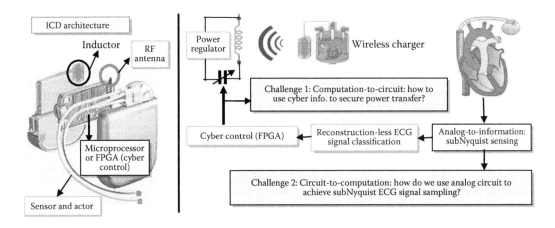

**Figure 13.1** Magnetic coupling for wireless IMD charge. ICD, Implantable cardiovascular Defibrillator; FPGA, Field Programmable Gate Array.

worn at all times. Because the implantable device has no power supply of its own, it must always be connected to the external device. Many patients find this undesirable. Second, magnetic coupling can often only provide data rates of less than 50 kbps [3]. This data rate limits the number of possible IMD applications. Finally, magnetic coupling often has a very short range. Because of these (and other) limitations, the MICS communication band was developed. The next section will detail MICS technology.

## 13.3.2 Medical Implant Communication Service

The MICS communication band was established in 1999 to allow for communication with IMDs. It consists of the frequencies from 402 to 405 MHz, and it allows for longer transmission range (typically about 2 m) and faster transmission rates than magnetic coupling [3]. This frequency band was chosen for several reasons. First, it allows for good signal propagation through the human body. Second, the MICS band was previously used to support meteorological aids such as weather balloons [3]. Because IMDs require only short transmission distances, and because IMDs are not typically used in close proximity to weather balloons, these two types of technology can share the MICS band. Finally, the MICS band was also available in many European countries, which allowed for international compatibility.

MICS does not only specify a frequency band for wireless IMD communication, but it also specifies several device requirements. Any MICS compliant device [3] must

- Use as little power as possible during 400-MHz communications. This is because implant battery power is often very limited. Thus, current draw during communication sessions should be limited to less than 6 mA for most implantable devices.
- Make use of a lower power mode when asleep and periodically "sniff" for a wake-up signal. This requirement also reduces power use.
- Allow for data rates of at least 20 kbps.
- Allow for a transmission range of at least 2 m.

These four requirements encompass the most important MICS specifications. Obviously, they present chip designers with many challenges. The most difficult challenge is power consumption.

**Table 13.2 MICS Requirements**

| | Requirement | Challenges |
|---|---|---|
| Frequency band | 402–405 MHz | N/A |
| Maximum current draw during communication | 6 mA | Requires low power components. Often results in weak transmission strength. |
| Minimum data rate | 20 kbps | Can only use 3 MHz of bandwidth. |
| Minimum transmission range | 2 m | Difficult to achieve with low power components. |

Two of the four listed requirements (current limit and "sniffing" capability) are directly related to power consumption, and the other two (data rate and transmission range) are indirectly related to power consumption. Thus, every component in a MICS compliant device must be extremely power efficient. Also, MICS compliant devices require very sensitive antennas because of the low-power requirements during communication [3]. Table 13.2 summarizes MICS requirements [4]. Devices that meet these requirements have been designed.

MICS has truly revolutionized wireless communication with IMDs. While magnetic coupling was sufficient for many older IMDs, it did not provide the data rates or the functional structure of MICS. With MICS now firmly established as a communication standard, IMD technology will surely advance. Doctors have envisioned IMD systems that monitor and control many signals in the human body, and MICs might allow for some of these systems to become feasible [3]. As low-power technology evolves, look for more MICs applications.

## 13.3.3 Security

As was mentioned previously, security is incredibly important when communicating wirelessly with IMDs. There are many people who might wish to interfere with IMD operation for a variety of reasons. These people are often referred to as adversaries. Adversaries can be categorized using four groups: passive, active, coordinated, and insiders [2]. Passive adversaries monitor signals transmitted by IMDs and by different sources communicating with the IMD. Active adversaries interfere with communications between IMDs and other equipment. Coordinated adversaries work in a group to achieve their goals. For example, one adversary could be close to the patient while the other is near an IMD programmer. Insider adversaries are the hardest to detect because they are adversaries that pose as health care professionals, hardware engineers, or even patients. Much more time will be spent on IMD communication security later in this chapter.

Data privacy is not the only IMD communication concern. IMD manufacturers must also be concerned with data tampering [4]. No person (including the patient) should be able to tamper with IMD files or interfere with IMD behavior by providing erroneous inputs. All information stored in an IMD should be editable only by an authorized health care professional. This includes data that are present from the beginning of IMD operation (such as serial number) and data that are stored after the IMD has commenced operation (such as log files).

With these ideas in mind, several requirements for IMD security have been developed. These requirements are

■ Authorized clinical access: Allows the patient's doctor and his/her staff to be able to change the settings on a patient's IMD when the patient is in the clinic.

- Emergency access: Allows the medical staff at other hospitals and clinics to be able to change the patient's IMD settings when the patient is taken there for an emergency situation.
- Security: There shall be no unauthorized person(s) to be able to change the settings or view the information dealing with a patient's IMD [4].

Achieving authorized clinical access is fairly simple, and there are a number of ways that this can be done. Most of the methods involve some sort of password security system. That is, the doctor at a clinic must have a password in order to access the device. However, it would be insufficient for only the doctor at the clinic to have the password for the device, because that would not allow for emergency care by other doctors. It is important for emergency care doctors to be able to access the device without using a method that would compromise security.

Several methods have been developed to allow for emergency access to IMDs. The primary approaches are passwords, additional patient body modifications, patient behavior changes, and other methods that are passive with respect to the patient [4]. Examples of these types of systems include medical bracelets with a password engraved on them, tattoos with a password in barcode form, ultraviolet tattoos with a barcode, and proximity bootstrapping devices. One of the most interesting of these solutions is the password tattoo, and there have been many individuals in the medical community who have suggested that this method would be best for all involved. The main problem with using tattoos, however, is that it requires more body modifications for patients besides the IMD that is implanted. People have invented a tattoo that is only visible by ultraviolet light. This is not visible to the naked eye, and this would be sufficient for patients who do not like the extra body modifications. The ultraviolet tattoo is clever and is the most secure way to ensure that the patient's IMD settings remain secure.

Some security methods require patients to allow some kind of modification to their behavior. For example, some patients have been asked to wear a wristwatch or wristband that contains all of the patient's IMD information. This approach allows the patient's private information to be concealed for the most part. However, given the chance, the information on the wristband may still be accessed by unauthorized individuals. Another downfall to this approach is that the wristband could become lost or damaged from every day wear and tear. If the wristband becomes damaged, then the IMD information would not be accessible to the proper authorities and this could result in a severe emergency. Also, this approach is not ideal for all patients because some patients refuse to wear the wristwatch for various personal reasons. Some patients may already wear medical alert bracelets, and this would provide contrast to the wristband systems [4].

As was mentioned earlier, some security systems are passive with respect to the patient. These systems are entirely built into an IMD and do not require any additional body modifications, behavior changes, or extra equipment [4]. These types of IMDs track various metrics such as the location of the patient, whether the patient is sitting or standing, the patient's heartbeat, and the patient's blood pressure. If the device determines that the patient is in an emergency situation, it contacts the appropriate medical authorities. This approach is very efficient in dealing with lower security failures and safety failures.

The proximity bootstrapping security method makes use of an external device (called a bootstrapping device) that can temporarily access an IMD key when it becomes in contact with a patient. This allows medical staff to obtain access to the IMD. This approach is only available through short and medium range. The downfall of this approach is that it is not always available for a patient during a time of an emergency unless the patient is near a bootstrapping device.

Recently, research has been conducted to determine how patients feel about various types of IMD security systems. The research showed that patients have two main preferences regarding

IMD security. First, it is important for patients to feel like their information is secure. Because of this, patients tend not to like password bracelets, because they feel like this type of system allows anyone to access their medical data. Second, patients do not like for their IMD security system to interfere with their normal behavior or change their physical appearance. Thus, patients tend to dislike visible tattoos. Ultraviolet tattoos, however, are fairly popular with patients because they are not visible to the naked eye.

## 13.4 Circuit Design

Recently, much research has been performed in the area of IMD circuit design. In this section, we will discuss one particular IMD circuit [11]. A high-level block diagram for the circuit that they designed can be seen in Figure 13.2. The system consists of two parts: an internal part and an external part. The external part has a class-E driver, which includes a high-efficiency RF transmitter. The external part is inductively coupled to the internal part. The internal part consists of a power regulation circuit, a signal processing circuit, and an actuator. Each of these components will now be discussed in more detail.

The power regulation circuit is the component that is responsible for providing power to the rest of the internal part of the system. It consists of a full-wave rectifier and a voltage regulator. It receives RF power signals from the external system, rectifies the signals, and then regulates them to the appropriate level. This process results in a stable DC power source that can be used by the rest of the electric components. It is important to note that the regulation is just as important as the rectification in this system because the RF power signals usually fluctuate within a certain range.

The signal processing unit is responsible for processing the data that are received through the magnetic coupling. This process is achieved through the use of an amplitude shift keying (ASK) demodulator. ASK modulators are fairly low power and easy to implement. They consist of an inverting amplifier, a low-pass filter, and a comparator.

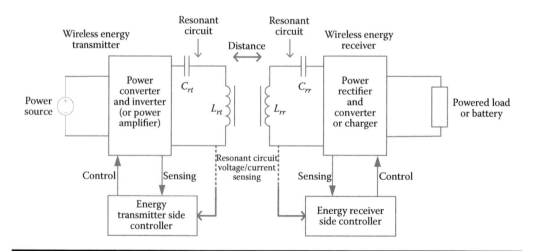

**Figure 13.2 Illustrative block diagram of wireless energy transfer via resonant inductive coupling.**

After the data signals are demodulated, they are passed along to a controller. The controller uses the data to determine an appropriate setting for the current driver. The current driver then drives the actuator, which is the component that actually applies the therapy to the patient. The therapy varies based on the setting that is chosen by the controller. In this way, the therapy is able to adapt almost instantaneously to the needs of the patient.

Li, Hu, and Zhang [11] were able to implement this design on a chip using a 0.5-μm 2P3M standard CMOS process. The chip was tested in the human body temperature range, and the results were successful. This circuit is a good representation of modern IMD design techniques. Of course, it is interesting to discuss the possibilities of future IMD technology.

## 13.5  The Future of IMD Technology

Now that we have seen an example of modern IMD technology, let us discuss where IMD technology will go in the future. Many of the challenges associated with IMD design stem from the necessity to achieve reliable low-power wireless communication. As has already been mentioned, battery life is a huge impediment to IMD design. A couple of things will likely happen in the future to ease this challenge. First, more power-efficient methods of wireless transmission will be developed. These methods could include new antenna technology and new modulation methods. Second, electric components, in general, will become more power efficient. As controllers, rectifiers, and other electric components become more efficient, the strain on IMD batteries will be lessened by a good amount.

So what will IMD designers do when they achieve more power efficient devices? First, patient stress will be reduced because battery maintenance will be reduced. Second, IMD designers and medical professionals will likely develop new, advanced uses for IMDs. Some medical professionals have already envisioned a future in which nearly everyone will have some sort of IMD. Athletes could make use of IMDs to inject adrenaline into their bloodstream, and soldiers could make use of IMDs to enhance their sensory awareness on the battlefield. Even average (and healthy) people could make use of IMDs to constantly monitor their vital signs. The potential medical applications of IMD technology are almost endless.

Of course, as with any new technology, new developments in the IMD field will almost certainly lead to ethical debates in the future. Consider the previously given example of athletes using IMDs to automatically inject adrenaline into their bloodstream as they become tired during a sporting event. Would this type of activity be allowed by professional sports organizations? Should it be allowed? These questions (and many more) are a taste of what will face medical professionals as IMD technology advances. Humans are often apprehensive of human–machine combinations (sometimes referred to colloquially as "cyborgs"). These ethical issues and apprehensions would have to be discussed.

Another major issue in the future of IMD technology is primarily related to the business world. Before we can discuss IMD business in more detail, however, we must first discuss the government regulations that govern the design and sale of IMDs in the United States. The US Food and Drug Administration (FDA) regulates all medical devices. According to the FDA, medical devices include essentially any instrument or apparatus that is used to diagnose disease, cure disease, or affect the function of the human body through nonchemical means. As one might guess, IMDs fall into this category. In fact, most IMDs are classified as class 3 medical devices, meaning that they are subject to general controls (misbranding prohibitions, oversight of manufacturing practices, etc.), special controls (labeling requirements, mandatory performance standards, etc.), and premarket approval by the FDA [12].

With the large number of regulations regarding the manufacture, marketing, and sale of IMDs, any company that wishes to enter the IMD field must be concerned with not only the needs of the customer but also the requirements of the government [12]. This makes for a relatively high barrier of entry into the IMD market, which, in turn, has resulted in a small number of companies that sell IMDs [2]. Having fewer companies in a market tends to lead to less innovation and higher prices in that market. Obviously, that situation is not ideal.

So what can be done to lower the barriers of entry into the IMD market? One of the most obvious solutions would be to reduce government regulation. Of course, this solution should be pursued with caution; less government regulation could lead to IMDs that are not safe. Another possible solution is for the government to subsidize IMD companies. While this might be an attractive solution because of its simplicity, it is unlikely in the United States because of high government budget deficits. Perhaps the most likely solution is for private capital to enter the IMD market to assist start-up businesses and entrepreneurial endeavors. With the incredible potential in the IMD field, it is quite likely that venture capitalists will begin to look for opportunities to invest in new companies in the field.

Regardless of the challenges on the business side of IMD design, the future does look bright for IMD technology. The potential applications are almost endless, and it is a popular topic in scientific research. These two advantages will almost certainly lead to more innovation in the IMD market in the future.

## 13.6 Conclusions

This chapter has provided an overview of IMDs. We discussed the definition of IMDs and also briefly discussed some of their possible uses and classifications. We then discussed two specific types of IMDs: the pacemaker and the neurostimulator. Both of these IMDs are active IMDs, meaning that they provide a constant therapy. While discussing the pacemaker, we learned of challenges regarding battery life and reliability. While discussing the neurostimulator, we learned about the differences between IPG systems (which are fully implanted into the patient) and RF systems (which make use of both an internal component and an external component).

We then discussed wireless communication. Almost all IMDs make use of wireless communication to send and receive data, and many IMDs are also powered wirelessly. There are two main methods for achieving wireless communication with IMDs: magnetic coupling and MICS. Many older IMDs make use of magnetic coupling. Newer devices, on the other hand, make use of the MICS communication protocol to transmit data. MICS also allows for higher data rates than old-fashioned magnetic coupling. Also, MICS has been important in standardizing the method of communicating with IMDs.

We also discussed security as it relates to wireless communication with IMDs. Of course, it is incredibly important that all medical data be secured while communicating with a patient's IMD. There are several popular methods for securing these data. These methods include password bracelets, password tattoos (visible and ultraviolet), and proximity bootstrapping. For obvious reasons, patients tend to prefer security methods that do not require a change in either their behavior or their appearance.

After concluding our discussion of wireless communication, we discussed a particular IMD circuit design. This particular circuit made use of magnetic coupling to pass power and data from an external IMD component to an internal IMD component. The internal component made use of an ASK demodulator to receive data from the external component. These data were then used to drive the therapy process.

Finally, we discussed the future of IMD technology. The future holds mainly two things for IMDs: lower power consumption and more medical applications. As more medical applications are devised, we will likely see an increase in the number of IMD users. We could even see a time when most people use IMDs to enhance bodily functions.

IMDs are powerful devices indeed. With current IMD technology, we can regulate heartbeats, treat chronic pain, and deliver pharmaceutical products. Without IMDs, many patients would be in much worse conditions. The continuing advancement of IMD technology is one of the most interesting and promising trends in medical science. As our understanding of low-power technology and wireless communication technology increases, we could see great advances in the field of IMDs. The future of IMDs is bright.

# References

1. Wikimedia Foundation, Inc. 2012. Implant (medicine). Available at http://en.wikipedia.org/wiki/Implant_%28medicine%29 (Accessed on April 26, 2012).
2. M. Guo. 2007. Implantable medical devices. Class report. University of Alabama, Tuscaloosa.
3. P. D. Bradley. 2007. Implantable ultralow-power radio chip facilitates in-body communications, *RF Design*, 20–24, June.
4. T. Denning, A. Borning, B. Friedman, B. T. Gill, T. Kohno, and W. H. Maisel. 2010. Patients, pacemakers, and implantable defibrillators: Human values and security for wireless implantable medical devices. In *Conference on Human Factors in Computing Systems*, ACM, New York.
5. National Heart Lung and Blood Institute. 2012. What is a pacemaker? [online]. Available at http://www.nhlbi.nih.gov/health/health-topics/topics/pace/ (Accessed on April 26, 2012).
6. Wikimedia Foundation, Inc. 2012. Artificial cardiac pacemaker. Available at http://en.wikipedia.org/wiki/Artificial_cardiac_pacemaker (Accessed on April 26, 2012).
7. American Pain Foundation. Implantable pain therapies are a treatment option [online]. Available at http://www.painfoundation.org/painsafe/person-with-pain/implantable-pain-therapies/implantable-treatment-option.html (Accessed on April 26, 2012).
8. American Pain Foundation. 2006. Treatment options: A guide for people living with pain [online]. Available at http://www.painfoundation.org/learn/publications/files/TreatmentOptions2006.pdf (Accessed on April 26, 2012).
9. American Pain Foundation. Neurostimulation can be used safely [online]. Available at http://www.painfoundation.org/painsafe/person-with-pain/implantable-pain-therapies/neurostimulation-safety.html (Accessed on April 26, 2012).
10. C.-H. Kao and K.-T. Tang. 2009. Wireless power and data transmission with ASK demodulator and power regulator for a biomedical implantable SOC. Paper presented at IEEE/NIH Life Science Systems and Applications Workshop, Bethesda, MD.
11. X. Li, J. Hu, and H. Zhang. 2010. Wireless energy and signal transmission for micro implantable medical system. Paper presented at IEEE International Conference on Nano/Micro Engineered and Molecular Systems, Xiamen.
12. U.S. Food and Drug Administration. 2010. Medical devices [online]. Available at http://www.fda.gov/medicaldevices/deviceregulationandguidance/overview/classifyyourdevice/ucm051512.htm (Accessed on April 26, 2012).

# Chapter 14

# The Application of Machine Learning in Monitoring Physical Activity with Shoe Sensors

Wenlong Tang, Ting Zhang, and Edward Sazonov

## Contents

# 14.1 Introduction

Cyber-physical systems (CPSs) are integrations of computation with physical processes. Embedded computers and networks monitor and control the physical processes, which usually affect computations and vice versa [1]. CPSs have been widely found in such fields as aerospace, automation, chemical engineering, civil infrastructure, energy, health care, manufacturing, and transportation.

In this chapter, a CPS for monitoring human physical activity will be introduced. The system consists of a computational part that includes machine learning algorithms, computers, and the physical elements that are shoe sensors and humans. The computational part acquires signals from the human subjects through the shoe sensors and calculates the time spent in various activities during the day. The subjects are then provided feedback on how to adjust the levels of physical activity to live a healthier lifestyle. The computation and the subjects interact with each other tightly in this feedback cycle. In the rest of this section, machine learning and the shoe sensors will be introduced. Section 14.2 will briefly describe the three machine learning approaches: support vector machine (SVM), artificial neural network (ANN), and decision trees (DT). Section 14.3 will present and compare the results from the three machine learning approaches. Some discussions and conclusions are provided in Sections 14.4 and 14.5, respectively.

## 14.1.1 Machine Learning

Machine learning has had several definitions since the late 1950s. One of the earliest definitions was from Arthur Samuel, an American pioneer in the field of artificial intelligence, in 1959. He defined machine learning as a "field of study that gives computers the ability to learn without being explicitly programmed" [2]. In 1983, Herbert Simon made an alternative definition as "learning denotes changes in the system that are adaptive in the sense that they enable the system to do the same task or tasks drawn from the same population more efficiently and more effectively the next time" [3]. Another definition was provided by Tom M. Mitchell in 1997: "A computer program is said to learn from experience $E$ with respect to some class of tasks $T$ and performance measure $P$, if its performance at tasks in $T$, as measured by $P$, improves with experience $E$" [4]. In summary, machine learning methods build general models based on the data from particular samples.

According to different learning types, machine learning methods can be classified into three categories: supervised learning, unsupervised learning, and reinforcement learning. Supervised learning (pattern recognition) refers to when the models have to learn a projection between inputs and outputs in order to predict new information (a training data set is required in this case), with representative methods such as SVM, ANN, regressions, and DT. Unsupervised learning, such as clustering, principle component analysis, and independent component analysis, refers to when the models have to extract information from the distribution of the input data. Reinforcement learning refers to when the models have to learn a regularity that projects states to actions that leads to a maximal reward, such as Monte Carlo methods [5].

Recently, machine learning has been widely used in computer science, engineering, bioinformatics, and biostatistics. For example, machine learning approaches are employed to teach machines in human tracking and identification [6]. In bioinformatics and biostatistics, great challenges are brought by the continuous arrival of new data that requires researchers to quickly translate raw data into scientific knowledge. Machine learning is a powerful tool in the integration of high-throughput and high-dimensional data, such as the analysis of gene expression data [7] and copy number variation data [8]. In this chapter, we will talk about the applications of three widely utilized machine learning methods—SVM, ANN, and DT—in a shoe sensor system for monitoring people's daily physical activity.

### 14.1.2 Shoe Sensors

Monitoring of posture allocations and physical activities is important in many areas of biomedical research [9]. It can help to live a healthier lifestyle. For instance, prostate cancer has been proven to be directly related to extensive sitting [10]. Obesity could be caused by insufficient physical activity and prolonged car driving [11]. Higher physical activity is associated with lower risk of osteoporosis [12]. Moreover, abnormal patterns of daily activities are symptoms of many diseases. It is reported that children with autism have weak muscles that may result in fewer daily activities than healthy children [13].

The shoe sensor system introduced here for monitoring physical activities, called "SmartShoe," collects heel acceleration and plantar pressure data (as shown in Figure 14.1a). Each shoe has five force sensitive resistors (FSRs) [14] integrated in a flexible insole. The five FSRs are positioned under the five critical points of contact: heel, heads of metatarsal bones, and the hallux (as shown in Figure 14.1b). These locations allow for differentiation of the most important parts of a gait cycle, such as swing phase, stance phase, heel strike, and toe-off as well as accounting for differences in loading of anterior and posterior areas of the foot in ascending/descending stairs and cycling. A three-dimensional (3-D) accelerometer [15] located on the back of each shoe provides the motion information in three dimensions. The accelerometer is constantly powered on, with a power of 2.55 mW. To save power, voltage supply to pressure sensors is turned on only for the duration of the measurement (128 µS), resulting in the worst-case consumption of 6 µJ/sample or 150 µW at 25 Hz. The wireless board, the battery, and the power switch are installed on a rigid circuit board glued to the back of each shoe (as shown in Figure 14.1a). The tail of the flexible insole is fed through a narrow cut in the shoe and connected to the same circuit board. The sensor system is very lightweight (less than 35 g) and creates minimum observable interference with motion patterns and little perceived changes in tactile sense or rigidity of the shoe.

SmartShoe is based on a combination of two sensor modalities: acceleration and pressure. The hypothesis that drove the selection and placement of the sensors is that the combination of these two modalities uniquely identifies many different major postures and activities and thus would

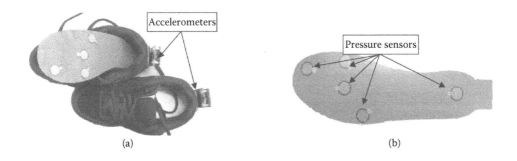

(a)                                                          (b)

**Figure 14.1** (a) A pair of shoes with the wearable sensors, the wireless transmitter, and the batteries. The accelerometers on the back of the shoes can measure anterior–posterior ($A_{AP}$), medial–lateral ($A_{ML}$), and superior–inferior ($A_{SI}$) axes of accelerations. (b) Pressure sensitive insoles are equipped in the shoes. $P_H$ is the heel pressure sensor, $P_{5M}$, $P_{3M}$, and $P_{1M}$ are the fifth, the third, and the first metatarsal head sensors, respectively, and $P_{HX}$ is the hallux sensor.

possibly need no feature extraction from the time series. The sensors can be fully embedded into a shoe and thus can be applied in one (or two) location(s) on the body to minimize the interference with activities of daily life. Moreover, because the monitoring system is integrated into conventional footwear, it requires minimal extra effort to wear the sensors, thus reducing the conspicuousness and burden associated with activity monitoring and facilitating everyday use.

The acceleration and pressure signals are sampled at 25 Hz by a 12-bit A/D converter and sent over a wireless link to a base computer. The wireless system used for data acquisition is based on the Wireless Intelligent Sensor and Actuator Network [16], which allows for data sampling simultaneously (with a difference of no more than 10 μS) from both shoes. The total power consumption of the wireless sensor system is approximately 15 mW, which enables up to 20 h of continuous monitoring on a 100-mAh battery. The sensor data are streamed to a portable computer with a LabVIEW data acquisition program and stored on the hard drive for further processing.

A single sample of data from a shoe is represented by a vector $S = \{A_{AP}, A_{ML}, A_{SI}, P_{H}, P_{5M}, P_{3M}, P_{1M}, P_{HX}\}$, where $A_{AP}$ is the anterior–posterior acceleration, $A_{ML}$ is the medial–lateral acceleration, $A_{SI}$ is the superior–inferior acceleration, $P_{H}$ is the heel pressure, $P_{5M}$, $P_{3M}$, and $P_{1M}$ are the pressures from the fifth, the third, and the first metatarsal head sensors, respectively, and $P_{HX}$ is the pressure from the hallux sensor. Our previous study has shown that a higher accuracy of classifying activities could be achieved by discarding signals from $P_{5M}$ and $P_{3M}$ [9]. Thus, those signals were discarded in the analysis from the beginning.

## 14.2 Machine Learning Approaches

After the data acquisition, the time series of data from both shoes were combined as $f_i = \{S_{L}, S_{R}\}_i$, $i = \{1,…, M\}$, where $S_{L}$, $S_{R}$ were the data samples from the left and right shoe, respectively, and $M$ was the length of the time series. Because different sensors may generate signals in different scales, a normalization procedure was performed on the sensor data as follows:

$$x_i = [X_i - \min(F)]/[\max(F) - \min(F)] \tag{14.1}$$

where $X_i$ is the sensor signal for the $i$th epoch, $x_i$ is the normalized value of $X_i$, and $F$ is the entire sensor signal. After the normalization, all the elements follow $x_i \in [0, 1]$. The data set after the normalization is a raw sensor data set, and classifiers can be trained and validated with the raw sensor data without further processing. After the above simple signal processing, machine learning approaches, such as SVM, ANN, and DT, are employed to classify different physical activities. In this section, the fundamental knowledge of the three machine learning methods will be delivered.

### 14.2.1 SVM

SVM was first established in the late 1970s and is widely used for pattern classification. It was developed from the theory of structural risk minimization [17]. SVM supports classification both in binary [18] and multiclass formulations [19]. The binary case is introduced in this section first. The notation in this section will mostly follow that of Vapnick's book [17].

Suppose we have $l$ subjects, and each subject has a label associated "truth" $y_i$ and a vector $x_i$ $\in R^n$, where $i = 1,…,1$. Now we assume the data are "*iid*" (independently drawn and identically distributed) and there is some unknown probability distribution $P(x, y)$ from which the data are drawn. Our task is to build machines that are able to learn the mapping $x_i \mapsto y_i$. The machines are actually functions $f(\boldsymbol{x}, \alpha)$, which are dependent on the parameters $\alpha$ for an input $\boldsymbol{x}$. The $\alpha$ values can be determined from the training procedure.

The expectation of the test error for a trained machine is

$$R(\alpha) = \int \frac{1}{2}|y - f(\boldsymbol{x}, \alpha)| dP(\boldsymbol{x}, y), \tag{14.2}$$

where $R(\alpha)$ is called the risk. The empirical risk $R_{emp}(\alpha)$ is defined as measuring the mean error rate on the training data set:

$$R_{emp}(\alpha) = \frac{1}{2l} \sum_{i=1}^{l} |y_i - f(\boldsymbol{x}_i, \alpha)|, \tag{14.3}$$

where $R_{emp}(\alpha)$ is a fixed number for a particular set of $\alpha$ and $\{\boldsymbol{x}_i, y_i\}$. For the binary case, $y_i$ can only be two values. We use 1 and –1 to label binary classes. Thus, the value of $\frac{1}{2}|y_i - f(\boldsymbol{x}_i, \alpha)|$ can only be 0 and 1. This value is called loss. If we choose a $\eta$ and make $0 < \eta < 1$, then for losses taking these values, with probability $1 - \eta$, the following boundary holds [17]:

$$R(\alpha) < R_{emp}(\alpha) + \sqrt{\frac{h[\log(2l/h) + 1] - \log(\eta/4)}{l}}, \tag{14.4}$$

where $h$ is a nonnegative integer named as the Vapnik Chervonenkis (VC) dimension. The VC dimension is a property of a set of functions $\{f(\alpha)\}$.

In a two-dimensional space, for example, the set $\{f(\alpha)\}$ consists of oriented straight lines that can separate the two classes of subjects (as shown in Figure 14.2).

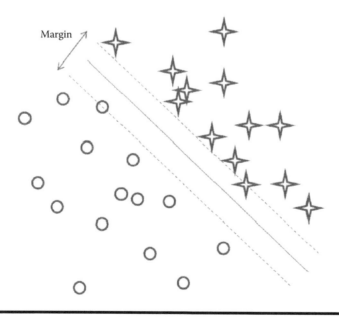

**Figure 14.2    A straight line to separate the two classes with margin marked.**

In a multidimensional space, the line in Figure 14.2 will be a hyperplane. The points **x** on the hyperplane satisfy **wx** + $b$ = 0, where **w** is normal to the hyperplane. If ‖**w**‖ is the Euclidean norm of **w**, |$b$|/‖**w**‖ is the perpendicular distance from the hyperplane to the origin. Suppose $d_+$ and $d_-$ are the shortest distance from the hyperplane to the nearest positive and negative subjects, the "margin" of a hyperplane can be defined as $d_+ + d_-$. The linear SVMs are looking for the separating hyperplane with the largest margin, which can be expressed [20] as

$$x_i\mathbf{w} + b \geq +1 \text{ for } y_i = +1, \tag{14.5}$$

$$x_i\mathbf{w} + b \leq -1 \text{ for } y_i = -1, \tag{14.6}$$

and can also be written as

$$y_i(x_i\mathbf{w} + b) - 1 \geq 0 \cdot \forall i \tag{14.7}$$

Then, the equation can be switched to a Lagrangian formulation:

$$Lp \equiv \frac{1}{2}\|\mathbf{w}\|^2 - \sum_{i=1}^{l}\alpha_i y_i(x_i\mathbf{w} + b) + \sum_{i=1}^{l}\alpha_i. \tag{14.8}$$

Now the problem has been changed to minimize $Lp$ with respect to **w**, $b$, and simultaneously require that the derivatives of $Lp$ with respect to all the $\alpha_i$ vanish. It is a convex quadratic programming problem that is simpler to handle.

For a multiclass classification problem such as the problem of monitoring human's physical activities, several binary SVM classifiers need to be utilized. For example, one-against-one classification method trains $k(k-1)/2$ binary classifiers (where $k$ is the number of classes) to solve $k$-class classification problems. During the prediction, the class with most votes becomes the winner. One against one compares favorably with other multiclass SVM classification methods [21] and is a default choice in the libSVM package [22] that has been utilized in this study. The pairwise coupling procedure for one-against-one classification can be obtained by a variety of methods [23].

### 14.2.2 ANN

ANN is a part of computational intelligence and a computational model that attempts to simulate processes in the human brain [24]. The model of a neuron is shown in Figure 14.3. The model has $N$ inputs that are denoted by $x_0, x_1,\ldots, x_{N-1}$. Each input has a corresponding weight denoted by

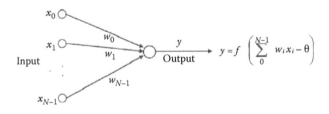

**Figure 14.3    An artificial neural network model with one neuron.**

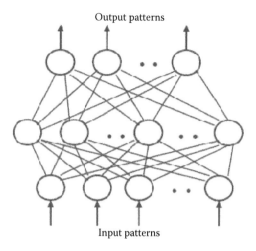

Output patterns

Input patterns

**Figure 14.4    MLP with a three-layer structure.**

$w_0$, $w_1$,..., $w_{N-1}$. $y$ is the output. $\theta$ is the bias. Perceptron algorithm, developed by F. Rosenblatt in 1958 [25], is one of the broadly used ANNs. If $d$ is used to denote the desired output, the learning procedure of the perceptron, also called perceptron training, is to minimize $d - y$. The weights keep updating during the learning procedure.

A multilayer perceptron (MLP), as shown in Figure 14.4, is usually employed to solve problems requiring a nonlinear decision boundary. The layer on the top that provides output patterns is called the output layer, the layer in the middle is called the hidden layer, and the bottom layer with input patterns is called the input layer. This three-layer structure is sufficient to solve most of the modeling problems. MLP was applied in monitoring human posture allocations and physical activities by using SmartShoe. MLP is powerful in information processing systems to solve pattern analysis problems or feature vector classifications [26].

## 14.2.3 DT

DT [9] is a hierarchical model that recursively separates the input space into class regions. The final decision-making model is a tree-like structure that is composed of decision nodes (conjunctions of features) and leaves (class labels). Ross Quinlan invented algorithms ID3, C4.5, and C5.0 to implement DT. In the study of monitoring people's major physical activities, C5.0 was applied on features computed from the sensor signals. A DT classifier can be trained with either raw data or computed features. However, in real-life applications, DTs are usually used along with feature computation. The features calculated in this study are mean, standard deviation, entropy, variance, maximum value, number of mean crossings, and mean absolute deviation [27]. In this way, the dimension of the feature data is much smaller than that of the raw data, which results in smaller size trees; thus, the computation can be less complicated. Also, the tree model can be easily accessible for analysis and interpretation.

Boosting algorithms are usually applied together with DT classification to reach higher accuracy. Boosting algorithms are actually iteratively learning procedures. Advanced boosting option is originally built in Quinlan's C5.0 algorithm to increase accuracy. The C5.0 algorithm generates 10 different trees as simple classifiers, and then integrates them and improves the performance through boosting.

## 14.3 Experimental Study by Using SVM, ANN, and DT

### 14.3.1 Experimental Design

The data collection for the study of monitoring people's major physical activities was performed on a group of nine human subjects—three males and six females. The study was approved by the Institutional Review Board at Clarkson University. Each subject was required to provide informed consent. All subjects were healthy, nonsmokers, and sedentary to moderately active (less than two to three bouts of exercise per week or participation in any sporting activities less than 3 h per week). Pregnant women and those who had impairments that prevented physical activity were excluded. Subjects were 23.7 ± 4.3 (mean, standard deviation) years old with a range from 18 to 31, had a weight of 70.5 ± 15.8 kg with a range from 55.6 to 100.9 kg, and had a body mass index of 25.2 ± 6.5 kg/m² with a range from 18.1 to 39.4 kg/m². The shoe sizes (US) ranged from 9.5 to 11 for men and from 7 to 9 for women.

The data collection for each subject was performed during a single 2.5–3 h visit. The subjects wore the SmartShoe sensors for the duration of the visit. The subjects were not restricted in the ways they performed activities. Sitting represented sitting in a chair with a rigid back; standing represented free standing without any specialized equipment; walking and jogging were combined as one activity and performed on the Biodex Gait Trainer 1 treadmill [28]; the subjects used stairs between different floors for ascending and descending with a railing available; and cycling was performed on an Ergomedic 828E bicycle exerciser [29]. The signals collected from the SmartShoe sensors were divided into segments (epochs) of 2-s duration that were used as an atomic time interval for activity classification. Table 14.1 shows counts of all the epochs recorded for the six classes of activities, totaling in over 11.5 h of data (20,890 epochs). All data were used in the analysis and approximately equally distributed across the nine subjects. Recognizing these six classes from the SmartShoe data was the major goal of this study.

**Table 14.1  Description and the Sample Size of the Six Posture/Activity Classes**

| Activity/Class | Description (Total Duration) | Number of Epochs |
|---|---|---|
| Sitting | Including sitting motionless and with fidgeting (1 h 47 min) | 3218 |
| Standing | Including standing motionless and with fidgeting (1 h 47 min) | 3207 |
| Walking/jogging | Includes several speeds, slopes, and load conditions (5 h 57 min) | 10721 |
| Ascending stairs | (18 min) | 550 |
| Descending stairs | (17 min) | 506 |
| Cycling | Includes two load conditions: 50 and 75 rpm (1 h 30 min) | 2688 |

## 14.3.2 Validation

The leave-one-out [30] cross-validation method was used to evaluate the performance of the classifiers described above in this study. All the activity epochs from one subject were taken as the testing data, and the epochs from the remaining subjects were used to train the classifier. This procedure was repeated nine times until all subjects were considered for testing. All the test results were then combined into a cumulative confusion matrix, allowing calculation of the class-specific recall (the proportion of a class instances that are correctly identified) and class-specific precision (the proportion of the predicted class cases that are correct) for each activity.

## 14.3.3 Results

Table 14.2 shows the cumulative confusion matrix when the libSVM was applied on the raw sensor data. Radial basis function kernel was used here because it usually provides a better performance in the classification than other kernels [31]. The intersection of the class-specific recall and the class-specific precision is the overall accuracy, which is 97.3%.

A three-layer MLP implemented using MATLAB®'s neural network toolbox was also employed to perform the classification. The hidden layer of the MLP contained 240 neurons for the classification with the raw sensor data. Optimized Levenberg–Marquardt with adaptive momentum was picked as the training algorithm. Table 14.3 shows the cumulative confusion matrix when the MLP classifier was used on the raw sensor data. The overall accuracy is 97.6%.

Figure 14.5 shows a DT structure built for subject 2. DT was applied on the feature data calculated from the raw sensor data for each of the 2-s epochs. The tree leaves are the activities and the connecting nodes are values that are the thresholds for the computed features. The representations of the sensor readings in the DT algorithms are shown in Table 14.4.

In Figure 14.5, the connection nodes are displayed as feature names followed by sensor names. Each branch of the tree shows the attribute usages in the algorithm. It is obvious that not all features are involved or all sensors are used. Figure 14.5 shows that the maximum value of the pressure sensor 2, which is the pressure signal under the first metatarsal head, is used for all the vectors

**Table 14.2  Cumulative Confusion Matrix When the SVM Classifier Was Used on the Raw Sensor Data**

| | Predict | | | | | | |
|---|---|---|---|---|---|---|---|
| *Actual* | *Sit* | *Stand* | *Walk* | *Ascend Stairs* | *Descend Stairs* | *Cycle* | *Recall (%)* |
| Sit | 3174 | 2 | 0 | 0 | 0 | 42 | 98.6 |
| Stand | 21 | 3140 | 2 | 44 | 0 | 0 | 97.9 |
| Walk | 2 | 1 | 10634 | 74 | 10 | 0 | 99.2 |
| Ascend Stairs | 9 | 7 | 33 | 471 | 29 | 1 | 85.6 |
| Descend Stairs | 0 | 2 | 39 | 87 | 378 | 0 | 74.7 |
| Cycle | 152 | 4 | 0 | 0 | 0 | 2532 | 94.2 |
| Precision (%) | 94.5 | 99.5 | 99.3 | 69.7 | 90.6 | 98.3 | 97.3 |

**Table 14.3  Cumulative Confusion Matrix When the MLP Classifier Was Used on the Raw Sensor Data**

| | | | Predict | | | | |
|---|---|---|---|---|---|---|---|
| Actual | Sit | Stand | Walk | Ascend Stairs | Descend Stairs | Cycle | Recall (%) |
| Sit | 3214 | 2 | 0 | 0 | 0 | 2 | 99.9 |
| Stand | 17 | 3187 | 1 | 2 | 0 | 0 | 99.4 |
| Walk | 9 | 8 | 10668 | 23 | 9 | 4 | 99.5 |
| Ascend Stairs | 12 | 11 | 75 | 394 | 24 | 34 | 71.6 |
| Descend Stairs | 14 | 6 | 84 | 32 | 360 | 10 | 71.1 |
| Cycle | 115 | 0 | 0 | 0 | 0 | 2573 | 95.7 |
| Precision (%) | 95.1 | 99.2 | 98.5 | 87.4 | 91.6 | 98.1 | 97.6 |

among all the features, while the mean and maximum value from the acceleration signal 3, which is the acceleration signal in the vertical direction, is used as the second most used feature. All other features that are not listed in Figure 14.5 are not used during the classification of subject 2.

Table 14.5 is the cumulative confusion matrix when DT is used on the feature data. The overall classification accuracy is 88.6%. The classification accuracy is not as high as the individual model [27]; however, it is still applicable in a lot of circumstances.

```
Decision tree:
max_pre2 <=2914:
:... mean_acc3 <= 1889: cycle(146/1)
:    mean_acc3 > 1889:
:    :... max_pre1 > 2171: stand (175)
:        max_pre1 <= 2171:
:        entropy_pre3 <= 8795: sit (166)
:        entropy_pre3 >8795: stand (5)
max_pre2 > 2914:
:... max_acc3 > 3886: stairs_down (24/1)
    max_acc3 <= 3886:
    :... mean_acc1 <= 804: stairs_up (21/1)
        mean_acc1 > 804:
        :... mean_acc2 <= 2015: stairs_up (11/1)
            mean_acc2 >2015:
            :...mean_acc1 <=1220:
            :... mean_acc2 <= 2135: walk (621)
            :    mean_acc2 >2135:
            :    :...mean_acc2<= 2178: walk (8)
            :        mean_acc2 > 2178: stairs_down (3)
            mean_acc1>1220:
            :...variance_acc2 > 62622: stairs_up (2/1)
                variance_acc2 <= 62622:
                :...variance_acc3 <= 34976: stand (4)
                    variance_acc3 > 34976:walk (5)
```

**Figure 14.5  Decision tree generated for the classification of subject 2.**

**Table 14.4  Sensor Representations**

| Sensor Signals | Descriptions | Notations in Decision Trees |
|---|---|---|
| $A_{AP}$ | Acceleration in the anterior–posterior direction | acc1 |
| $A_{ML}$ | Acceleration in the medial–lateral direction | acc2 |
| $A_{SI}$ | Acceleration in the superior–inferior direction | acc3 |
| $P_H$ | Pressure applied on the force sensitive resistor (FSR) under the heel | pre1 |
| $P_{1M}$ | Pressure applied on the FSR under the first metatarsal head | pre2 |
| $P_{HX}$ | Pressure applied on the FSR under the hallux | pre3 |

**Table 14.5  Cumulative Confusion Matrix When the DT Classifier Was Used on the Feature Data**

| | Predict | | | | | | |
|---|---|---|---|---|---|---|---|
| Actual | Sit | Stand | Walk | Ascend Stairs | Descend Stairs | Cycle | Recall (%) |
| Sit | 3012 | 156 | 0 | 50 | 0 | 0 | 93.6 |
| Stand | 171 | 3013 | 9 | 6 | 8 | 0 | 93.9 |
| Walk | 27 | 525 | 9273 | 4 | 413 | 479 | 86.5 |
| Ascend Stairs | 0 | 11 | 186 | 314 | 24 | 15 | 57.1 |
| Descend Stairs | 0 | 1 | 62 | 10 | 366 | 67 | 72.3 |
| Cycle | 2 | 0 | 76 | 78 | 1 | 2531 | 94.2 |
| Precision (%) | 93.8 | 81.3 | 96.5 | 68.0 | 45.1 | 81.9 | 88.6 |

# 14.4  Discussions

In this chapter, three different machine learning approaches were applied to a CPS problem of recognition of postures and activities. The SVM and MLP were applied to the raw sensor data. The DT classification was used on the features extracted from the raw sensor signals. The SmartShoe system demonstrated high-activity classification accuracy with all of the machine learning methods used.

It can be observed from the confusion matrices that SVM and MLP achieved high overall accuracy, 97.3% and 97.6%, in classifying these six activities. Because both methods produced comparable results, the accuracy most likely represents the actual performance that can be achieved in physical activity recognition by SmartShoe. However, the DT classification had a lower overall accuracy, 88.6%, compared with SVM and MLP. SVM and MLP also held higher mean values of class-specific recalls and precisions for the six activities than that of DT, indicating better performance in classifying these six activities. There are a couple of possible reasons to explain these results. First, the learning algorithms of the DT have many limitations [32] such as instability and unwieldiness. Second, DT classification was applied on the feature data. Relative values of momentary pressure and acceleration contained information important for classification of activity (e.g., ascending stairs is somewhat similar to walking; however, the forefoot is loaded substantially more than the heel), and some of this information might be lost when the features were computed. It may be possible to find an improved feature set that would preserve such information.

The advantage of DT is that it requires fewer computational resources in the decision part because the dimension of the data set was much decreased by the feature extraction and thus is more likely to be embedded in a cell phone or even sensors for computation and can be executed in real life. The results of the DT algorithm suggest direct and effective ways to simplify the feature set and thus reduce the total computational time.

## 14.5 Conclusions

This chapter introduces a CPS, SmartShoe, which can be employed to monitor people's major physical activities. Experimental studies demonstrate that the SmartShoe system has a capability of accurate monitoring postures and activities conveniently. The concepts of machine learning and the three popular machine learning approaches employed in this study were described and compared. The users can decide what method is more appropriate with regard to their particular situation. The biomedical applications that involve recognizing people's activities could take advantage of the CPS described in this chapter.

## References

1. E. Lee. 2008. Cyber physical systems: Design challenges. Paper presented at International Symposium on Object/Component/Service-Oriented Real-Time Distributed Computing (ISORC).
2. A. Ng. 2011. 01 and 02: Introduction, regression analysis, and gradient descent. Available at http://www.holehouse.org/mlclass/01_02_Introduction_regression_analysis_and_gr.html (Accessed on November 21, 2012).
3. J. J. R. Anderson, R. S. Michalski, J. G. Carbonell, and T. M. Mitchell. 1983. *Machine Learning: An Artificial Intelligence Approach*. New York: Morgan Kaufman Publishers.
4. T. Mitchell. 1997. *Machine Learning*. New York: McGraw Hill.
5. J. Vitay. 2011. Machine Learning 01-Introduction. Available at http://www.tu-chemnitz.de/informatik/KI/edu/ml/ws2011/ML01-Introduction.pdf.
6. Q. Hao, F. Hu, and Y. Xiao. 2009. Multiple human tracking and identification with wireless distributed pyroelectric sensor systems. *IEEE Syst. J.* 3:4, 428–439.
7. W. Tang, H. Cao, J. Duan, and Y.-P. Wang. 2011. A compressed sensing based approach for subtyping of leukemia from gene expression data. *J. Bioinformatics Comput. Biol.* 9:5, 631–645.
8. W. Tang, H. Cao, J.-G. Zhang, J. Duan, D. Lin, and Y.-P. Wang. 2012. Subtyping of glioma by combining gene expression and CNVs data based on a compressive sensing approach. *Med. Adv. Genetic Eng.* 1, 101.
9. E. Sazonov, G. Fulk, J. Hill, Y. Schutz, and R. Browning. 2011. Monitoring of posture allocations and activities by a shoe-based wearable sensor. *IEEE Trans. Biomed. Eng.* 58:4, 983–990.
10. T. Kubo, K. Ozasa, K. Mikami, K. Wakai, Y. Fujino, Y. Watanabe, T. Miki et al. 2006. Prospective cohort study of the risk of prostate cancer among rotating-shift workers: Findings from the Japan Collaborative Cohort Study. *Am. J. Epidemiol.* 164:6, 549–555.

11. L. D. Frank, M. A. Andresen, and T. L. Schmid. 2004. Obesity relationships with community design, physical activity, and time spent in cars. *Am. J. Prev. Med.* 27:2, 87–96.

12. T. V. Nguyen, P. N. Sambrook, and J. A. Eisman. 1998. Bone loss, physical activity, and weight change in elderly women: The Dubbo Osteoporosis Epidemiology Study. *J. Bone Miner. Res.* 13:9, 1458–1467.

13. S. J. Rogers, S. Hepburn, and E. Wehner. 2003. Parent reports of sensory symptoms in toddlers with autism and those with other developmental disorders. *J. Autism Dev. Disord.* 33:6, 631–642.

14. Interlink Electronics. 2012. Interlink Electronics: Sensor Technologies [online]. Available at http://www.interlinkelectronics.com/.

15. STMicroelectronics | MEMS inertial sensor: 3-axis – ±; 2g/±; 6g linear accelerometer—LIS3L02AS4, vol. 2012.

16. E. Sazonov, K. Janoyan, and R. Jha. 2004. Wireless intelligent sensor network for autonomous structural health monitoring. In *Proceedings of SPIE Annual International Symposium on Smart Structures and Materials*. International Society for Optics and Photonics, Bellingham, WA, pp. 305–314.

17. V. Vapnick. 1995. *The Nature of Statistical Learning Theory*. Springer-Verlag, New York.

18. M. Wegkamp and M. Yuan. 2011. Support vector machines with a reject option. *Bernoulli 2011* 17, 1368–1385.

19. R. Zhang and D. N. Metaxas. 2006. *Support Vector Machine with Reject Option for Image Categorization*. BMVA Press.

20. C. J. C. Burges. 1998. A tutorial on support vector machines for pattern recognition. *Data Min. Knowl. Discov.* 2:2, 121–167.

21. H. Chih-Wei and L. Chih-Jen. 2002. A comparison of methods for multiclass support vector machines. *IEEE Trans. Neural Networks* 13:2, 415–425.

22. C.-C. Chang and C.-J. Lin. 2011. LIBSVM: A library for support vector machines. *ACM Trans. Intell. Syst. Technol.* 2:3, 1–27.

23. T.-F. Wu, C.-J. Lin, and R. C. Weng. 2004. Probability estimates for multi-class classification by pairwise coupling. *J. Mach. Learning Res.* 5, 975–1005.

24. E. Sazonov. 2012. *Neural Networks*. University of Alabama.

25. F. Rosenblatt. 1958. The perceptron: A probabilistic model for information storage and organization in the brain. *Psychol. Rev. Cornell Aeronaut. Lab.* 65:6, 386–408.

26. R. Nayak, L. Jain, and B. Ting. 2001. Artificial neural networks in biomedical engineering: A review. In *Asia-Pacific Conference on Advance Computation*.

27. T. Zhang, W. Tang, and E. S. Sazonov. 2012. Classification of posture and activities by using decision trees. *Conf. Proc. IEEE Eng. Med. Biol. Soc.* 18:31, 27.3.

28. "Biodex | Physical Medicine and Rehabilitation, Nuclear Medicine Supplies and Accessories, Medical Imaging Tables and Accessories." [Online]. Available: http://www.biodex.com/. (Accessed on December 18, 2012).

29. Monark. 2012. Monark Exercise AB [online]. Available at http://www.monarkexercise.se/ (Accessed on December 18, 2012).

30. B. Efron and R. J. Tibshirani. 1993. *An Introduction to the Bootstrap*. Chapman & Hall, New York.

31. S. S. Keerthi and C. J. Lin. 2003. Asymptotic behaviors of support vector machines with Gaussian kernel. *Neural Comput.* 15:7, 1667–1689.

32. Bright Hub PM. 2012. A review of decision tree disadvantages [online]. Available at http://www.brighthubpm.com/project-planning/106005-disadvantages-to-using-decision-trees/ (Accessed on January 14, 2013).

# CIVILIAN CYBER-PHYSICAL SYSTEM APPLICATIONS

## Chapter 15

# Energy Efficient Building

Preston Arnett, Ian Wolfe, and Fei Hu

## Contents

## 15.1 Cyber-Physical Systems for Building Energy Saving

With ever-rising energy costs driving research into alternative energy sources, a seemingly untapped area of energy research comes in the form of finding more and more ways to reduce already present energy burdens. Buildings consume around 70% of the energy used daily in the United States. Heating, ventilation, and air conditioning (HVAC) systems, in particular, consume a large amount of energy. The heating and cooling of buildings and homes consume nearly three quarters of all of the energy in the United States annually; yet there have been no significant pushes to reduce this figure recently. The approaches presented throughout this chapter propose simple ways to model, control, and monitor energy usage in ways that could eventually lead to noteworthy reductions to the energy usage in buildings of all kinds.

We can add inputs to the HVAC controllers in order to increase the systems operating efficiency [1]. A key input that should be considered is the current occupancy of the building. The main obstacle to producing this input is the inaccuracy and limited usage of sensors for detecting occupancy. A good occupancy detection system is proposed in [1]. It involves using a Reed switch along with passive infrared (PIR) sensor to provide increased occupant detection. PIR sensors detect movement and by themselves are not accurate enough to evaluate the occupancy of a building. The Reed switch detects whenever a door to the room is opened. In conjunction with the PIR sensors, the system can mark a room as occupied when a door is opened then closed and the PIR detects movement in the room. This way, if an occupant stops moving after entering the room, the system will keep the area as occupied. The one fault with this system is when one occupant leaves a room while another stays still inside the room. For this circumstance, the authors added an interrupt to their algorithm, so that whenever the PIR sensors detect movement, the room will automatically be marked as occupied.

The sensors were deployed in a wing of the Computer Science and Engineering (CSE) building at the University of California for testing. The authors observed that many rooms were left unoccupied for long stretches of time, supporting the idea that occupancy could be a key input for reducing HVAC energy usage. The sensors were tested over time on their accuracy to detect occupancy and the results showed that they performed accurately.

The sensors were equipped with low cost antenna in order to make the system wireless, to improve the ease of installation. It was tested to determine how many base stations were needed in order to consistently collect data from all of the sensors. It was found that three to four base stations are needed to cover a floor of the CSE building. The battery life of all the sensors and base stations were also tested, and it was found that the sensors would have a battery lifetime of over 7 years if they detected up to 50 events a day.

To estimate the efficiency the proposed system could have on a building, the energy simulation program EnergyPlus was used. The program is used by the US Department of Energy and is superior in accuracy to programs such as BLAST. In EnergyPlus, a single story building was created and occupancy data were input from the data collected using the sensors in the CSE building in the previous experiment. The energy savings were calculated if the system kept occupied areas at 22.9°C and unoccupied zones at 26.1°C. The energy savings were found to be between 10% and 15% over a four-month period during the summer varying with the outside temperature. The wireless detection system is low cost, is easy to implement, and still provides accurate detection of whether or not a room is occupied. The sensing system described can reduce building energy consumption by 10% to 15% [1].

The system is indeed low cost, as neither PIR sensors nor Reed switches are expensive to purchase. The implementation was also confirmed to be easy because both sensors could be installed without the addition of new wiring to the building. The sensors were also found to be highly accurate at determining occupancy in the CSE building experiment. The EnergyPlus simulation found the energy savings to be around 10% to 15%; however, this is an ideal simulation so real results might deviate [1].

## 15.2 Occupancy-Based Demand Response HVAC Control Strategy

Every year, an extreme portion of the total amount of energy produced is used to provide ventilation, heating, and cooling solutions. Unfortunately, much of this energy is wasted by efficient systems that heat, cool, and ventilate rooms at full occupancy settings even when they have little to no occupancy. In [2] they proposed a HVAC control scheme that is centered on real-time occupancy monitoring and prediction methods to curbed energy use in these systems.

They have used the concept of a moving widow Markov chain. A Markov chain consists of a number of known states where these known states can change probabilistically over time. In the context of a HVAC system, one can use the occupancy numbers as the various states. One of the proposed methods of collecting occupancy data for this model is collected by using the wireless sensors. These wireless sensors detect movement and in conjunction with defined hourly state changes, a transition matrix for each hour can be constructed. One of the biggest limitations to this method is the sheer number of states that are required to fully characterize an entire building (the paper cites close to 4 million states for a simple four-room scenario).

When the moving window Markov chain is computed, this information is then implemented into a control strategy for both ventilation and temperature. The American Society of Heating, Refrigerating, and Air-Conditioning Engineers (ASHRAE) Standard 62.1 uses the following formula to determine the ventilation rate needed for a given room:

$$V_{bz} = R_p P_z + R_a A_z$$

Here, $V_{bz}$ is the ventilation rate, $R_p$ is the minimum CFM/person, $P_z$ is the number of people, $R_a$ is the minimum CFM/ft$^2$, and $A_z$ is the floor area.

One can observe that the only variable in the equation above that is unknown at any given time is the number of people, $P_z$. In conjunction with the real-time data and the predictive Markov chain, this variable can be determined on a per need basis during any given time slot.

Heating and cooling are handled in a way similar to a ventilation system. First, minimum values are selected that meet ASHRAE standards. These values will be used for operating hours in which there would be less than 10 min of use. In hours in which the probability of more than 10 min of use has been predicted, temperatures are set for more comfortable targets than those of the minimums. The probabilities of occupancy are all drawn from the moving window Markov chain.

After simulating and testing the model, it is found that there is possibly 8.8% energy savings in typical office space. Buildings that feature areas that are frequently unused saw savings approaching almost 20%. These savings increase to 26.5% in the colder months, which would suggest that colder areas would benefit from this system even more than usual.

In summary, [2] presents a very practical and implementable HVAC system that could potentially save a large amount of energy. The use of a Markov chain and ground data to predict hourly room occupancy is an interesting solution. The SCOPES sensor system that was employed has a large amount of detection error associated with its use. Overall, the results seem promising despite there being some areas in which the technology can become more efficient.

## 15.3 ARIMA-Based Electricity Use Forecasts

ARIMAX, also known as Auto Regressive Integrated Moving Average with eXternal (or eXogenous) input, was developed for forecasting in economics and other fields [3]. Because building energy usage is a time series, it can be analyzed using an ARIMAX model. The authors created and tested such a model using occupancy of the building as an independent variable. The expected use of this is to predict energy usage ahead of time and to use that information to respond to energy costs from utility companies.

To test the model devised, several types of wireless sensors were deployed in a test building in order to collect data. Contact sensors were placed on exterior doors to track heat lost from the building and on interior doors to help track occupancy flow. Sensors were placed on the exterior of the building to provide climate data and on the interior to track occupancy flow and carbon dioxide concentration. These sensors will help to calculate energy used in the HVAC and other systems.

The data set constructed using data from the sensors only takes weekdays into account, as peak demand energy control is the main expected use of the model. The variables are all in hours to make computations more efficient. By using data of total building energy usage and the variations of occupancy taken from the sensors, the authors were able to determine how energy usage fluctuates with changes in occupancy.

From the results gathered by the authors several conclusions were drawn. One example is that motion sensor tracking for controlling the occupancy variable was detrimental to the ARIMA model, and logins to the buildings computer network made a better variable for occupant energy usage. The authors also indicated several improvements that could be made to the model to improve on its prediction capabilities. These improvements include adding logoffs as well as logins as a variable and using the model in traditional office buildings, because the lab building the authors tested had a large portion of energy used by lab equipment.

Overall, [3] explains how an ARIMAX model can be used to predict energy usage of buildings by taking several variables into account, a main one being occupancy. The model performed well in the experiment conducted by the author. There were some issues with the performance, some of which were pointed out by the authors in their concluding remarks.

One issue with the model is incorporating occupancy data into the ARIMA model effectively, because most variables used, including motion sensor data, do not change the model enough to be useful. The authors got around this by using network logins by occupants to their computers. The biggest downside to this input is the fact that if a user logs in and then leaves the building, their logoff is not recorded and the energy usage assigned to them will be higher than it should be. Another issue with this article is simplicity of the ARIMAX model itself. The model was not able to reflect occupancy well with most variables because of how simple the mathematical formulae for the model are. If the model was more complex, it might be able to predict energy usage in a building more effectively. The final problem with this article [3] is with the data collected by the authors. Several days' worth of sensor data was lost, or unsuccessfully collected; this may have been a factor in how difficult implementing an occupancy variable into the model was.

## 15.4 Circuit-Level Energy Monitoring

Device-level energy use is important in understanding how to more efficiently use energy during building operation. The main issue with conventional ways of measuring device energy usage is that it does not take into account the user activity that operates certain devices. In [4] it combines circuit level energy monitoring with Granger causality analysis to effectively link user activity with energy use, while keeping the cost of implementation and intrusiveness low. By linking user activity to device power use, plans could then be implemented to reduce building energy usage through reduced or altered occupant activity.

It focused on energy usage in office settings, mainly due to office occupants' increased likelihood to leave appliances on when not in use. Residential occupants tend to worry more about energy use since they pay the utility bill. The authors used IP network traffic as an indicator of whether a particular occupant is present or not, and can measure energy use in small zones by using the circuit level energy monitoring. The Granger causality is then used to estimate correlation between user presence and the resultant energy draw. This type of system might be less accurate than other systems measuring at an appliance level, but is more cost effective and less intrusive to users.

For the Granger causal model utilized in this experiment, $X_t$ and $Y_t$ were used as time series data, with $X_t$ being traffic of IP addresses and $Y_t$ being circuit branch power usage. The causal model was then defined using the following equations, with $\varepsilon_t$ and $v_t$ being white noise series and $m$ being the number of data points being used to define the current data point at time $t$.

$$X_t = \sum_{j=1}^{m} a_j X_{t-j} + \sum_{j=1}^{m} b_j X_{t-j} + \varepsilon_t$$

$$Y_t = \sum_{j=1}^{m} c_j X_{t-j} + \sum_{j=1}^{m} d_j Y_{t-j} + v_t$$

Veris E-30 is the software the authors chose to monitor power usage at the different branches of a circuit breaker. Traffic from occupants' personal IP addresses was used to monitor user presence in the building. The Granger causal model was able to successfully correlate users with energy usage under certain assumptions; however, the model does have limitations. One limitation is that the system cannot distinguish devices with constant power use regardless of whether they are in use or not. The system also can have issues when several IP addresses are registered in an area near a device drawing energy. The biggest limitation to the system is the fact that if a user does not register their IP while in the area, any corresponding energy use will go without an identified user.

There are several limitations to the proposed system that cannot be easily remedied without the system becoming more expensive or more intrusive on occupants. One of these is the fact that appliance usage is not monitored. So if one device in an area is used by a different occupant than all the other devices in that area, the energy usage could be linked to the wrong user. Measuring power draw at an appliance level would require a more costly system. This same issue applies with devices that draw constant power. Without the ability to separate those devices from the other appliances in an area, users can be linked to power use they do not cause.

The other limit is reducing a building's power use via this model would require the cooperation of occupants. People forget to turn devices off and are not suitable for efficient energy control. There is also the issue of intrusion of privacy, and whether companies should track user's energy use and impose rules on them to correct any abnormalities in device utilization.

## 15.5 A Limited-Data Model of Building Energy Consumption

In [5] it proposes a model that breaks down building energy consumption for practical widespread use in determining energy efficiency. The model is shown to be accurate to true energy consumption numbers with up to 10% error. This model is then used to predict the energy efficiency of several different scenarios. The model uses data that can be obtained easily and cheaply, and is used only for prediction purposes.

They used software named EnergyPlus, which combines the calculation of object energy use with the energy used to heat areas. Since small objects tend to use insignificant power individually, the model used had a small inventory that does not contain every object found in modern buildings. The model instead combines all of the small objects together, since combined they have a more significant energy usage. To estimate the energy usage of objects, three types of uses were recorded: constant energy use, timed energy use, and submetered energy use.

Besides object energy use, the energy used to heat and cool buildings had to be measured as well. The equations below were used to calculate the power required to heat a building. The first of these equations represents the HVAC load or the energy needed to maintain a fixed temperature $T_i$ (21°C in this case). The integral of the second equation tracks the temperature of the area's wall over time. Using these equations the energy needed to heat and cool buildings can be calculated.

$$\frac{dQ_i}{dt} = -H_i(T_i - T_w(t)) + P(t)$$

$$\frac{dQ_w}{dt} = C_w \frac{dT_w}{dt} = -H_e(T_w(t) - T_e(t)) + H_e(T_i - T_w(t))$$

The last piece needed to get a good estimate of energy usage is occupancy. Some areas or buildings might be powered off when there are no occupants. Fewer occupants also mean fewer objects running and using energy. Combining all of these different variables, they were able to calculate the energy saved in several different scenarios. The energy savings can be seen in Table 15.1.

The metered scenario shows the energy usage recorded from the meters, and the current conditions scenario shows what the calculated energy use is normally. The current conditions scenario was used to calculate the saved energy. The normal computing scenario shows what the energy use would be if the building did not house a large server and the network supporting it. The PC off scenario shows the energy use if occupants turn off computers when not working, and the LED lighting scenario shows the energy that could be saved by using LED lights.

Overall, [5] presents a semiaccurate model for calculating energy efficiency of buildings while considering several variables. The model was found to be only within 10% error in regard to estimating efficiency, so there are many ways it could be improved. It does not take oil heating into account versus electricity used for HVAC. Fixing this could improve the efficiency greatly because HVAC is the biggest portion of energy usage in a building.

**Table 15.1  Predicted Yearly Energy Savings**

| Scenario | Average Power (kW) | Change (kW) | Savings (£) |
|---|---|---|---|
| Metered | 275 | | |
| Current conditions | 213 | | |
| Normal conditions | 118 | 95 | 83,000 |
| PCs off | 206 | 7 | 6100 |
| LED lighting | 192 | 21 | 18,000 |

## 15.6 Reducing and Monitoring Energy Usage in Residential Settings

### 15.6.1 TinyEARS: Spying on House Appliances with Audio Sensor Nodes

In [6] the authors proposed a way to install an inexpensive and easily installed system for monitoring energy usage in residential locations. The system used is Tiny Energy Accounting and Reporting System (TinyEARS). It uses acoustic signatures of household appliances to track energy use, and their system achieves results within a 10% error margin. Because the authors rely primarily on acoustics, the installation and cost of the sensors used are better than in other residential energy monitoring systems.

The system is able to obtain adequate energy usage results from measuring the acoustic signatures of household appliances. Such a feature was checked against real-time results obtained from a household power meter. TinyEARS system can be easily deployed owing to the limited number of sensors required, the system only requires one sensor node per room, and TinyEARS can recognize appliances with a 94% success rate with simple coding.

TinyEARS consists of three different layers: event detection, device detection, and time correlation (Figure 15.1). The event detection layer tells the device detection layer when the power meter experiences an increase in energy demanded. The device layer can then use acoustic sensors to find out which appliance is using the extra energy. The time layer helps decide which acoustic detections to match with which increases in energy demand on the power meter. To utilize these three layers, TinyEARS requires a power meter, acoustic sensors, and other units. Using Mel frequency cepstral coefficients (MFCC), the system is able to recognize up to 94% of the devices and match them with the correct energy usage spike.

In the test run performed, a 99% success rate at identifying devices was achieved, but only four devices were tested, and all of them had distinct acoustic signatures. To fully test the capabilities of the system, it should be integrated into a larger, busier household. Some kitchens can have as many as 10 devices running at busy hours. The test run seemed to be too simple to get a grasp of how effective the system really is.

The biggest issue with the TinyEARS system is that it has a 10% error in detecting device power consumption. This percentage could make a huge difference if the device it misses is pulling a majority of the power used. Also, there is no discussion as to how this system can be used to

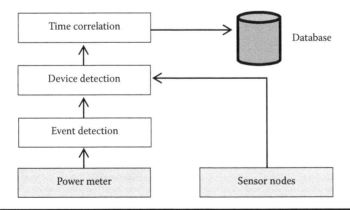

**Figure 15.1  Decision architecture of TinyEARS system.**

save energy. The easiest application would be to use TinyEARS to alert homeowners as to which devices are using the most power at what times. However, unless the owners are willing to use devices left or are forgetting to turn devices off, this will not lead to less energy consumption [6].

### 15.6.2 Private Memoirs of a Smart Meter

As the smart grid initiative moves forward, smart meters will begin to gain more and more prevalence in households across the country. These smart, high-granularity meters will allow very precise billing data and usage statistics to be collected. This will facilitate more accurate peak-time billing and load balancing for power providers. Unfortunately, this drastic increase in data accuracy can be used in the wrong way to glean unintended personal data from users.

Without prior knowledge of user activity, in [7] they were able to discern interesting private data about individuals. Sometimes this power activity based information on answers and questions that, on the surface, are harmless. Things like whether or not there is a newborn in your family are only harmless until you realize that third parties pay good money for information like that.

Data collection for this test case was facilitated by instrumenting each homes circuit with The Energy Detective (TED) energy monitor. The architecture used (see Figure 15.2) logs household power consumption every second and transmits the data to a remote server via the Web. Every transmitted entry contains a time stamp and the average power consumption in kilowatts. These two pieces of information form what is referred to as a power tuple. Figure 15.2 shows such an architecture.

The scheme is performed as follows: preprocess power traces using an off-the-shelf clustering algorithm to identify and label similar types of power events, tag each power event with the one or more defining characteristics, filter out automated appliances by observing their signatures during periods of low power activity, and map opaque labels to real-life events using a small amount of externally gathered knowledge. For labeling power events, power traces are processed

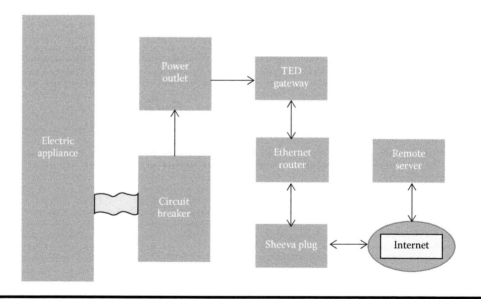

**Figure 15.2  Building monitoring architecture.**

through a density-based clustering algorithm that will group the power traces into power segments. With this information, similar shapes can easily be identified and relevant labels can be applied. Next, the power segments are joined with a couple more pieces of information. The most important pieces of information tagged to the power segments consist of its duration and the power increase/decrease at the beginning of the segment, also known as its power step. The researchers also label power segments that have similar shapes. At the end of this process, a 6-tuple consisting of the segment's label, start time, average power, duration, beginning power step, and a shape label is generated. The 6-tuples are filtered using these data to narrow down certain events. These data are also used to isolate and filter out automated appliances as well. When all of this analysis has been performed, prior knowledge can be applied to map the data to real-life events.

The goal of a privacy enhancing architecture would be to obscure the information that was able to find while still maintaining the advantages that smart meters offer. This means that the following criteria must still be in place: critical peak billing and dynamic pricing, support tamper and energy theft alarms, support power failure and restoration notifications, and support demand response for home notifications. One way to obfuscate this information would be to implement a zero-knowledge billing protocol. This protocol has three phases in every billing cycle: registration, tuple gathering, and reconciliation.

For the first phase of the protocol, registration, smart meters would cryptographically generate $N$ number of tags and a set of pseudorandom keys. The $N$ tags would correspond to the number of power tuples needed to compute power consumption for a billing interval. The overall security of this stage can be improved by increasing the number of items in the set of pseudorandom keys available; around 10 has been proven to provide adequate protection. This then moves to the tuple gathering stage. Here, the smart meters generate tuples ($[r_i, t_i, \rho_i]$). In this stage, $r_i$ corresponds to the pseudorandom tag that was discussed in the registration phase. The remaining parameters relate to timestamp $t_i$ and power usage reported $\rho_i$. Gateways will not reveal what random tuples belong to which smart meter. Finally, reconciliation is imitated and the client computes the bill using variable pricing, $E = \sum_{i=1}^{N} \text{Cost}(t_i, \rho_i)$ [7].

Table 15.2 shows the energy monitoring time granularity.

**Table 15.2  How to Set Up Time Resolution**

| Question | Pattern | Granularity |
|---|---|---|
| Were you home during your sick leave? | Yes: Power activities during the day<br>No: Low power usage during the day | Hour/minute |
| Did you get a good night's sleep? | Yes: No power events during night<br>No: Unexpected power events overnight | Hour/minute |
| Did you leave your child at home? | Yes: One person activity patterns<br>No: Normal, no-person power patterns | Minute/second |
| Did you eat a hot or cold breakfast? | Cold: No power events matching hot appliances<br>Hot: Power signature matching a kitchen appliance | Second |

### 15.6.3 *Wireless, Collaborative Virtual Sensors for Thermal Comfort*

One of the most popular methods of monitoring building energy efficiency is to employ wireless sensor networks throughout a building [8]. These methods are, unfortunately, limited by the absolute cost of implementing this across a large area. What Ploennigs et al. propose are methods that both maximize building efficiency and minimize the cost to prospective implementations.

The overall efficiency of a building ultimately comes down to two main things: an energy efficient design and intelligent building operation. The best building solutions take an ideal combination of operational costs, personal comfort, and energy efficiency. For companies, the operational cost is of the utmost importance. Minimizing this cost can save companies a significant amount of money in long-term situations.

For thermal comfort, one very popular model comes in the form of Fanger's Predicted Mean Vote (PMV) model. This model is contingent on the following factors: air temperature, radiant temperature, humidity, occupant's clothing, and activity. The result of the calculation resolves into a PMV value that ranges from –3.5 to 3.5 (see Table 15.3). The activity and clothing variables are often predetermined, so the overall calculation simplification can occur by keeping these values fixed. This calculation simplification can be taken even further by using a virtual sensor (monitoring simple air temperature data) to simulate the more complex sensors needed.

Figure 15.3 shows the sensor data collection topologies.

Using this method, the virtual sensors receive the more complex information needed for the generation of a PMV calculation from a centrally located reference sensor. This allows much cheaper implementation across a large area and greatly decreases the complexity of implementation.

### 15.6.4 *Zero-Configuration Sensor Network Architecture for Smart Buildings*

In [9] it is not concerned with reducing the energy usage of buildings directly. The purpose is to design and prototype a system that can connect and coordinate sensors distributed throughout a building. The authors developed their software with the intended use of collecting sensor data in a building in order for the data to be sent to another, separate system that could use the data to make energy saving changes to the building operations. The system presented integrates sensors and actuators into the IP network of a building.

**Table 15.3  PMV Model Ranges**

| Fanger's PMV Model Classifications | PMV Values |
| --- | --- |
| Hot | PMV ≥ 2.5 |
| Warm | 2.5 > PMV ≥ 1.5 |
| Slightly warm | 1.5 > PMV ≥ 0.5 |
| Neutral | 0.5 > PMV ≥ –0.5 |
| Slightly cool | –0.05 > PMV ≥ –1.5 |
| Cool | –1.5 > PMV ≥ –2.5 |
| Cold | –2.5 > PMV |

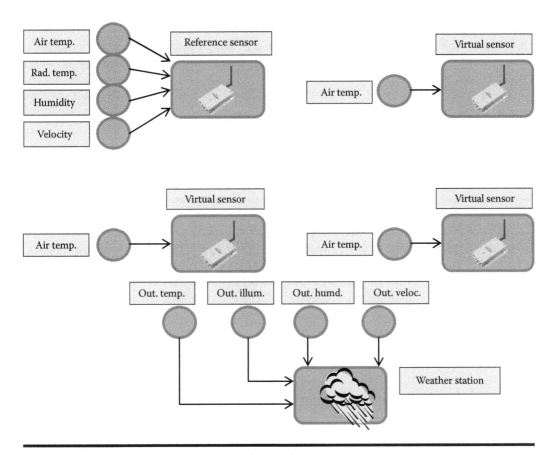

**Figure 15.3   Different sensor data collection methods.**

The authors decided to integrate the sensors and actuators into an IP network because of its ability to connect to more types of devices as compared to specialized protocols. The sensors were connected using wireless LAN in order to reduce installation time and costs, and nodes are not powered by batteries because of the high power draw of the transceiver.

Nodes are able to be autodiscovered by the system, configuring themselves into the network by what information they gather and can provide. The sensor data can stay up to date by either programming the nodes to send data to the network at intervals or getting the server to gather sensor data according to demand. The authors used a Representational State Transfer (REST) protocol.

The RESTful API has three different objects it differentiates between. The first is a member that refers to the function of any corresponding sensor and actuator. The other two objects are a list of members (collection) and a list of these collections (root collection). After the system was designed, the authors performed a test implementation. The system was implemented using TinyOS on Pixie nodes. The system was able to perform correctly but had some lag due to the hardware restrictions and computing power.

It only deals with integration of the sensors into a building existing IP network. It does not deal with the bigger issues of gathering reliable information from a sensor network that is not too costly, or converting this sensed information and utilizing it to somehow regulate energy flow in the building. It is trying to solve the potential problem of linking sensors to the building network in order to utilize the data collected.

Overall, the authors in [9] have done a good job of explaining how all the different parts of the system work and what guidelines were used in their design and operation. Some issues with the system include how the operation is limited by lag produced by the hardware and software of the system.

### 15.6.5 Using Circuit-Level Power Measurements in Household Energy Management Systems

The information gathering part of energy management has been explained in [10]. The authors design a system for making power measurements at the circuit level in order to more accurately determine power usage. This system is made for households as opposed to office settings, because with offices, area power consumption is normally all that is required, but with the sheer number and variety of devices in a household setting, circuit level measurement is desired.

The cheapest way to measure power consumption in residential areas would be to just measure total power consumption with a meter. This does not give a picture of the energy consumption of devices, however. To get the best picture of energy consumption in a household, a meter could be installed on every device in the residence. The problem with this is the high cost and amount of installation required for such a system. The authors of this paper decided to go to the middle route and design a system that measures circuit level power consumption and then estimates device power usage from the gathered data. This would theoretically provide data that can be used by an energy management system to control power usage at a device level, while requiring the low cost of a simpler setup.

The authors in [10] used circuit-level power measurements because each circuit should only have a couple of devices associated with it, so the occurrence of indecipherable devices should be relatively low. Because there are fewer devices to differentiate between on each circuit, a level-based disaggregation algorithm can be utilized. This makes it easier to determine devices with constant power usages over algorithms that use edge-based algorithms, which look for power draw changes to find devices. The algorithm is also able to determine when devices are on or off through training. This can be done by the user turning each device on and off or by the system doing it through control systems.

For testing the system, meters were set up on circuits each containing three different devices. WattNodes with a 1% accuracy and power usage of 2 W were used. The system was able to differentiate between devices fairly well and had an average error of less than 6%.

Overall, it proposes an interesting and logical solution to the problem of power measurement in residential areas. In theory the system proposed makes a lot of sense, as it is low cost and should allow for fairly accurate device power usage tracking. The problems come with actually implementing such a system. With three devices the system had errors under 6%. However, in households, there might be more than three devices attached to a single circuit. The test also only included a PC, an LCD, and a lamp, all with very different power usage signatures. In actual residences, the power usage of devices might be very similar or even the same, making differentiation much more difficult.

Another outstanding issue with the proposed system is the training period required. Not all household devices will have load-controlling devices attached, so used interaction would be required for training. This would require either extra work or the inclusion of control devices on the end devices in order for automated training. This does not sound difficult but may turn off the casual residential client and will lessen the cost difference between this system and a system with meters attached to each separate end device.

## 15.7 Human–Building–Computer Interaction

With phones and other electronic devices becoming more mature, the possibilities for a building–computer interface providing occupants with energy usage data becomes more feasible. These data could then be used by the occupant to control and reduce a building's energy usage. This chapter outlines a system that includes sensors, several cloud services, and a smartphone interface, which provide building data to a user to control and minimize energy usage.

The flow diagram in Figure 15.4 shows the basic workings of a human–building–computer interaction (HBCI) system [11]. Sensors in the physical environment (buildings) gather data and send them to electronic devices to be controlled. Occupants can then view the data and provide feedback to maximize energy usage.

In the experiments used for this article, several distinctions were made in order to streamline and improve the simple flow of the HBCI system. Objects in the physical environment, whether referring to rooms, computers, or sensors, are tagged so that the cloud can distinguish between all of them easily and without error. Each object has different HBCI services associated with it, which are given to the cloud with the aforementioned tagging system. The cloud can then perform different actions using the data gathered from the different objects. In HBCI, an Android smartphone was used to view and manipulate data gathered from the environment.

Smartphones gather data from different objects by either scanning a QR code or using localization. After an object is selected, different actions, data, and object properties are shown on the smartphone. User verification is required to perform certain actions and is obtained by using the object identifying code and the occupant's credentials. An issue brought up with this type of interaction is the potential for security breaches and unauthorized use of the system.

It also outlines the basics of the application used for the system, called mPAD. The mPAD has three different types of access—public, private, and root access—with varying degrees of control of the system. Root access is required to view user history of an object, and private access is required to perform actions on an object. Each object has different services. One service outlined in [11] is the energy service. This service shows the energy usage of different objects in a space. Another service is occupancy, which allows users to check into spaces, allowing the cloud to track

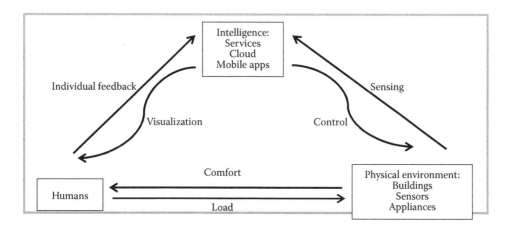

**Figure 15.4** **Human–building–computer interaction flow diagram.**

use of areas. Users interact with each service to provide data and control actions performed based on gathered data.

The biggest downfall to this system is the need for user input. If the system instead used sensors and determined automatically what actions to perform to conserve energy, not only would the system be less labor intensive but also the risk of false authorizations would be removed.

The system also assumes competence of all users involved in the system. This might work in the small environment tested in this scenario, but in bigger buildings it might become an issue. With the introduction of more users into the system also comes an increased risk of misuse and false authorization. To improve this scheme, a larger system should be implemented and tested. There was also no numerical estimate of the amount of energy this system could save. In order for this system to be improved on, a basis of energy efficiency must first be determined.

## 15.8 Conclusions

With ever-increasing energy costs, decreasing one of the most substantial sources of energy use is both prudent and worthwhile. While there is potentially game-changing energy research out there, these studies show that significant reductions can be made with relatively cheap technology. Cheaper sensor networks have pushed researchers into developing models and methods that can control HVAC systems in a manner that decreases energy use while also maintaining comfortable living conditions. Advancements in technology also bring about other concerns like privacy that must be taken into account. Overall, the field of building energy research is providing exciting new developments and still has plenty of room for further improvement.

## References

1. V. L. Erickson and A. E. Cerpa. 2010. Occupancy based demand response HVAC control strategy. In *Proceedings of the 2nd ACM Workshop on Embedded Sensing Systems for Energy-Efficiency in Building (BuildSys '10)*. ACM, New York, pp. 7–12.
2. G. R. Newsham and B. J. Birt. 2010. Building-level occupancy data to improve ARIMA-based electricity use forecasts. In *Proceedings of the 2nd ACM Workshop on Embedded Sensing Systems for Energy-Efficiency in Building (BuildSys '10)*. ACM, New York, pp. 13–18.
3. Y. Agarwal, B. Balaji, R. Gupta, J. Lyles, M. Wei, and T. Weng. 2010. Occupancy-driven energy management for smart building automation. In *Proceedings of the 2nd ACM Workshop on Embedded Sensing Systems for Energy-Efficiency in Building (BuildSys '10)*. ACM, New York, pp. 1–6.
4. Y. Kim, R. Balani, H. Zhao, and M. B. Srivastava. 2010. Granger causality analysis on IP traffic and circuit-level energy monitoring. In *Proceedings of the 2nd ACM Workshop on Embedded Sensing Systems for Energy-Efficiency in Building (BuildSys '10)*. ACM, New York, pp. 43–48.
5. A. Rice, S. Hay, and D. Ryder-Cook. 2010. A limited-data model of building energy consumption. In *Proceedings of the 2nd ACM Workshop on Embedded Sensing Systems for Energy-Efficiency in Building (BuildSys '10)*. ACM, New York, pp. 67–72.
6. Z. C. Taysi, M. A. Guvensan, and T. Melodia. 2010. TinyEARS: Spying on house appliances with audio sensor nodes. In *Proceedings of the 2nd ACM Workshop on Embedded Sensing Systems for Energy-Efficiency in Building (BuildSys '10)*. ACM, New York, pp. 31–36.
7. A. Molina-Markham, P. Shenoy, K. Fu, E. Cecchet, and D. Irwin. 2010. Private memoirs of a smart meter. In *Proceedings of the 2nd ACM Workshop on Embedded Sensing Systems for Energy-Efficiency in Building (BuildSys '10)*. ACM, New York, pp. 61–66.

8. J. Ploennigs, B. Hensel, and K. Kabitzsch. 2010. Wireless, collaborative virtual sensors for thermal comfort. In *Proceedings of the 2nd ACM Workshop on Embedded Sensing Systems for Energy-Efficiency in Building* (*BuildSys '10*). ACM, New York, pp. 79–84.

9. L. Schor, P. Sommer, and R. Wattenhofer. 2009. Towards a zero-configuration wireless sensor network architecture for smart buildings. In *Proceedings of the First ACM Workshop on Embedded Sensing Systems for Energy-Efficiency in Buildings* (*BuildSys '09*). ACM, New York, pp. 31–36.

10. A. Marchiori and Q. Han. 2009. Using circuit-level power measurements in household energy management systems. In *Proceedings of the First ACM Workshop on Embedded Sensing Systems for Energy-Efficiency in Buildings* (*BuildSys '09*). ACM, New York, pp. 7–12.

11. J. Hsu, P. Mohan, X. Jiang, J. Ortiz, S. Shankar, S. Dawson-Haggerty, and D. Culler. 2010. HBCI: Human-building-computer interaction. In *Proceedings of the 2nd ACM Workshop on Embedded Sensing Systems for Energy-Efficiency in Building* (*BuildSys '10*). ACM, New York, pp. 55–60.

*Chapter 16*

# Cyber-Physical System for Smart Grid Applications

Huiying Zhen and Fei Hu

## Contents

## 16.1 Introduction

New technologies and innovations continue to boost industry. However, a significant improvement in the electric power system has rarely been made for decades. The Northeast Blackout of 2003 was a widespread power outage in the United States, which impacted many places including Cleveland, Akron, Toledo, New York City, Westchester, Orange and Rockland, Baltimore, Buffalo, Rochester, Binghamton, Albany, Detroit, and parts of New Jersey. This event put the reliability of the US power grid into question. Meanwhile, the increase in electric load and consumption demands enlarges power complications, such as overloads and voltage sags. In America the power grid contributes to 40% of all nationwide carbon dioxide emissions. In view of both economic and environmental perspectives, extensive changes must be made to improve the current power system that has poor stability and efficiency. Therefore many nations are now modernizing their power grids. American Electric Power Research Institute (EPRI) started a research on "Intelligrid" in 2001; the European Commission set up the European Technology Platform (ETP) in 2005 to create a vision of 20/20/2020; and State Grid in China started to construct more practical and fundamental digital grid. They believe that the power grid not only requires reliability, scalability, manageability, and extensibility but also should be secure, cooperative, and cost-effective. Such an electric infrastructure is called a "smart grid." On the basis of the US Department of Energy's definition, a smart grid integrates advanced sensing technologies, control methods, and integrated communications into the current electricity grid. Through advanced sensing technologies and control methods, it can capture and analyze data regarding power usage, delivery, and generation in near real time. Figure 16.1 shows the traditional structure of an electric grid with power generation, power distribution, and so on. It also shows an "intelligence" infrastructure where the Global Positioning System (GPS) sends signals to make sure that the entire system is synchronized. The smart meters communicate bidirectionally with the neighbors via multihop routing, and the collectors are connected directly to a utility provider. Communication and optimizing of the grid make the demand have intelligent operations.

As we can see, a smart grid is a typical cyber-physical system (CPS). It has both cyber units (information networks, sensors, actuators, controllers, control software, etc.) and physical units (such as converters and generators). Table 16.1 shows some examples of CPS sensors and controllers in a smart grid system.

The smart grid can provide predictive information and corresponding recommendations to all stakeholders (e.g., utilities, suppliers, and consumers) regarding the optimization of their power utilization [1]. It may achieve demand response and dynamic price regulations, real-time and online equipment monitoring and measurement, self-healing wide-area protection, distributed generation and alternate energy resources, real-time simulation, and contingency analysis. Apparently, it can be regarded as a "system of systems" that involves both information technology (IT) and electricity system operations and governance. Such a complex CPS certainly bears many challenges, especially in cybersecurity and privacy aspects.

There are many wireless networks available for the "intelligence" infrastructure. However, which one best fits the smart grid requirements depends on the power system architecture and varieties of communication modules and wireless connections. In a smart grid, multihop wireless networking is necessary because electric equipment that needs to exchange information is out of communication range of each other.

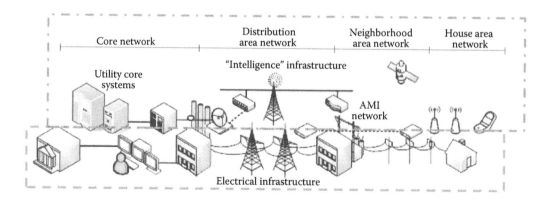

**Figure 16.1 Electrical infrastructure and the intelligence infrastructure.**

**Table 16.1   Examples of Smart Grid Sensors/Controllers**

| Location | Sensor/Controller Names | Functions |
|---|---|---|
| Storage | Alternating current (AC) sensor | Measure AC frequency oscillation |
| | Power level sensor | Indicate the current energy storage |
| | Power release controller | Controls release/storage of electricity |
| Transmission line | Impedance sensor | Detect power line breakage |
| | Inverter controller | Stop/allow the local AC transmission |
| Wind turbine | Pressure/wind speed sensors | Shows the energy generation strength |
| | Wind blade controller | Stop the wind fans if hardware fault detected |
| | Wind generator controller | Control for wind power generation |
| Solar panel | Solar PV voltage sensor | Shows the energy generation strength |
| | Solar PV controller | Control for solar power generation/integration |
| MG/PG interface | Switch controller | Switch between islanded/connected modes |
| Generator | Temperature/vibration sensor | Indicates generator's working status |

To simplify network organization and maintenance, the entire network needs to be self-organized [2]. For example, some communication modules are for wireless sensing and control and run at a low duty cycle, but other communication modules may need to constantly forward data traffic. In a smart grid, communication modules associated with electric devices are usually stationary, but mobile connections need to be supported at the customer side or on some handheld devices.

From the security viewpoint the grid may be subject to physical attacks. Malicious software can harm the control system or use up the system resources to perform the attacker's tasks. Any of these disruptions can be dangerous. Threats such as tampering with billing information of particular users can cause a major economic disturbance if they are not monitored carefully. The power grids are a major resource to national defense. Thus, any form of attacks can cause havoc. Furthermore, increased connectivity of the grids may invade the consumers' privacy. Failure to eliminate these threats will hinder the modernization of the existing power industry. Although contemporary security technologies, such as virtual private networks (VPNs), intrusion detection systems (IDSs), public key infrastructure (PKI), antivirus software, and firewalls, have well protected the IT infrastructure, they still cannot be directly used for the smart grids due to their inherent differences, as described in Table 16.2.

As an example, intruders may utilize VPN to hack the power grid. The North American Equipment Council reported the effects of a slammer worm on the power utilities used over in North America [3]. The worm migrated through a VPN connection to a company's corporate network until it finally reached the critical supervisory control and data acquisition (SCADA) network. It infected a server on the control-center LAN that was running MS-SQL. The worm traffic blocked SCADA traffic. We may transplant some IT security techniques into the smart grid to

**Table 16.2    Differences between the IT Network and Smart Grid**

|  | *IT Network* | *Smart Grid* |
|---|---|---|
| Security objectives | Confidentiality, integrity, availability | Availability, integrity, confidentiality [4] |
| Architecture | (1) Flexible and dynamic topology<br>(2) Center server requires more protection than periphery hosts [4] | (1) Relatively stable treelike hierarchy topology<br>(2) Some field devices require the same security level as the central server [4] |
| Quality of service | (1) Transmission delay and occasional failures are tolerated<br>(2) Allow rebooting | (1) High restrictions on transmission delay and failures<br>(2) Rebooting is not acceptable [4] |
| Technology | (1) Diverse operating systems<br>(2) Public networks<br>(3) IP-based communication protocols | (1) Proprietary operating systems<br>(2) Private networks<br>(3) IEC61850-and Distributed Network Protocol (DNP)-based communication protocols |

meet its security and privacy requirements. However, while choosing any of the possible security measures, there always exists a trade-off among security, cost, and performance.

## 16.2  Communications in Smart Grid

Figure 16.1 shows the concept of a typical power generation, transmission, and distribution system. The power plant generates electricity and increases the voltage to a high level. The high-voltage electricity is transmitted to a distant transmission substation where the voltage is reduced and the electricity is further transmitted to lower-level transmission substations. Finally, the electricity is transmitted to local distribution substations, which will distribute it to consumers. Several distribution lines emanate from each distribution substation, and each line supplies a number of consumers such as residential houses and other local loads.

The safe operation of a power grid requires that critical electricity parameters (e.g., voltage and frequency) always stay within their operating ranges. However, unexpected events like short circuits and the imbalance between power demand and supply may affect those electricity parameters that may be out of their operating ranges. For safe operation, the parameters are monitored and controlled in real time. In some extreme cases, protection schemes are triggered to separate a problematic grid component. The smart grid will bring new features into the power grid such as renewable-based generation, demand response, wide-area protection, and smart metering. The core of the smart grid is an intelligent communication system that links all components together in an efficient and secure manner. Smart grid communications can be unicast or multicast. Unicast means one-to-one communications and has many applications such as sending a measurement report from a field device to the control center or sending a control command in the reverse direction. Multicast means one-to-many communications such as broadcasting a command to a specific group of receivers.

When the grid topology suddenly changes owing to unexpected loss of a large generator, transmission line, or load, some protective actions may be triggered to disconnect other generators or transmission lines, resulting in a cascaded failure. Such cascaded failures caused the well-known Northeast Blackout in 2003 [5]. Wide-area protection schemes are being deployed to prevent cascaded failures. In these schemes, phasor measurement units (PMUs) measure system parameters such as current and voltage at precisely synchronized times and multicast the phasor data to the control centers. For example, in the North American SynchroPhasor Initiative, each PMU multicasts measured system parameters to the control centers at a rate of 30 times per second. From the phasor data the control centers detect problems like electricity frequency drop and issue control commands to open or close appropriate switches.

During periods of peak energy consumption, utility companies send alerts to ask consumers to reduce their power consumption by temporarily turning off nonessential appliances. If enough consumers comply with the requests, the power reduction could be enough to avoid building an additional expensive power plant. With support from variable-rate pricing and smart metering technologies that measure energy usage at different times of a day, demand response is expected to be widely deployed in the smart grid [6]. Because millions of appliances may be involved, unicast may not be a good solution to transmit the alert message. Thus, for fast demand response, the control center generally multicasts alert messages to a large number of remote appliances.

For safe operations, some control actions such as opening a circuit switch will be sent from the control center to remote field devices. The control commands are delivered through the supervisory control and data acquisition (SCADA) system (see Figure 16.2). In many cases, a control command should be sent to a large number of field devices, and it should be executed immediately and hence multicast is the right choice. For example, in case of emergency the control center multicasts important messages such as "emergency shutdown" to all or a large fraction of substations and field devices.

To date, the architectural framework and implementation standards of the smart grid are still under investigation by the academic [7,8], industrial [4,9,10], and government sectors [11,12]. Although there are various designs for the grid architecture, almost every case follows the common reference model [10] proposed by the US National Institute of Standards and Technology (NIST).

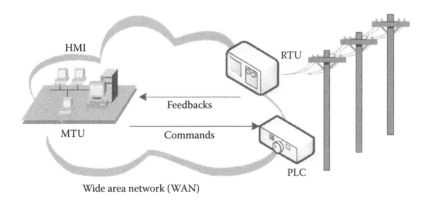

**Figure 16.2 A typical SCADA architecture. (From National Communications System, Supervisory control and data acquisition (SCADA) systems. Technical Report. Available at http://www.ncs.gov/library/tech bulletins/2004/tib 04-1.pdf, 2004.)**

The SCADA system is responsible for the real-time monitoring and control of the power delivery network [9]. Through intelligent remote control and distributed automation (DA) management at medium voltage substations, it can help the grid reduce operation and maintenance costs, and ensure the reliability of the power supply [9,13]. Two related subsystems are the energy management system (EMS) and the distribution management system (DMS) [10].

Basically, SCADA systems consist of four parts (as shown in Figure 16.3) [14]: (1) field data interface devices such as remote terminal units (RTUs) and programmable logic controllers (PLCs), (2) a communication system (e.g., telephone, radio, cable, and satellite), (3) a central master terminal unit (MTU), and (4) human–machine interface (HMI) software or systems. By using RTUs and PLCs, most control actions can be performed automatically and remotely [9,10]. The Idaho National Laboratory (INL) report [15] claimed that the current SCADA system has lots of vulnerabilities. Creery and Byres [16] discussed a few realistic situations of attacks on physical SCADA systems that caused a major stir in the industry. To secure a SCADA network, a variety of technologies are involved, including user and device authentication, firewalls, Internet Protocol Security (IPSec), VPN, IDSs, etc. [17].

Many studies [11,12] have found that the plug-in hybrid electric vehicle (PHEV), in addition to reducing carbon emissions and reliance on fossil fuels, could also provide a means to support DER in the smart grid. Because most PHEV batteries are designed to speed up rapidly for fast discharge, parked PHEVs can supply electric power to the grid. This vehicle-to-grid concept may improve the efficiency and increase the reliability of the power grid. However, it is still under development and the trade-off between costs and benefits is still uncertain [18].

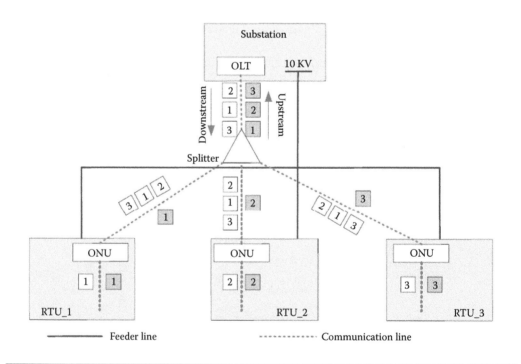

**Figure 16.3 Typical tree-based EPON system for the power grid. (Based on Sun, Z. et al., *The 2nd International Conference on Advanced Computer Control (ICACC 2010)*, vol. 3. IEEE, New Brunswick, NJ, 2010, pp. 246–250.)**

The communication standards for the power industry were developed by five leading organizations including the IEEE, the International Electro-technical Commission (IEC), and the Distributed Network Protocol (DNP3) Users Group [14]. The most prevalent protocols for SCADA communication systems are IEC 60870-5 and DNP3 [14]. The IEC protocol is typically used in Europe for communication between MTU and RTUs in SCADA systems [19,20]. The DNP3, which is derived from IEC 60870-5 and recognized by the IEEE 1379 standard, is widely used in Asia and North America. IEC 61850 has now been released to support more enhanced capabilities including a peer-to-peer communication mode for field devices. It can be regarded as a successor to the DNP3.

IEC 62351 [21] is a standard that specifies security constraints and concerns of the above communication protocols and standards. It consists of eight parts. The first two parts present an introduction to its background and a glossary of terms. Part 3 specifies the security requirements for TCP/IP profiles in IEC 60870 and IEC 61850. In particular, it describes the transport level security (TLS) configuration for secure interactions [21]. Part 4 addresses manufacturing message specification (MMS) ISO 9506 protocol security in the IEC 61850 standard. Specifically, the MMS will work with the TLS to secure communications [19]. Not all components are required to adopt this secure mechanism [19]. Part 5 focuses on the security of serial communication in IEC 60870 and DNP3. It suggests that the TLS encryption mechanism can be utilized for serial communication to enable confidentiality and integrity. As for the authentication, the serial version can only address replay, spoofing, modification, and some DoS attacks. It cannot prevent eavesdropping, traffic analysis, or repudiation due to its limited computing capability. However, it could be protected by alternate methods, such as VPNs or "bump-in-the-wire," a scheme that uses an IPSec device as a firewall to filter unwanted packages from the Internet technologies, depending on the capabilities of the devices and communications involved [19]. Relevant key management measures are also described in this part.

## 16.3 Cybersecurity Issues on Smart Grid

The traditional power infrastructure focuses on developing equipment to improve availability, integrity, and confidentiality. Because the concept of smart grid comes up, contemporary communication technologies and equipment have been typically regarded as a support of the power industry's reliability. However, increased connectivity is becoming more critical to the cybersecurity of the power system. In a broad sense, the cybersecurity of the power industry covers all IT and communications issues that affect the operation of power delivery systems and the management of the utilities [22]. The development of a secure smart grid would encounter the following four challenges [4]:

1. The power delivery system has new communication requirements in terms of protocols, delay, bandwidth, and cost. Avoiding early obsolescence is essential in smart grid security development.
2. Many legacy devices have been used in power automation systems for decades. Most of them only focus on a certain functionality and thus lack sufficient memory space or computational capability to deal with security problems. Integrating the existing legacy equipment into the smart grid without weakening their control performance is a challenge.
3. Networking in the current power grid uses heterogeneous technologies and protocols such as ModBus [23], ModBus+ [23], ProfiBus (Process Field Bus) [24], Inter-control Center Communication Protocol (ICCP), and DNP3 [14]. Nevertheless, most of them were designed for connectivity without cybersecurity.
4. Current power systems are usually proprietary systems that provide specific performances and functionalities but not security.

Many organizations are currently involved with the development of smart grid security requirements, including North American Electrical Reliability Corporation—Critical Infrastructure Protection (NERC CIP), International Society of Automation (ISA), National Infrastructure Protection Plan (NIPP), IEEE (1402), and NIST. One prominent set of requirements has been reported by the NIST Cyber Security Coordination Task Group (CSCTG) [22]. After reviewing the NIST CSCTG report [9] and existing research [9,13,19,20] on cybersecurity, we have categorized the relevant issues into five groups (shown in Table 16.2). Notice that general security problems such as software engineering practices, firewalls, circuit designs, and patch management are not included in the table.

### 16.3.1 Device Issues

Devices like programmable logical controllers (PLCs), RTUs, and IEDs are widely deployed in power delivery systems to allow administrators to perform maintenance or to dispatch functionalities from a remote location [4]. However, they also enable malicious users to manipulate the device and disrupt normal operations of the grid, such as shutting down running devices to disconnect power services or tampering with sensing data to misguide the decisions of the operators [4]. In [25] they discussed such a cyber vulnerability, in which an attacker could switch off hundreds of or even millions of smart meters with remote, off switches. Although no agreed solutions are proposed in present standards and regulations, some recommended countermeasures in [25] may be considered for further discussions. For the devices, the IEEE 1686–2007 standard has specified security requirements. However, experience shows that typical IEDs are far from complying with this standard. As described in Table 16.2, potential security problems may exist in the applications of smart meter, customer interfaces, and PHEVs. As for the meter device, a conventional physical meter can be modified by reversing the internal usage counter (aka meter inversion) or be manipulated to control the calculation of the electric flow [26]. Addressing this problem may require hardware support. Several algorithms [27,28] have been proposed to prevent the meter data from being compromised. They analyzed the trade-off between security and efficiency and designed two algorithms for per-hop and end-to-end communication protocol, respectively. They used Advanced Encryption Standard (AES) with 128-bit shared key to encrypt the line between the meter and the gateway, which showed that their protocol is reliable and energy efficient (according to their experiment results).

As for the customer interfaces and PHEVs, not many works are conducted to address potential security problems. Ongoing relevant research mainly focuses on issues of malware attacks and fast encryption. Metke and Ekl [29] proposed some suggestions for malware protection on embedded systems and general purpose computer systems. For embedded systems, manufacturers should take full responsibility for securing software development and upgrade procedures. To meet this requirement, three possible approaches have been proposed in [29]. First, the manufacturer may issue a public key to each device and encrypt all new software with the corresponding private key. The device can then validate the source of the updated patch and thus secure the system. The second method is called the high assurance boot (HAB). The embedded system will be validated once it boots up. The validation script is safely coded into its hardware by the manufacturer. Because not all devices can be rebooted very often, secure validation software is considered as the third solution. By using a device attestation technique, devices can be validated while running. When it comes to general purpose computer systems, the authors in [29] argue that current antivirus software cannot prevent the system from suffering malware attacks. Although there is currently no solution, one thing is recommended: all mobile codes (e.g., ActiveX, JavaScript, and Flash animation) in the smart grid should be strictly controlled from suppliers to operators. As we know, tens of millions of sensors or RTUs are deployed in the grid for DA. These devices have limited bandwidth, power (battery or long sleep cycles), storage, memory, and intermittent connections. Because of these constraints, applications like key management should require less centralization and more persistent connectivity than current approaches; it should also retain a certain level of trust and security for the entire infrastructure. NIST requirements suggest that each device has unique key and credential materials such that, if one has been compromised, others will not be affected.

Zhang et al. [30] proposed a 256-bit AES-based solution to secure the traffic between two smart grid devices in Ethernet networks. AES algorithms have inherent requirements for the smart grid: it should need only small memory spaces and must be able to be used for wireless sensor networks (WSNs). In their design, all data packets in Ethernet networks consist of four fields: one header and three data fields. Specifically, the header contains the destination IP address. All other nodes except the recipient cannot read the data payload and will simply discard it. The data payload includes three fields. Each of them is 16 bytes, because AES will only process 16-byte sized data. To indicate whether a message is encrypted or not, the header adds an extra AES status flag; thus this message may be transmitted through other networks. By using the Altera Cyclone-2 Field Programmable Gate Array (FPGA) based platform, they have successfully implemented their design into the hardware. Experiment results indicate that the data transmission is secure only if no eavesdroppers exist on the Ethernet network and that the throughput (bytes encrypted per second) can be 1202 bps [30].

## 16.3.2 Networking Issues

Potential security problems of networking in smart grids mainly exist in the Internet, wireless networks, and sensor networks. Just like the Internet, multiple networking technologies can be utilized for the smart grid, including fiber optics, land mobile radio, 3G/4G (WiMax), RS-232/ RS-485 serial links, WiFi, and so on [30]. Which one will be used depends on the requirements of the grid environment and is an open issue in the development of smart grid communication standards (Table 16.3).

For wired networks, Sun et al. [31] claimed that Ethernet Passive Optical Networks (EPONs) would be a promising solution for the smart grid broadband access networks due to the following metrics: (1) backward compatibility, (2) low-cost fiber deployment and maintenance, and (3) minimal protocol overhead. EPON also has been regarded as next-generation Gigabit-Ethernet by IEEE 802.3ah standard. In [31], a tree-based EPON broadcasts messages to every optical network unit (ONU), all of which share one common channel to deliver data to an optical line termination (OLT). In this case, every ONU is able to capture all downstream traffic from the OLT and will vie with other ONUs for limited upload bandwidth. Therefore, EPON can be easily attacked by methods such as spoofing, DoS, and eavesdropping. By using identity-based cryptography (IBC) and challenge response technology, the authors then proposed a secure communication protocol for the EPON. Unlike traditional asymmetric cryptographic approaches (e.g., PKI), the IBC generates a public key by using an arbitrary data string, and the corresponding private key binds this information, which is signed by a trusted key distribution center (KDC) [31]. In their scheme, the OLT and ONUs periodically perform mutual authentication. First, the OLT challenges an ONU, $i$, with a message, $n$, encrypted with $i$'s public key. After verifying this message, ONU $i$ will respond with $n$ and a random number $m$ encrypted with OLT's public key. After getting this response message, the OLT can verify the identity of $i$. Finally, the OLT will send $m$ back to $i$.

**Table 16.3 Smart Grid Security Issues**

| Topics | Keywords | Cybersecurity Issues | |
| --- | --- | --- | --- |
| | | *Potential Problems* | *Possible Solutions* |
| Devices | Smart meter | 1. Customer tariff varies on individuals, and thus breaches of the metering database may lead to alternate bills.<br>2. Meters may suffer physical attacks such as battery change, removal, and modification [4].<br>3. Functions like remote connect/disconnect meters and outage reporting may be used by unwarranted third parties. | 1. Ensure the integrity of meter data.<br>2. Secure meter maintenance.<br>3. Detect unauthorized changes on meter.<br>4. Authorize all accesses to/from AMI networks [22]. |
| | Customer interface | 1. Home appliances can interact with service providers or other AMI devices. When manipulated by malicious intruders, they could be unsafe factors in residential areas [32].<br>2. Energy-related information can be revealed on IEDs or on the Internet. Unwarranted data may misguide users' decisions. | 1. Access control to all customer interfaces.<br>2. Validate notified information.<br>3. Improve security of hardware and software upgrades [22]. |
| | PHEV | PHEV can be charged at different locations. Inaccurate billing or unwarranted service will disrupt operations of the market [22]. | Establish electric vehicle standards. |
| Networking | Internet | Certain applications may be built on the Internet. Inherent problems like malicious malware and denial of service (DoS) attacks are threats to the grid operations [22]. | 1. Adopt TCP/IP for smart grid networks [22].<br>2. VPN (IPSec), SSH, SSL/TLS [9].<br>3. Intrusion detection and firewalls. |

*(continued)*

**Table 16.3   (Continued) Smart Grid Security Issues**

| Topics | Keywords | Cybersecurity Issues | | |
|---|---|---|---|---|
| | | *Potential Problems* | *Possible Solutions* | |
| | Wireless network | In wireless networks, layer 2/3 can be easily attacked by traffic injection and modification. Without routing security, traffic on these layers is not reliable [9,22]. | 1. Protect routing protocols in layer 2/3 networks [3].<br><br>2. Security capabilities in 802.11i, 802.16e, and 3GPP LTE [29]. | |
| | Sensor network | Sensor data are critical for the grid. Intercepting, tampering, misrepresenting, or forging these data can damage the grid. | AES encryption [28]. | |
| Dispatching and management | Asset management | 1. When assets need to be replaced, unplanned outages and equipment damage could occur [22].<br><br>2. Compatibility problems could emerge while integrating legacy devices into the grid, which may cause the system to fail or malfunction. | 1. Maximize the lifetime of assets through cooperation among relevant operators [22].<br><br>2. Backup data.<br><br>3. Enabling backward compatibility. | |
| | Cipher key management | 1. Data encryption and digital signatures are required in sensors to secure communications. Most of existing cryptographic schemes lack efficiency under limited space and computation [22].<br><br>2. Access and communication may occur across different domains [33]. To manage their own credential keys in different areas is difficult, especially in a national-wide scenario [22].<br><br>3. Device or system may be "locked out" when an emergency occurs. | 1. PKI [29].<br><br>2. Identity-based encryption (IBE) [33].<br><br>3. Hierarchical, decentralized, and delegated schemes and their hybridization [22].<br><br>4. Design a bypass or "cold boot" method for emergency while remaining secure in daily operations [22]. | |

*(continued)*

**Table 16.3 (Continued) Smart Grid Security Issues**

| Topics | Cybersecurity Issues | | |
| --- | --- | --- | --- |
| | Keywords | Potential Problems | Possible Solutions |
| | Real-time operation | Some applications (e.g., real-time process) must meet limited time constraints. Increasing interoperability may cause unbounded and uncontrollable delays of the power system. | Minimize and make predictable timing impacts of security protections [3]. |
| Anomaly detection | Temporal information | 1. Unsecured time information may be used for replay attacks and revoked access, which has a significant impact on many security protocols [4,23].<br>2. Timestamps in event logs may be tampered with by malicious people [22]. | 1. Use PMUs to ensure accurate time information [23].<br>2. Adopt existing forensic technologies to ensure temporal logs are accurate [34]. |
| | Data and service | RTUs may be damaged in various ways. The accuracy of transmitted data and the quality of services therefore cannot be guaranteed [35]. | Utilize fraud detection algorithms and models used in credit card transaction monitoring [36]. |

Thus, *i* is able to verify the authority of the OLT. This mechanism establishes a secure channel between two devices. In fact, it is also adopted by the DNP standard for secure communication (e.g., challenge response mode and aggressive mode) [20]. However, the authors did not give a simulation or an experiment to evaluate its performance in terms of time delay, package overhead, scalability, etc. Moreover, they had not discussed how to set up a KDC but rather assumed that one existed.

## 16.4 Privacy Issues on Smart Grid

Intelligent control and economic management of energy consumption require good interoperability between consumers and service providers. Unprotected energy-related data can cause invasions of privacy in the smart grid. In particular, radio waves in AMI may disclose information about when and where people are and what they are doing [37]. Failure to address privacy issues in the smart grid will not be accepted by regulators and customers. In this section, we will give a brief overview of current studies on privacy issues in the smart grid.

### 16.4.1 Personal Information

Personal information is any recorded information that can identify an individual directly or indirectly [22]. Besides name, biographical, and contact information, it may also involve personal choices, social activities, health problems, or any economic, physical, or mental information derived from the above, and information about other relatives. Considering the smart grid context, any type of energy use data that links to personal information should be secured and monitored in a proper way. NIST guidelines [22] have provided a list of personal information that may be available through the smart grid as follows:

1. Name: responsible for the account
2. Address: location to which service is being taken
3. Account number: unique identifier for the account
4. Meter reading: kW and kWh consumption recorded at 15–60 min intervals during the current billing cycle
5. Current bill: current amount due on the account
6. Billing history: past meter readings and bills, including history of late payments/failure to pay, if any
7. HAN: in-home electrical appliances
8. Lifestyle: when the home is occupied and it is unoccupied, when occupants are awake and when they are asleep, how many various appliances are used, etc.
9. DER: the presence of on-site generation and/or storage devices, operational status, net supply to or consumption from the grid, usage patterns
10. Meter IP: the IP address for the meter, if applicable
11. Service provider: identity of the party supplying this account, relevant only in retail access markets

## 16.4.2 Privacy Concerns

In the context of the smart grid, energy consumption data obtained by a third party may disclose personal information without one's permission. Besides laws and regulations to protect personal information in the smart grid, we also require a secure mechanism to prevent privacy violations from breaching local data and remote copies. According to the NIST study [38], four typical areas of privacy concern in the smart grid are presented as follows.

First, fraud should be considered, especially when energy consumption is attributed to a different location (e.g., in PHEVs' case) [22]. The metering system (either physical recording or electronic metering systems) should not allow any personnel abuse or modifications of the collected data [22]. In particular, NIST's report [38] has analyzed two relevant privacy use cases in detail. One case is about a landlord with tenants who have PHEVs that require being charged separately. For the purpose of preserving the privacy of the tenants, a utility is used to authenticate communications between the smart meter and PHEVs through a secure line and energy services communication interface (ESCI) provided by the utility and/or vehicle manufacturer. Another case regards the PHEV general registration and enrollment process. In order to complete the initial setup for PHEVs, NIST believes that utilities should offer the following services to customers: (1) enrollment, (2) registration, (3) initial connection, (4) ability to repeatedly reestablish connection between a utility and PHEV, (5) ability to provide a PHEV tariff or charging status information to customer interfaces, and (6) correct bill.

Second, data in the smart meter and HAN could reveal certain activities of home smart appliances [22]. In addition, it can be used for tracking specific times and locations of energy consumption in specific areas of the home, which may further indicate the appliances used and/or types of activities. For example, appliance vendors may want this kind of data to know both how and why individuals used their products in certain ways. Such information therefore could impact appliance warranties. Meanwhile, other entities may need these data to conduct target marketing. In [38] a system is designed that utilized a power router and a rechargeable battery to hide or obscure load signatures in a home area. In this system, they assume that the home will have several energy storage and generation devices in the future. Through a power router, appliance load signature or usage pattern will be moderated and thus cannot be recognized and tracked by a malicious intruder. They have further improved this model in [39] and named it as ElecPrivacy. Besides, a number of privacy measurement approaches are provided for this model in [39].

Third, obtaining near-real-time data on energy consumption may help to infer whether a residence or facility is occupied, where people are in the structure, what they are doing, and so on [22]. Authors in [40] proposed a data aggregation approach for meters based on spanning tree topology. By using a homomorphic encryption method, data are secured all the way from home meters to the data center. It can well protect the privacy of individual power usage according to their analysis and evaluation in [41]. In [41], researchers pointed out that customers would possibly deploy a separate measurement device at home to better monitor their power usage. The redundant meter data, if transmitted in an unsecured wireless line, could leak customer's information to an eavesdropper. By compressing the data to a rate below its entropy, the authors in [41] proposed a coding method that well addressed this problem.

Fourth, personal lifestyle information derived from energy use data could be valuable to some vendors or parties [22]. For instance, vendors may use this information for targeted marketing. The beneficial information may be revealed by new technologies like smart meters, time of use and demand rates, and direct load control of equipment. They could be further sold and used for energy management analysis and peer comparisons. Cavoukian et al. [42] proposed an escrow-based anonymization scheme to prevent personal information from being tracked by unauthorized third parties. They categorized metering data into two parts: high frequency and low frequency. Corresponding setup and communication procedures were then designed for each type of data. These procedures are both regular PKI authentication approaches. Because the anonymity degree of the system depends on the size of the anonymity set, the wide deployment of such a scheme requires further investigation.

In addition, two aspects of the smart grid data need to be considered in terms of existing laws and regulatory policies to ensure that new types of data are addressed [22]: (1) granular and available data on use of individual appliances by time and location, and (2) public awareness of contractual agreements about data ownership and what may be revealed about people's daily activities.

### 16.4.3 Recommendations

NIST has delivered a report [22] on the consumer-to-utility privacy impact assessment (PIA) of the smart grid. Ten potential design principles are proposed to address privacy issues in the smart grid:

1. An organization should ensure that information security and privacy policies and practices exist and are documented and followed. Audit functions should be present to monitor all data accesses and modifications.
2. Before collecting and sharing personal information and energy use data, a clearly specified notice should be announced.

3. Available choices should be presented to all users. Organizations need to obtain users' consent or implied consent if it is not feasible, with respect to the collection, use, and disclosure of their personal information.

4. Only personal information that is required to fulfill the stated purpose should be collected from individuals. Treatment of the information should conform to these privacy principles.

5. Information should only be divulged to those parties authorized to receive it. Personal information should be aggregated or anonymized wherever possible to limit the potential for computer matching of records. Personal information should only be kept as long as it is necessary to fulfill the purposes for which it was collected.

6. The organization should allow individuals to check their corresponding personal information and to request the correction of perceived inaccuracies. Personal information should be notified about parties with whom personal information has been shared.

7. Personal information should be used only for the purposes for which it was collected. Personal information should not be disclosed to any other parties except for those identified in the notice or with the explicit consent of the service recipient.

8. Personal information in all forms should be protected from unauthorized modification, copying, disclosure, access, use, loss, or theft.

9. Organizations should ensure the data usage information is complete, accurate, and relevant for the purposes identified in the notice.

10. Privacy policies should be made available to service recipients. These service recipients should be given the ability and process to challenge an organization's compliance with their state privacy regulations and organizational privacy policies as well as their actual privacy practices.

Cavoukian et al. [42] presented the conceptual model "SmartPrivacy" to prevent potential invasions of privacy while ensuring full functionality of the smart grid. Specifically, in the case of utilities providing personal information to a third party with the express consent of an individual, the following are examples of SmartPrivacy defaults that offer greater protection of privacy:

■ The information provided to third parties should be minimized such that it only fulfills the purpose of relevant services. For example, partial location data (e.g., the first few digits of a zip code) may be sufficient for services that allow for comparison of neighborhood averages and other features such as weather statistics.

■ When data are transmitted, the risk of interception arises. Appropriate and secure channels of transmission between different communication protocols are required to ensure strong privacy protection in the smart grid.

■ Anonymize identity if possible. When sharing data with a third party, consider using a pseudonym such as a unique number that the individual would be permitted to reset at any time.

■ Third parties should not request information from the utility about consumers, or consumers must be able to maintain control over the type of information that is disclosed to third parties by the utility.

■ Third parties should agree not to correlate data with data obtained from other sources or the individual, without the consent of the individual.

## 16.5 Future Research Directions

Generally speaking, three areas should be further studied to enhance the security level of the grid: (1) integrity and confidentiality of the transmitted data, (2) building a robust and efficient dispatching and management model for a SCADA system, and (3) establishing a universal policy and standard for secure communication technology. We have also examined the privacy concerns in the smart grid. To eliminate personal information leakage problems, we believe that state-of-the-art techniques like anonymity, access control, and accountability might provide solutions. Possible future research directions may include the following aspects:

1. *Control system security*: Industrial control has not paid enough attention to security issues. In recent years, people have paid more attention to control system security to protect power generation, transmission, and distribution. Codesigns of control and security in smart grids will be interesting topics in the future.

2. *Power system security*: Besides cybersecurity, vulnerabilities in the physical power grid should also be further explored and studied. Because new devices will be largely deployed, no one can guarantee that the power line itself is 100% secure. Standards and regulations for those new components and their compatibilities need to be modified accordingly.

3. *Accountability*: As we know, advanced cybersecurity technologies may well protect every level of the current network infrastructure. However, new vulnerabilities and risks continue to emerge under the particular framework of the smart grid. As a complement, accountability is required to further secure the smart grid in terms of privacy, integrity, and confidentiality. Even if a security issue presents itself, the built-in accountability mechanism will determine who is responsible for it. When detected, some problems can be fixed automatically through the predefined program, while others may provide valuable information to experts for evaluation.

    Generally speaking, accountability means that the system is recordable and traceable. Every single change in a local host or network traffic, which may be the most important or most desirable information, can be used as evidence in future judgment. Under such a circumstance, no one can deny their actions, not even the administrators or other users with high privileges. Together with some suitable punishments or laws in the real world, this will prevent many attacks.

    One case study for accountability is the monthly electricity bill of every homeowner. Although the cost of electricity could be determined by the smart meter, we still doubt its reliability. The utility or the smart meter itself may alter transmitted data to suit someone's interests or for some other reasons (e.g., because they are under attack). As a consequence, homeowners could have two different electric bills: one from the utility and one from the smart meter. We have proposed an approach to address this issue in [42]. With accountability, the false party can be detected by provable evidences. To extend the accountability concept to the whole-grid area, regional and nationwide power management systems should be involved.

4. *Integrity and confidentiality*: Integrity and confidentiality are two main aspects for computer and network security design. Naturally, they are still essential for securing the smart grids. For example, integrating with huge numbers of DERs may incorporate with distributed database management system and cloud computing technologies. Whether or not we could adopt current solutions to provide integrity and confidentiality for smart grid is a future research direction.

5. *Privacy*: Privacy issues in cybersecurity may be addressed by adopting newly anonymous communication technologies. Current approaches to anonymize traffic in general networks will cause overhead problems or delay issues. For some time-critical operations, limited bandwidth and less connectivity features in the smart grid may hinder the implementation of anonymity. Some pilot works are presented in [39] and [43]. In addition, a network traffic camouflage technique could be considered to hide critical entities (e.g., database or control center) in the grid.

## 16.6 Conclusions

This chapter mainly gives an overview of CPS issues (especially security) in the smart grid. According to existing research, almost every aspect related to IT technology in the smart grid has potential vulnerabilities due to inherent security risks in the general IT environment. This chapter also provides future research directions. Cybersecurity and privacy issues in the smart grid are new challenging topics across the fields of power industry, electrical engineering, and computer science. More in-depth research is required to develop such a promising power grid in the near future.

## References

1. NaturalGas.org. 2010. Natural gas and the environment. Available at http://www.naturalgas.org/environment/naturalgas.asp.
2. Cisco Systems, Inc. 2009. Internet protocol architecture for the smart grid. White paper.
3. A. Faruqui, R. Hledik, and S. Sergici. 2009. Piloting the smart grid. *Electr. J.* 22:7, 55–69.
4. D. Wei, Y. Lu, M. Jafari, P. Skare, and K. Rohde. 2010. An integrated security system of protecting smart grid against cyber attacks. In *Innovative Smart Grid Technologies* (*ISGT 2010*), IEEE, New Brunswick, NJ, pp. 1–7.
5. Q. Li and G. Cao. 2011. Multicast authentication in the smart grid with one-time signature. *IEEE Trans. Smart Grid* 2:4, 686–696.
6. Y. Simmhan, B. Cao, M. Giakkoupis, and V. K. Prasanna. 2011. Adaptive rate stream processing for smart grid applications on clouds. In *Proceedings of the 2nd International Workshop on Scientific Cloud Computing* (*ScienceCloud '11*). ACM, New York, pp. 33–38.
7. A. Clark and C. J. Pavlovski. 2010. Wireless networks for the smart energy grid: Application aware networks. In *Proceedings of the International MultiConference of Engineers and Computer Scientists*, 2010 Vol. II (IMECS 2010). International Association of Engineers, Hong Kong, 1243–1248.
8. J. Gadze. 2009. Control-aware wireless sensor network platform for the smart electric grid. *Int. J. Comput. Sci. Network Security* 9:1, 16–26.
9. A. R. Metke and R. L. Ekl. 2010. Smart grid security technology. In *Innovative Smart Grid Technologies* (*ISGT 2010*). IEEE, New Brunswick, NJ, pp. 1–7.
10. W. Y. Chu and Dennis J. H. Lin. 2009. Communication strategies in enabling smart grid development. In *The 8th International Conference on Advances in Power System Control, Operation and Management* (*APSCOM 2009*). IEEE, New Brunswick, NJ, pp. 1–6.
11. U.S. National Institute of Standards and Technology. 2010. NIST framework and roadmap for smart grid interoperability standards, release 1.0. *NIST Spec. Publ.* 1108. Available at http://www.smartgrid.gov/standards/roadmap.
12. U.S. NETL. 2007. A systems view of the modern grid. White paper. Available at http://www.smartgrid.gov/white papers.
13. G. N. Ericsson. 2010. Cyber security and power system communication—Essential parts of a smart grid infrastructure. *IEEE Trans. Power Delivery* 25:3, 1501–1507.

14. National Communications System. 2004. Supervisory control and data acquisition (SCADA) systems. Technical report. Available at http://www.ncs.gov/library/tech bulletins/2004/tib 04-1.pdf.

15. Idaho National Laboratory. 2008. Common cyber security vulnerabilities observed in control system assessments by the INL NSTB program. Technical report INL/EXT-08-13979. Available at http://www.inl.gov/scada/publications.

16. A. Creery and E. J. Byres. 2005. Industrial cybersecurity for power system and SCADA networks. In *Industry Applications Society 52nd Annual Petroleum and Chemical Industry Conference*. IEEE, New Brunswick, NJ, pp. 303–309.

17. F. Alsiherov and T. Kim. 2010. Secure SCADA network technology and methods. In *Proceedings of the 12th WSEAS International Conference on Automatic Control, Modelling and Simulation*. World Scientific and Engineering Academy and Society, Stevens Point, WI, pp. 434–438.

18. U.S. Department of Energy. 2009. Smart grid system report. White paper. Available at http://www.oe.energy.gov/SGSRMain090707lowres.pdf.

19. S. Ward, J. O'Brien, B. Beresh, G. Benmouyal, D. Holstein, J. T. Tengdin, K. Fodero et al. 2007. Cyber security issues for protective relays. In *IEEE Power Engineering Society General Meeting*. IEEE, Brunswick, NJ, pp. 1–27.

20. S. Hong and M. Lee. 2010. Challenges and direction toward secure communication in the SCADA system. In *Proceedings of the 8th Annual Communication Networks and Services Research Conference*. IEEE, New Brunswick, NJ, pp. 381–386.

21. IEC TC57. 2010. Power system control and associated communications data and communication security, IEC 62351 Part 1 to 8, Technical specification and draft.

22. U.S. National Institute of Standards and Technology. 2010. Guidelines for smart grid cyber security (vol. 1 to 3), NIST IR-7628. Available at http://csrc.nist.gov/publications/PubsNISTIRs.html#NIST-IR-7628.

23. ModBus. ModBus specifications and implementation guides. ModBus Protocol Specification v1.1b.

24. ProfiBus. ProfiBus standard. Specifications and Standards. Available at http://www.profibus.com/downloads/specifications-standards/.

25. R. Anderson and S. Fuloria. 2010. Who controls the off switch? In *Proceedings of the 1st IEEE SmartGridComm 2010*. IEEE, New Brunswick, NJ, pp. 96–101.

26. P. McDaniel and S. McLaughlin. 2009. Security and privacy challenges in the smart grid. *IEEE Security Privacy* 7:3, 75–77.

27. F. Li, B. Luo, and P. Liu. 2010. Secure information aggregation for smart grids using homomorphic encryption. In *Proceedings of the 1st IEEE SmartGridComm 2010*. IEEE, New Brunswick, NJ, pp. 327–332.

28. A. Bartoli, J. Hernandez-Serrano, M. Soriano, M. Dohler, A. Kountouris, and D. Barthel. 2010. Secure lossless aggregation for smart grid M2M networks. In *Proceedings of the 1st IEEE SmartGridComm 2010*. IEEE, New Brunswick, NJ, pp. 333–338.

29. A. R. Metke and R. L. Ekl. 2010. Security technology for smart grid networks. *IEEE Trans. Smart Grid* 1:1, 99–107.

30. P. Zhang, O. Elkeelany, and L. McDaniel. 2010. An implementation of secured smart grid ethernet communications using AES. In *Proceedings of the IEEE SoutheastCon (SoutheastCon 2010)*. IEEE, New Brunswick, NJ, 394–397.

31. Z. Sun, S. Huo, Y. Ma, and F. Sun. 2010. Security mechanism for smart distribution grid using ethernet passive optical network. In *The 2nd International Conference on Advanced Computer Control (ICACC 2010)*, vol. 3. IEEE, New Brunswick, NJ, pp. 246–250.

32. P. McDaniel and S. McLaughlin. 2009. Security and privacy challenges in the smart grid. *IEEE Security Privacy* 7:3, 75–77.

33. H. K.-H. So, S. H. M. Kwok, E. Y. Lam, and K.-S. Lui. 2010. Zero-configuration identity-based signcryption scheme for smart grid. In *The First IEEE International Conference on Smart Grid Communications (SmartGridComm)*. IEEE, New Brunswick, NJ, pp. 321–326.

34. Q. Pang, H. Gao, and M. Xiang. 2010. Multi-agent based fault location algorithm for smart distribution grid. In *The 10th IET International Conference on Development in Power System Protection (DPSP 2010)*. The Institution of Engineering and Technology, Hertfordshire, U.K., pp. 1–5.

35. S. Fries, H. J. Hof, and M. Seewald. 2010. Enhancing IEC 62351 to improve security for energy automation in smart grid environments. In *The 5th International Conference on Internet and Web Applications and Services (ICIW 2010)*. IEEE, New Brunswick, NJ, pp. 135–142.

36. Z. Fan, G. Kalogridis, C. Efthymiou, M. Sooriyabandara, M. Serizawa, and J. McGeehan. 2010. The new frontier of communications research: Smart grid and smart metering. In *Proceedings of the 1st International Conference on Energy-Efficient Computing and Networking* ACM, New York, pp. 115–118.

37. G. Kalogridis, C. Efthymiou, S. Z. Denic, T. A. Lewis, and R. Cepeda. 2010. Privacy for smart meters: towards undetectable appliance load signatures. In *The First IEEE International Conference on Smart Grid Communications (SmartGridComm)*. IEEE, New Brunswick, NJ, pp. 232–237.

38. NIST Interagency or Internal Reports 7628. 2010. *Guidelines for Smart Grid Cyber Security*, vol. 2, *Privacy and the Smart Grid*. The Smart Grid Interoperability Panel—Cyber Security Working Group. Available at http://csrc.nist.gov/publications/nistir/ir7628/nistir-7628_vol2.pdf.

39. F. Li, B. Luo, and P. Liu. 2010. Secure information aggregation for smart grids using homomorphic encryption. In *Proceedings of the 1st IEEE SmartGridComm 2010*. IEEE, New Brunswick, NJ, pp. 327–332.

40. D. P. Varodayan and G. X. Gao. 2010. Redundant metering for integrity with information-theoretic confidentiality. In *Proceedings of the 1st IEEE SmartGridComm 2010*. IEEE, New Brunswick, NJ, pp. 345–349.

41. C. Efthymiou and G. Kalogridis. 2010. Smart grid privacy via anonymization of smart metering data. In *The First IEEE International Conference on Smart Grid Communications (SmartGridComm)*. IEEE, New Brunswick, NJ, pp. 238–243.

42. A. Cavoukian, J. Polonetsky, and C. Wolf. 2010. SmartPrivacy for the smart grid: embedding privacy into the design of electricity conservation. In *Identity in the Information Society*. Springer, New York, pp. 275–294.

43. G. Kalogridis, S. Z. Denic, T. Lewis, and R. Cepeda. 2011. Privacy protection system and metrics for hiding electrical events. *Int. J. Security Networks* 6:1, 14–27.

## Chapter 17

# Cyber-Physical System for Transportation Applications

Matthew Bell, Loilin Muirhead, and Fei Hu

## Contents

## 17.1 Networked Automotive Cyber-Physical Systems

First, we will introduce the next generation automotive cyber-physical systems (CPSs) and how they can be incorporated into traffic conditions, battlefield operations, and medical areas. These systems have the ability to sense, actuate, and communicate in a wireless network through the cyber world. Using CPS by use of sensors in automobiles would give drivers the ability to find information on things like road conditions, traffic congestion, and wind conditions. This information would be shared between the driver and vehicle. However, with every newly emerging technology comes some drawbacks. The main concern is the delay of information and the way to allocate resources in such a way to minimize this from happening, and the possible delivery of unique environments in a desired time frame [1].

This system also comes in handy when vehicle owners have to deal with vehicle theft or unauthorized usage, where sensors can be used to detect fingerprints. Unauthorized usage could be monitored by access patterns. An example would be battlefield operations where an officer wants to know the movements of, say, a specific tank.

The above sensor-based CPS applications can be seen in the medical field, which could use CPS to monitor patients who did not necessarily have to be in a hospital but needed a little attention. The sensor would monitor heart rate, oxygen, pressure, and so on. Usage of this technology in the medical field would also help medical professionals communicate quicker with patients and coworkers. This technology faces a few challenges in the medical field, and one of them is in the midst of disaster/terrorism. In times of disaster, being able to coordinate the movements of health providers would provide a more efficient patient care but would be difficult to implement. With any environment that supports a wireless infrastructure, there are obviously some vulnerabilities. There are concerns of identity disclosure, denial of service, and the reliability of communication between health providers and pharmacies while patient confidentiality is upheld.

CPS technology overall could revolutionize patient care, revamp automobile usage, and take battlefield operations to a level never seen before in history, causing a new shift in the technological era.

## 17.2 Arterial Traffic Condition Estimation

Here we will discuss how incomplete or sparse data can be gathered through the use of a CPS that can estimate and predict the travel times on arterial (or secondary) roads in an urban environment. In [2] the CPS used was that of the Global Positioning System (GPS) placed in a fleet of roughly 500 taxis in the San Francisco, California, area. Using the data that were acquired through GPS data, the authors then tested the model that they had developed and found that it increased the accuracy by 35% over that of traditional vehicular probe data.

Because traffic congestion can have serious impacts on both economic and societal activity throughout the country, the authors of this article believed that creating a system capable of both monitoring and predicting traffic patterns in real time would allow a system to be built that could then minimize the traffic congestion. Because the majority of arterial roads in the country lack the traffic sensing sensors (radars, video cameras, and loop detectors) that are found on primary roads, this system had to be able to make these predictions using the sparse amount of data available. The data from the GPS units are limited, along with the unknown cycles of the traffic lights; the models and mathematical algorithms presented in the article were designed to convert the sparse data into data that can be used to predict traffic congestion [3].

The method the authors proposed on using is based on a probabilistic graphical model called the *coupled hidden Markov model* (CHMM). These models have been widely used to interpret sensor data from roadways [4]. One downside to the CHMM is that it is based on a model that is expecting a measurement every 30 s from a fixed location. So to improve on these limitations, the model that the authors proposed would be able to make traffic estimations from random locations at random times. This was accomplished by first creating traffic model assumptions, then constructing a graphical model and particle filter, and finally applying the model to a real-world scenario.

### 17.2.1 Traffic Model Assumptions

To model traffic flow through an intersection, they made six key assumptions to determine the distribution of the travel time between any two lights [3]. These assumptions are a triangular fundamental diagram, stationarity of traffic, first-in first-out, discrete congestion states, conditional independence of link travel, and conditional independence of state transitions. After making these assumptions, the authors were then able to derive the following formula to represent the probability distribution of vehicle locations:

$$\propto_{x_1,x_2} = \int_{x_1}^{x_2} \wp_x(x)dx$$

with $\propto_{x_1,x_2}$ being the ratio between the distance $|x_2 - x_1|$ and $\wp_x(x)$ and $Z$ being equal to

$$\wp_x(x) = \frac{1}{Z}dx$$

$$Z = \int_0^L d(u)du$$

The probability distribution $\wp_x(x)$ was then normalized so that is was equal to

$$
\begin{cases}
\tilde{\rho}_a & x \in [0, L-(\iota_r + \iota_{max})] \\
\tilde{\rho}_a + \tilde{\rho}_b \dfrac{x-(\iota_r + \iota_{max})}{\iota_{max}}, & x \in [L-(\iota_r + \iota_{max}), L - \iota_r] \\
\tilde{\rho}_a + \tilde{\rho}_b & x \in [L-\iota_r, L]
\end{cases}
$$

with $\tilde{\rho}_a = \dfrac{\rho_a}{Z}$, $Z = \rho_a L + \dfrac{1}{2}\iota_{max}\rho_b + \iota_r \rho_b$, and $\tilde{\rho}_b = \dfrac{\rho_b}{Z} = 2\dfrac{1-\rho_a L}{\iota_{max} + 2\iota_r}$. These formulas provided the authors with a mathematical formula so that they could determine how the vehicles were distributed and where backups occurred along a stretch of road by only using the parameters $\tilde{\rho}_a, \iota_r, \iota_{max}$. Using experimental results, the authors were able to verify this model by using only a small amount of data.

## 17.2.2 Graphical Model

To create the graphical model, one obstacle that had to be overcome was that the GPS units did not directly report the amount of time it required to travel from one road segment to the next, or in other words, a driver could drive through several blocks in between GPS readings. To overcome this problem, the authors proposed using a method to break down the total time traveled into the amount of time traveled in each road segment. Owing to the size of the area that the authors were intending to cover, they decided to use the well-known problem solving *expectation maximization* (EM) algorithm to iterate between finding the probability of each step given the time interval (*E* step) and the probabilities of each state for each link and time interval (*M* step).

For the *E* step the authors had to make an approximation based on particle filtering. Each of these particles is used to determine the probability of the congestion of each road segment during

**Table 17.1 Comparison between the Graphical Model and the Baseline Model**

| Model | Root-Mean-Squared Error | Mean Percentage Error (%) |
|---|---|---|
| Graphical (with density) | 46 | 30.1 |
| Graphical (without density) | 50 | 34.3 |
| Baseline | 63 | 44.4 |

the time interval. They did not go into detail on who implemented a particle filter but instead consulted [5] for more information (Table 17.1).

For the $M$ step the authors had to maximize the log-likelihood given the expected probabilities of each link in the state on a given day. The log-likelihood equation was found to be

$$\Lambda(Y|\boldsymbol{z}, \boldsymbol{q}, \boldsymbol{P}, \boldsymbol{A}, \pi)$$

$$= \sum_{i=1}^{N}\sum_{s=1}^{S}\sum_{d=1}^{D}\sum_{t=1}^{T_d} z_{d}^{s},t,t\left(\sum_{i=1}^{I_{d,t,s}} \ln(g_{t,s}(y_i))\right) + \sum_{i=1}^{N}\sum_{d=1}^{D}\sum_{t=2}^{T_d}\sum_{s=1}^{S}\sum_{r=1}^{S^{N^I_n}} q_{d,t,s}^{s,r} \ln(A_t(r,s)) + \sum_{i=1}^{N}\sum_{d=1}^{D}\sum_{s=1}^{S} z_{d,0,t}^{s} \ln \pi_{t,s}$$

where $I_{d,t,l}$ is the travel time observed for a day of travel, $t$ is the time interval, and $l$ is the link. This equation was then optimized and the new equation is

$$\begin{array}{c} \max \\ \boldsymbol{P,A} \end{array} \Lambda\left(Y|\boldsymbol{z},\boldsymbol{q},\boldsymbol{P},\boldsymbol{A},\pi\right): \left\{ \begin{array}{l} \displaystyle\sum_{s=1}^{S} A_t(r,s) = 1, \forall t, r \\[2mm] A_t(r,s) \in [0,1], \forall t, r, s \\[2mm] \displaystyle\sum_{s=1}^{S} \pi_{t,s} = 1, \forall t \\[2mm] \pi_{t,s} \in [0,1], \forall t, s \end{array} \right.$$

Using this equation, the authors were able to create a highly scalable learning equation capable of analyzing real-time data.

## 17.2.3 Experimental Results

To confirm that the graphical model did provide a better model to predict traffic conditions, the authors applied their model to a fleet of about 500 taxis in San Francisco. Using the data received for each taxi and whether or not a taxi had a passenger, the authors were able to filter out the time periods when the taxi stopped for loading and unloading. To begin applying the model to the data collected from the taxis, the authors first had to set up a training period so that the historical traffic density parameters could be learned. When that parameter was established, the training data were then used again to set up the particle filter. Using the results gathered, the authors were able to increase the accuracy of travel times by 36.9% compared to the methods traditionally used.

Using the model the authors proposed, they were able to demonstrate a new method for estimating arterial traffic conditions. This method could also be adapted so that it could be used to optimize other CPS such as those used in health monitoring or structural testing.

# 17.3  Car Merging Assistant

A merging assistant can be implemented for mixed traffic situations in order to increase and make traffic flow more efficient around merging areas, thus minimizing changes in speed and conflict between vehicles [5]. The algorithm generated to accomplish this encourages smooth deceleration of the mainstream traffic so as to create a sizable gap for the ramp traffic or vehicles. Also, MATLAB® simulations can be done to show exactly how effective the merging assistant was on different traffic conditions and compositions. Then some of the pitfalls of the proposed algorithm are discussed, and some solutions are given to correct those pitfalls.

Merging areas have a rather big influence on the performance of highways around the world. A major area that causes a large amount of congestion and oscillation are the on-ramp vehicles that try to merge with the mainstream traffic or mainline traffic [4].

## 17.3.1  Merging Issues

We can define merging just as a special case of changing lanes or mandatory lane changes. There are three specific cases: (1) merging in between two vehicles, (2) merging in front of a group of vehicles, and (3) merging behind a group of vehicles.

The algorithm that was generated was based on case 1, because 2 and 3 are just special cases of 1. When designing the merging assistant, three main questions needed to be satisfied:

1. Out of the two traffic streams, which is to merge first?
2. In order to create a sizable gap for on-ramp vehicles wanting to merge at a certain time and place, what action must be taken by mainstream vehicles to allow this?
3. What action must be taken by the on-ramp vehicle to fit into the provided gap at that specific place and time with the needed acceleration and speed?

Manual merging provides that the mainline traffic be given precedence over the ramp traffic; this makes ramp traffic look for an appropriate gap in which to fit. Although with manual merging for case 1, there are no clear-cut rules due to scheduling problems, but the merging assistant can see the variation based on whether or not the algorithm is designed properly. To solve case 1, the algorithm can just give priority to mainstream traffic and have ramp vehicles wait until a gap is available.

Question 2 was considered in many different ways in how to be handled, but a prominent thought was that manual traffic be dealt with in a similar fashion, like in question 1, where if mainstream traffic priority had been breached, mainline vehicles would either change lanes or decelerate. This would be to avoid a potential crash.

When question 3 is considered, the authors look at several ideas posed by other authors. In aligning the ramp vehicles in the proper gap at the right place, time, speed, and acceleration, a ramp vehicle would cross a check-in point where it would then accelerate to the mainstream speed, of course, after a green signal is given. Another idea was instead of aligning to the specified gap, the vehicle would stay a safe distance from the vehicle in front and at the back, called a safety zone. Within all of the literature reviewed by these authors it was assumed that the vehicles are automated and knew the others' status. These assumptions are not valid in a mixed traffic situation because of the almost impossible task of scheduling the order in which merging takes place for manual vehicles [5].

### 17.3.2 Merging Assistant for Mixed Traffic

We now consider the merging assistant in a mixed traffic situation, where the mainline traffic is composed of cooperative adaptive cruise control (CACC) and manual vehicles. Perturbation is a component in determining the stability of traffic flow. So, one of the jobs of the merging assistant is to alleviate the amount of perturbations and their magnitude. An algorithm was proposed to create smooth traffic flow in such a way as to make ramp vehicles merge with mainstream traffic as efficiently as possible. Along with this it is assumed that there is a roadside unit (RSU), a detection device that can tell when a vehicle is approaching the acceleration lane. Also, a good point to note is that it is a single-lane highway example and vehicles can only decelerate; multiple highway lanes will be the subject of future research.

The logic of the merging assistant can be found in Figure 17.1. Vehicle A is the vehicle with CACC, and the idea is that it can decrease its speed so that when vehicle M starts to merge it stays within the specified safety zone. This answers questions 1 and 2 of the merging assistant's what to do list. Question 3 is then just making sure that there is a wide enough gap so that when the ramp vehicle reaches the acceleration lane, it can merge safely.

Dealing with the RSU, it should detect vehicle M at time $t_0$. The RSU determines the arrival of vehicle M at $t_1$ and tells the mainstream vehicles up ahead of the merging area. All vehicles that have the CACC feature and are within the RSU's range would be able to continuously know the position of a certain vehicle at the given time $t_1$ based on $v_A(t)$, its current position. If A comes within the safety zone of vehicle M, A will automatically decide whether to accelerate or decelerate [5]. In this case, acceleration is considered, and this is compared to a suitable deceleration rate

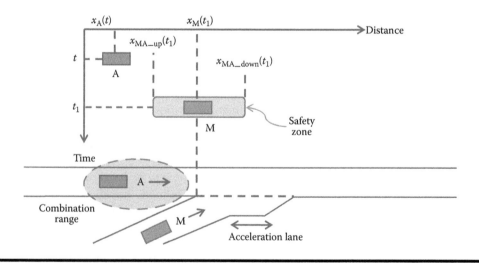

**Figure 17.1  Merging assistant logic.**

and to the acceleration rate of a given driver's comfortableness. Outside of the safety zone, vehicle A would just keep going at the driver's comfortable standards, as long as $t_1$ is not smaller than $t$.

The predicted position of A at $t_1$, $x_A(t_1)$, based on the present position $x_A(t)$ and also $v_A(t)$, the constant speed, is given by

$$x_A(t_1) = x_A(t) + v_A(t)(t_1 - t).$$

$x_{MA\_up}(t_1)$ and $x_{MA\_down}(t_1)$ are simply the upstream and downstream limits, respectively, of the safety zone for vehicle M with respect to A. If the position of vehicle A, $x_A(t_1)$, falls within $x_{MA\_up}(t_1)$ and $x_{MA\_down}(t_1)$ then vehicle A must either accelerate upstream or decelerate downstream all at $t_1$. Because option 1 (to accelerate) creates a problem because of the possible gap, then vehicle A communicates with the vehicle in front of it to say that it wants to accelerate. So option 2 is the only viable solution, determining that the acceleration of A can be done when vehicle A's position comes within the safety zone of M:

$$a(t) = \min\{\max(a_{merge\_asst}(t), d_{comfort}), a_{long\_norm}(t)\}$$

where $a_{long\_norm}$ represents the acceleration rate determined by the CACC controller, $d_{comfort}$ is the comfortable deceleration rate, and $a_{merge\_asst}$ is the acceleration determined by the merging assistant that is calculated from the equations of motion:

$$a_{merge\_asst}(t) = x_M(t_1) - x_A(t_1) - v_A(t)h_A, M/0.5(t_1 - t)2 + h_A, M(t_1 - t).$$

It is essential that $d_{comfort}$ be not close to zero because if it is, the merging assistant becomes less accurate.

## 17.3.3 Traffic Flow Simulations

Using the algorithm presented above the authors simulated traffic flow through MATLAB. Mainstream traffic was simulated at 1500 and 1800 vehicles $h^{-1}$ lane$^{-1}$ with 50% CACC. Also, the RSU's communication range is up to 400 m upstream. They simulate trajectories of vehicles that are just manual traffic and the other simulation is 50% CACC. The one with 50% CACC has smoother traffic than the manual, although these graphs look sort of similar. Vehicle kilometers traveled (VKT), average travel time on the network (Avg.TT), and the number of collisions are performance indicators that show the effectiveness of a system.

On the basis of the data gathered by the authors, CACC has a better VKT as well as Avg.TT. Although there are higher collisions with the CACC example, this is assuming that the driver did not override the CACC. The merging assistant seems to be effective only for high mainstream demand and higher penetration level.

The downside to the merging assistant is that when the mainline is rather blocked and the speed is low, vehicles with the merging assistant slowly move toward the merging point and wait on the ramp vehicle. This is a problem because there was a sizable gap already for the vehicle to merge into. This algorithm has yet to be corrected and further research has to be done in possible improvements of the merging assistant, such as only activating the merging assistant at moderately high speeds or using a CACC to create a larger gap.

In summary, the merging assistant was used in hopes of having smoother traffic and less conflict between vehicles near the merging area. In this situation, mainline traffic was composed of CACC and ramp vehicles were manually operated. The algorithm proposed by the authors was meant to have CACC decelerate in such a way that it made ramp traffic be able to merge in smoothly with the mainline also without being within the safety zone. Throughout the paper the fundamentals of the merging assistant were discussed as well as the three main questions in designing it. In a mixed traffic example, scheduling, traffic to be involved, and how to create a gap, are the problems in which solutions have been found. Simulation of the mixed traffic in MATLAB shows that the merging assistant is only effective in certain traffic situations. The main problem that was noticed during these simulations is that a very large gap was created when the mainline traffic waited for the ramp vehicles because of the safety zone specifications. Further research will be done on generating gaps, implementation of the merging assistant, and multiple highway lanes.

## 17.4 Arterial Traffic Prediction

In [6] a system called Mobile Millennium, which was designed to process and broadcast live arterial traffic conditions using traffic data captured through GPS enabled cell phones, was used. Using these data, the authors compared and analyzed both the *Logistic Regression* and *Spatiotemporal Auto Regressive Moving Average* (STARMA) algorithms. Using simulation data obtained from a New York field study, the authors tested both of the algorithms' learning and interference components.

Traffic can have a severe impact on both environmental and economic aspects of not only the United States but also the world. Because of this, it is essential that governments be able to reliably predict traffic conditions. One major challenge is the lack of infrastructure available to gather the required data to be able to predict such traffic conditions on nonmajor highways. Using GPS smartphones, the Mobile Millennium system collects traffic data, analyzes those data, and then sends back the traffic conditions to the end users, allowing them to choose less congested routes.

Because the data being collected are coming from nonstationary devices, one of the major challenges of modeling the Mobile Millennium system is the fact that the system does not have access to flow counts. Because of this, the authors used two different statistical learning models to overcome the flaws of other models like analytical flow or queuing-based models. The first of the models that they looked at was the *logistic regression* model. This model used a clustering algorithm between discrete traffic congestion states. The second model they examined was the STARMA [6] model, which is a time series model in which the previous travel time is used to predict the next travel time.

Because GPS smartphones are capable of recording locations every few seconds, a lot of travel information can be gathered this way. Some issues arise from this, however, for example, the strain that this constant sampling puts on mobile battery life and the privacy concerns that go with monitoring a cell phone even with the user's permission. To overcome this, the authors devised a sampling method called *virtual trip lines* (VTLs), which are specific coordinates that trigger the smartphone to collect and send its location. Each VTL has two GPS coordinates that make a virtual line that runs down roads of interest.

The authors created an arterial network with a total of $N$ pairs of VTLs with each pair having a unique ID $i \in \{1,\dots,N\}$, with each set of pairs denoted by $V = \{1,\dots,N\}$. Edge nodes ($e_{ij}$) were defined when both $i_d$ and $j_u$ are at the same geographical location with the subscripts denoting the flow of the traffic. They then defined the set of all upstream and downstream VTL pairs with

$$N^1(j) = \{j\} \bigcup \{i \in Y : e_{ij} \in \mathcal{E}\} \bigcup \{k \in V : e_{jk} \in \mathcal{E}\}$$

Aggregating the travel time data at time intervals $k = 0, t, 2t,\ldots$

$$A_{k,i=\{X_{t_{m,i}}|(k-1)t \le t_m < kt\}}$$

Each VTL pair had a defined binary mode of either *uncongested* or *congested*.

Using the above formulas along with the logistic regression model to estimate the contestation state of a VTL pair $i$ at a time interval $k$ to get the conditional probability equation gives

$$\hat{Q}_{k,i} = E_{h_i,g_i}\left[Q_{k,i}|Z_{k,i}^{r,s}\right] = \mathbb{P}_{h_i,g_i}\left[Q_{k,i} = 1|Z_{k,i}^{r,s}\right]$$

They then used several statistical equations to derive the logarithmic conditional likelihood equation that follows:

$$\mathcal{L}\left(\beta_i; \{Q_{k,i}\}_{k=0}^K, \left\{Z_{k,i}^{r,s}\right\}_{k=0}^K\right) = \sum_{k=0}^K \left(Q_{k,i}\beta_i^T Z_{k,i}^{r,s} - \log\left[1 + \exp\left(\beta_i^T Z_{k,i}^{r,s}\right)\right]\right)$$

This equation was called the *maximum likelihood estimate* (MLE). This equation was used by the authors to learn the parameters of a previously unseen data set and then to validate those parameters against those of a known data set.

The other model that the authors examined was the STARMA model, which is an extremely efficient estimation model. The STARMA model was vectorized for all VTL pairs $i \in V$ as

$$Z_k = \sum_{j=k-r}^{k-1}\sum_{n=0}^{s} \beta^{j,n}\Phi^{(n)}(Z_j) - \sum_{j=k-p}^{k-1}\sum_{n=0}^{q} \alpha^{j,n}\Phi^{(n)}(\epsilon_j) + \epsilon_k$$

where $\Phi^{(n)}(\cdot) = \left(\varphi_1^{(n)}(\cdot),\ldots,\varphi_N^{(n)}(\cdot)\right)^T$ and $\epsilon_k = (\epsilon_{k,1},\ldots,\epsilon_{k,N})^T$. The following logarithm was then maximized to obtain the MLE:

$$\mathbb{P}\left(\{Z_k\}_{k=0}^{K-1}; A,B,\sigma^2\right) = (2\pi)^{-\frac{KN}{2}} \left|\sigma^2 I_{KN \times KN}\right|^{-\frac{1}{2}} \exp\left(-\frac{S(A,B)}{2\sigma^2}\right)$$

To test both methods, the authors implemented and tested them using both simulations and real-life experimental data, testing both the accuracy and the effect the penetration rate had on it. The authors used two separate data sets to test the methods.

The first data set used was one that was generated using simulation software called Paramics. The simulation consisted of 1961 nodes, 4426 links, and 210 zones and was based from roads in

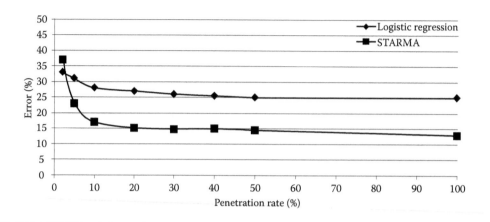

**Figure 17.2  Average forecast error versus penetration rate for all VTL pairs in the Paramics data set.**

Fresno, California. This subnetwork included a total of 380 links, with a total of 99 VTLs placed over them.

The second data set was gathered through the data collected via the Mobile Millennium launch project. This consisted of 20 drivers driving for 3 h on a 2.4-mi. loop in Manhattan. This experiment was then repeated two more times so that two of the runs could be used to train the models and the final run to then validate the results. To test the accuracy of both models, the data set was then divided into both a training set and a validation set. For the logistic regression model using a penetration of 5%, the study found that it could predict the average drive time forecast with an error percentage of approximately 30%. For the STARMA model with the same penetration percentage, they found that it could predict the travel time error percentage of approximately 22% (see Figure 17.2).

These data show that through the use of VTLs the authors were able to predict fairly accurately future travel times. This is important because it establishes a new method for data collection on arterial roads without the need of roadway sensors or loop detectors.

## 17.5  Road Traffic Delay Estimation

Traffic congestion is creating havoc in transportation networks in most of the civilized world, with congestion projected to double over the next decade. Tracking increased traffic on a GPS device could consume a lot of energy, which can deplete the battery on the device. There is also the issue associated with inaccurate position samples due to the limits in GPS technology when it comes to urban environments. This can be caused by what is called urban canyons, which are caused when being near tall buildings or tunnels, which prevent cell phones from having line of site with the satellites [7]. To overcome this, the authors presented a method that they called VTrack [8]. This model uses a *hidden Markov model* (HMM) to predict the travel of a vehicle over a city block level of the map.

The VTrack model does this by implementing a method that the authors dubbed *map matching*, where the system takes a position sample and matches it with the most likely position on the *map*. The system then uses the projected data of traffic flow to then relay to the user the best possible path. The authors used an extensive set of both GPS and WiFi location data that were gathered from 25 vehicles. They used a custom app along with in-vehicle computers to gather and transmit the data.

VTrack uses GPS and WiFi position sensors and runs a travel time estimation algorithm; how this works is it takes noisy position samples and uses those to determine the road the user is on, calculating the estimated travel time on roads. And the estimates made are then used to determine hotspots and route planning.

VTrack will not always be able to rely on GPS because of problems such as consuming too much cell phone battery power. With this known, VTrack can also operate using WiFi; point observations are converted into point estimations by a localization algorithm. The algorithm uses a database called wardrive that takes GPS coordinates and shows WiFi access points (APs). The sensors are given real-time data and fed to the estimation algorithm, which has two components: a map matcher and travel time estimator. The matcher determines what roads the user is driving on, and the estimator estimates travel times on road segments from the map-matched trajectory.

Hotspots are a piece of road where travel time is determined by the speed limit threshold. The goal is for users to view all hotspots within a given geographic range on a browser. This gives users the ability to avoid traffic at certain times in the day in which that road might be highly congested. In order to detect hotspots, two parameters have to be kept low: the miss rate and the false positive rate. The miss rate is the failed hotspots reported, and the false positive is the report of hotspots that actually are not hotspots. Real-time route planning looks to minimize travel time even through middrive. Errors ranging from about 10% to 15% are acceptable when dealing with route planning, basically at the high end 3–5 min error on about a 30-min drive.

To implement the VTrack method, the authors used a process called map matching. They chose this method because of its ability to use sparse or noisy samples and match them to a segment of roadway. When the cell phone is using GPS to capture the location data, it is able to report the position estimate directly to the system. But when the phone is using its WiFi sensor to determine its position, the sensor is only able to report back which APs it connected to, and then has to cross check the AP with a database of known AP location to determine where it is (Figure 17.3a).

To implement the map-matching algorithm, *outliers* were created to weed out data that were not physically possible to achieve. One such outlier would be if it broke a speed constraint that was set. The authors set a maximum speed threshold of $S_{outlier}$ = 200 mph, meaning that if it was found that a vehicle had traveled at a speed higher than that of $S_{outlier}$ from two set points, then the data would be thrown out because those speeds are just not possible.

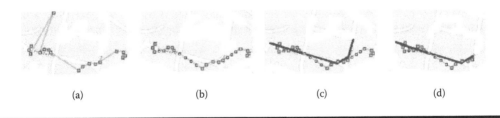

(a)          (b)          (c)          (d)

**Figure 17.3** **Process of map matching. (a) Raw trace. (b) Outlier removal and interpolation. (c) Viterbi matching. (d) Bad zone removal.** (Adapted from Thiagarajan, A. et al., VTrack: Accurate, energy-aware road traffic delay estimation using mobile phones. In *Proceedings of the 7th ACM Conference on Embedded Networked Sensor Systems (SenSys' 09)*. ACM, New York, p. 14, 2009.)

After removing the outliers the authors then focus on how to deal with outages in data. They defined an outage as a time period in which no usable data were collected. In [8] the authors used a scheme in which the system inserted interpolated points into the data table where outages were experienced. The algorithm gathered created these interpolated points by sampling at 1-s intervals along the segments that connected the last two known points, assuming that a constant speed was maintained during that time (Figure 17.3b).

When all the outliers had been removed and all the interpolation points applied, the authors then applied the Viterbi algorithm to then predict the most likely path the vehicle had traveled. The algorithm does this by computing the most likely sequence of road segments and outputting the predicted path that was taken (Figure 17.3c).

Finally, after the Viterbi algorithm had been applied, the authors then employed a method to remove all the zone that had been marked by the Viterbi algorithm as having a low probability of being the path taken by the vehicle (Figure 17.3d).

To guarantee that all road segments remain continuous, the authors developed a transition probability to reflect the following things: for a given road, there is a chance that the vehicle will still be on the road after a given time, a vehicle can only travel from one end of the road to the other by using the same intersections, and a vehicle cannot travel faster than $S_{outlier}$ on a road. To calculate this, the authors created the following probability distribution:

If $i = j$, $p = \varepsilon 2$. If $j$ does not start where $i$ ends, $p = 03$. If $j$ does start where $i$ ends, $p = \varepsilon$ or 0

with $p$ being the probability from a segment $I$ at sample $t$–1 to segment $j$ at sample $t$.

Overall, VTrack as presented in [8] used cell phones to accurately predict road travel time using different position samples, evaluated through hotspots and route planning. The issues with using VTrack were reducing energy consumption and travel time estimation.

## 17.6 Conclusions

Throughout this chapter we have discussed CPS technology that takes man to the next level of the cyber world, making people in different regions more connected than ever. CPSs were discussed where they can sense, actuate, and communicate in a wireless network, possibly having applications in the battlefield, traffic conditions, and medical areas. Another application for CPS is to predict travel times on arterial roads within urban environments.

A merging assistant was used for better traffic conditions owing to congestion near merging areas. Now there were some pitfalls due to dealing with the merging assistant because of all of the variables that need to be accounted for in traffic congestion. Overall, it seemed like a viable future solution to alleviating some of the traffic problems we currently have in the United States and abroad.

Engineers are trying to come up with solutions to try and solve this growing problem of traffic congestion. Not only have merging assistants been looked at but also phone apps that help direct drivers to spots with less congestion have been developed. Those new schemes can reduce the loss of life and improve the quality of living through reducing time traveled and improving air quality.

# References

1. Chellappan, S. and K. K. Madria. 2007. Networked automotive cyber physical systems: Applications, challenges and directions. *National Workshop on High-Confidence Automotive Cyber-Physical Systems*, Troy, MI, pp. 1–3.
2. Herring, R., A. Hofleitner, P. Abbeel, and A. Bayen. 2010. Estimating arterial traffic conditions using sparse probe data. *13th International IEEE Conference on Intelligent Transportation Systems*, Madeira Island, Portugal, pp. 1–8.
3. Herring, R., A. Holfleitner, S. Amin, T. A. Nasr, A. A. Khalek, and P. Abbeel. 2009. Using mobile phones to forecast arterial traffic through statistical learning. *89th Annual Meeting of the Transportation Research Board*, *59*, Washington, DC, pp. 1–22.
4. Kwon, J. and K. Murphy. 2000. Modeling freeway traffic with coupled HMMs. Technical report. University of California, Berkeley.
5. Pueboobpaphan, R., F. Liu, and B. Van Arem. 2010. The impacts of a communication based merging assistant on traffic flows of manual and equipped vehicles at an on-ramp using traffic flow simulation. *13th International IEEE Conference on Intelligent Transportation Systems*, Madeira Island, Portugal, pp. 1468–1473.
6. Russell, S. and P. Norvig. 1995. *Artificial Intelligence—A Modern Approach.* Prentice-Hall, Englewood Cliffs, NJ.
7. Sperling, D. and D. Gordon. 2009. *Two Billion Cars: Driving toward Sustainability.* Oxford University Press, New York.
8. Thiagarajan, A., L. Ravindranath, K. LaCurts, S. Madden, H. Balakrishnan, and S. Toledo. 2009. VTrack: Accurate, energy-aware road traffic delay estimation using mobile phones. In *Proceedings of the 7th ACM Conference on Embedded Networked Sensor Systems (SenSys' 09)*. ACM, New York, p. 14.

## Chapter 18

# Video Communications in Unmanned Aerial Vehicle-Based Cyber-Physical Systems

Mengcheng Guo, Fei Hu, Yang-ki Hong, Kenneth Ricks, and Jaber Abu-Qahouq

## Contents

## 18.1 Introduction

The unmanned aerial vehicle (UAV) is widely used in military and civilian applications. It can be used for field surveillance through the cameras deployed on the UAV body. However, when the UAVs are deployed in complicated radio signal conditions, it is a challenging issue to maintain the communication performance such as the throughput and time delays. Especially for the video transmission on UAVs, the task of maintaining high-quality video transmission is critical.

Before deploying a real UAV network, we built a UAV-based cyber-physical system (CPS) model via NS2. It has the following features: the UAVs can use spectrum sensors to detect any unused spectrum (i.e., unoccupied by the licensed primary users) in order to deliver their data via those idle spectrum bands, and the UAVs access the radio channels by following certain medium access control (MAC) protocols. Our goal is to avoid the interference with licensed users. We have also used the video evaluation metric (cyber part) to control the physical system's configuration and the performance as well.

We have designed a ferrite miniature antenna for the UAV system. Because video data are the major traffic in our UAV network, we have evaluated the video quality of experience (QoE) metrics under different antenna settings (by comparing our designed one with the existing ones). To detect the differences between the original and transmitted videos, we need to preprocess the video files to make them follow the identical format. We can then extract each video frame by following the time line. We have used the professional software measurement system unit (MSU) measurement tool. Our video analysis results show that our designed ferrite antenna has the best video performance in UAV systems.

## 18.2 UAV Video System Design

Today, UAVs have become a mature technology for environment monitoring [1]. To cover a large area, UAVs are often deployed to form a multihop wireless network. Our work targets high-quality

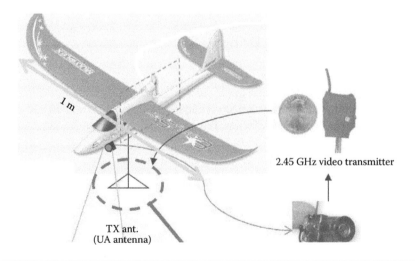

**Figure 18.1 Small UAV.**

video transmission in small UAVs (sUAVs) (Figure 18.1) that can send out video signals for 10 km away. However, such sUAVs (Figure 18.1) cannot send video for too long a distance owing to its limited antenna gain and power supply. Therefore, multihop (i.e., relaying data through hop-to-hop communications) radio frequency (RF) communications should be used to deliver video traffic. Each hop (between any two neighboring sUAVs) has a distance of less than 10 km. If two UAVs are more than 10 km away, one of them needs to increase antenna power in order to reach a long distance (which will shorten the battery life). Another motivation of using hop-to-hop relay communications is that mission-oriented sUAVs could be deployed in different heights in order to perform surveillance tasks with various viewing angles. Thus, a higher sUAV often needs to use lower sUAVs to help to relay its data (assuming video data).

The sUAV antenna needs characteristics such as circular or dual polarization, reduced visual signature, wide bandwidth, and directional polarization. These characteristics demand the development of novel magnetic materials and new antenna design, which would lead to reduction of antenna size and improvement of antenna performance. We thus developed a novel low-loss ferrite and designed and fabricated chip antenna (Figure 18.2) from ferrite. In order to test the antenna performance, we observe the video data transmission quality from an sUAV installed with our antenna. Especially, we target the following two types of measurement parameters:

1. *Video quality:* We investigate how the video QoE performance changes under the new designed antenna. We expect the video quality will be better than conventional sUAV antennas that do not have the same bandwidth and polarization directionality as ours. Regarding QoE metrics, there are two types of measurements: subjective quality measure and objective quality measure. Subjective quality measure is the best way to find quality of video because humans are the ultimate viewer. However, it requires a lot of manpower, which is costly and time consuming. It cannot be tested in real time and achieve automatic testing. Objective quality measure plays an important role in the video quality measure. Its mathematical models are easy to create and implement. There are three types of objective quality measures according to the availability of an original image with which a distorted video is to be compared with full reference that needs

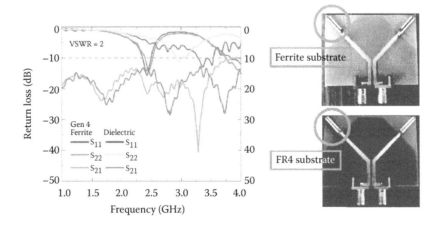

**Figure 18.2  Small UAV antenna.**

sender's video for comparisons, no reference (just relying on receiver's video data), and reduced reference that uses compressed source video as references. In our test system, we only focus on the full reference because it can offer offline analysis and it is the most accurate measure. We especially focus on the following six full reference quality metrics:

- Peak signal-to-noise ratio (PSNR): the largest SNR of a signal. A larger PSNR means less video distortion (compared to the original one).
- Modified sum of absolute differences (MSAD): the absolute difference between each pixel in the received video clip and the corresponding pixel in the original one. These differences are summed to create a simple metric of image similarity. Smaller MSAD means less distortion.
- Delta: the mean difference of the color components in the image points.
- Structural similarity index (SSIM): based on the measuring of three components (luminance, contrast, and structural similarity).
- Video quality metric (VQM): uses discrete cosine transform (DCT) to reflect human eye's perception level.
- MSU blurring: compares power of blurring of two images. If the value of the metric for the first picture is greater than for the second, it means the second picture is more blurred than the first picture.

2. *RF quality:* The impact of a ferrite antenna can be seen not only from the above mentioned video QoE measurements but also from the wireless communication performance measurement. This mainly includes network throughput (i.e., how many bits per second go through the entire sUAV network), packet loss rate (how many video packets are lost per second), network delay (how much time does it take to deliver a packet from a sender to a receiver?), and so on. We expect that the new antenna could improve packet arrival success rate and throughput.

Now the issue is, how do we measure the above two types of metrics through a set of feasible and cost-affordable methods, as well as through the corresponding hardware and software components? The easiest way is to wait for the ferrite antenna to be done and then test it in a real sUAV to see how video goes there. However, there are some realistic issues here: first, the antenna design is not a one-time business. It takes trial and error to come up with the best solution. For example, it is very difficult to lock the central frequency around 2.45 GHz and leave a ±0.25 G margin bandwidth. The antenna packaging cannot be done until everything is verified. Before antenna packaging and conductor insulation (to prevent energy leak), we need to treat the fragile antenna like a baby and should not risk installing it in the metal surface of the sUAV. It is necessary to test the antenna performance (including the above mentioned video/RF metrics) in a low-cost, easy-to-install ground device before we install it in the real sUAV. Second, the sUAV assembly and test is a time-consuming, high-cost activity. Each sUAV we use costs >$1000, not to mention hiring professionals to control its flight from the ground. It takes us significant time to select the right components to assemble an sUAV (if we do not use an assembled one, a ready-to-fly sUAV costs >$20,000 in many companies). While the antenna manufacturing takes generations to finalize, we cannot take the risk to always use sUAVs for each test because the sUAV can easily crash without delicate control. Also, some tests cannot be easily done in real sUAVs. For example, we do not have the budget to purchase dozens of sUAVs to test a multihop cognitive radio network.

Therefore, we propose to use a simulation–emulation–test bed (SET) based methodology to measure the sUAV video communication performance under our newly designed antenna. Here we define SET as follows:

1. *Simulation:* This refers to a purely software-based test without using any hardware units. Simulation perhaps has the lowest cost in our sUAV communication performance test. It avoids the risk of damaging expensive sUAVs, while providing accurate network protocol test. In our case, we choose NS2 as the simulation tool because of its well-modeled radio propagation conditions (such as multipath fading, Doppler effect, and radio diffraction). More importantly, it allows us to easily build any complex routing protocols above radio communications. This helps us to simulate a large-scale, multihop sUAV network.

   Note that here the simulation test also includes antenna software simulation before real antenna manufacturing. Antenna simulation could define antenna shape, printed circuit board for antenna control, and comprehensive antenna design parameters such as gain, central frequency, bandwidth, and so on.

2. *Emulation:* Instead of using software for all RF tests, emulation uses hardware in the physical layer test conditions. However, it is still not a real sUAV test. Emulation gives us a realistic radio communication environment due to its use of RF transceivers. In our case, we can easily replace the RF board's antenna with our designed one to see how our antenna enhances data communication performance (i.e., the above mentioned RF quality metrics). Because sUAVs are in a mobile environment, we could use remote control (RC) cars in a mobile communication scenario. We could also test antenna performance in different indoor/outdoor conditions.

   In our antenna test, besides the above mentioned RC car-based emulation, we also used embedded system boards (equipped with the new designed antenna) for the emulation in short-distance RF quality test.

3. *Test bed:* This test uses real sUAVs in the final video communication evaluation platform. Because the sUAV assembly (such as installing the antenna and video transceiver system in the sUAV) and pilot training all take a lot of time and expense, we wait until all the above simulation and emulation tests are done before conducting the sUAV test. The sUAV test bed gives us comprehensive video/RF quality measurement.

The above SET-based test methodology has the following advantages:

1. SET uses an easy-to-set up, "play-by-ear" test that fits our generation-by-generation antenna manufacturing style: We have deployed RC cars with real-time wireless video transmission system for video quality test. We also have ZigBee-compliant RF chips for a point-to-point communication test. Both of them allow quick antenna replacement in the wireless chip and thus can be conveniently taken to an indoor or indoor environment for tests.

2. SET uses simulations for proof-of-concept test of large-scale sUAV communications. We realize that it is too costly to use many sUAVs for cognitive radio test. An NS2-based accurate mobile ad hoc network (MANET) simulation could reflect the routing efficiency of our proposed ripple-cluster topology architecture. Such an architecture fits an sUAV network scenario when the sUAVs in a higher position send video data to lower sUAVs and then those video data eventually reach a ground station.

Figure 18.3 shows our SET-based sUAV video test principles and process.

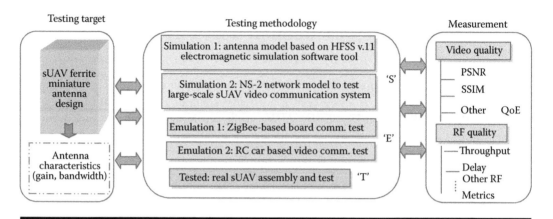

**Figure 18.3  SET-based methodology.**

## 18.3  Emulation and Measurement Design

Emulation uses emulated hardware (not real UAV) to test the antenna performance. It avoids the use of an expensive sUAV by emulating point-to-point, mobility-aware RF communications. We propose to use two approaches to emulate sUAV communications: one is based on ZigBee-compliant embedded system boards that allow the convenient replacement of the antenna; the other is to use an RC car for a longer distance RF link test.

### 18.3.1  Emulation Platform 1: Use ZigBee System as the Antenna Testing Platform

Our ultimate goal is to use a real sUAV to test long distance (>5 mi.) wireless communication performance. In the early stage, while we were assembling the UAV and wireless video system hardware components, we had already made an initial antenna prototype and aimed to test its performance to see if it is at least comparable or even better than the existing one in our ZigBee-compliant board that also works at 2.45 GHz. The reason of using a ZigBee board is it also uses 2.4-GHz central frequency. More importantly, its portability and embedded system architecture are very similar to the system installed in the real sUAVs.

The setup is shown in Figure 18.4. Its testing platform includes the following hardware components:

1. In the sender side, it has a laptop to generate source data, and a Texas Instruments (TI) RF transmitter with smartRF04EB and CC2430 board. They use a USB cable to connect each other. A laptop in the sender side allows the sender to move around in order to test the antenna performance in different distances.
2. The receiver side includes a TI RF receiver (the same board as the sender side) and a PC-based data display. They again use a USB interface. The reason for choosing a TI radio board is its antenna design represents one of the best radio sensitivity performances among ZigBee radio transceiver products.

During our antenna test, we first use TI's original antenna to send out data. Then we replace it with our designed antenna (see Figure 18.5) and test the receiver's signal quality. We focus on

ZigBee radio transceiver products

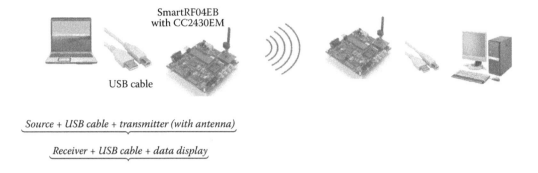

SmartRF04EB
with CC2430EM

USB cable

Source + USB cable + transmitter (with antenna)

Receiver + USB cable + data display

**Figure 18.4   ZigBee-based emulation setup.**

**Figure 18.5   Antenna replacement.**

the following two performance metrics during the wireless communication test under different antennas:

1. *Received signal strength indication (RSSI):* Typically, RSSI in the receiver side is determined by sender antenna power $P_S$, sender antenna gain $G_S$, receiver antenna gain $G_R$, sender–receiver distance $d$, and other factors (such as path loss ratio $\rho$), as follows:

$$P_R \propto \frac{P_S G_S G_R}{d^\rho}$$

$$RSSI = 10 \log(P_R) \text{ dB}$$

A better antenna can achieve higher gain and make a receiver's signal quality better.

2. *Packet error rate (PER):* This metric is the percentage of the number of damaged packets (due to wireless transmission) and the number of total received packets. A lower PER indicates a better antenna transmission performance.

We have also used C++ to build software to implement the real-time video stream system. It includes the following steps: (1) Capture and store the video frames with original data from web-cam, (2) use H.264 AVC to encode the captured video, (3) encapsulate the encoded H.264 video data with RTP packets, (4) transmit video data by different antennas over the wireless environment, (5) receive video data by different antennas, (6) depacketize the received data into video frames, (7) use H.264 AVC to decode the encoded video frames, (8) display and store the video frames, and (9) evaluate the received video quality with different antennas.

All these steps can be done in real time and in digital video format, which is more convenient than an analog UAV system in terms of evaluating the received video quality by different anten-nae. Figure 18.6 shows the framework of such a real-time video testing system. All parts of this system were programmed by C++ and Microsoft Foundation Class except the wireless network part that comes from the company setups. We can observe the video quality at the receiver with different antennas. We can easily replace the antenna in the wireless adapter with our antenna or other antennae. We can also collect the network statistics to evaluate the network quality of service. In our system, we can store both the original video data from the sender and the received video data in the receiver's memory because we know both video files. This benefit allows us to use the "full reference" to evaluate the video quality under different antennas.

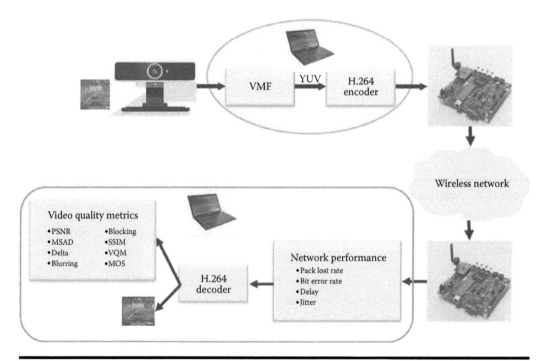

**Figure 18.6  Flow and framework of type one.**

### 18.3.2 Emulation 2: RC Car-Based Antenna Test

In order to emulate sUAV's mobility nature, we have designed an RC car-based video transmission system. The onboard and ground-station system integration is complete (they are platform independent). Our particular design utilizes linear power regulators that result in excessive heat generation. Version 2 of this board, which is in progress, will make use of switching regulators, thereby reducing the heat dissipation. A RC car platform to carry an onboard system for preliminary ground testing was built to support the onboard video system under test. Figure 18.7 shows our designed RC car platform.

On the basis of the software and hardware experiments we have developed the UAV video system in the physical world. The UAV platform that was chosen is a Raven-like UAV platform available from Nitroplanes (http://www.nitroplanes.com/projet-003-rq11.html), shown in Figure 18.8. The RQ-11 platform can be launched by hand, powered by an electric motor, and remote controlled via a handheld control console. The specifications of the RQ-11 are as follows: wingspan, 1300 mm; length, 1660 mm; weight, 900–1000 g; wing area, 30 $dm_2$; wing loading, 30–34 g $dm_2$. The biggest concern is its limit on the carried weight. We have reduced each possible component's weight by carefully choosing video systems.

In our video system, we use similar architecture as the ones in the RC car case. However, we have changed some parts to meet power and weight requirements. In the sender-side video system, because the battery voltage can vary in the range of 7–13.6 V and the required camera voltage is

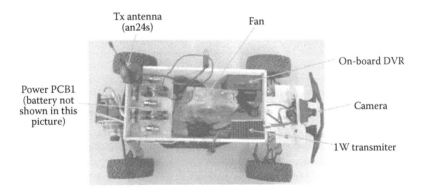

**Figure 18.7    RC car architecture and appearance.**

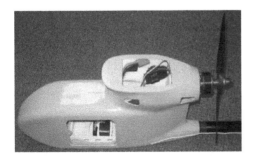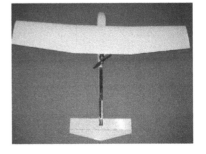

**Figure 18.8    (Left) RQ-11s UAV with installed video systems and (right) UAV appearance.**

12 V, a power converter that operates both step-up and step-down functions is needed. On the basis of this, we have designed a converter circuit because it is better than a buck-boost circuit for the system specifications. The transmitter we received requires 6 V, so we have designed a buck-type power converter circuit. The antenna will be swapped between the reference blade antenna and our new designed AL-Tenna. An optional video recorder may be used.

In order to plug the above antenna configuration into our UAV network simulation scenario, we have programmed the lower layer of the NS2 simulation tool to reflect the most important antenna specifications, such as central frequency 2.45 GHz, effective bandwidth 0.5 GHz around the central frequency, dual-polarization structure, antenna gain, and other parameters. Assume our antenna is equipped to hundreds of UAVs that form a MANET. Owing to different surveillance mission requirements, the UAVs may be deployed at different heights to capture real-time video in different viewing angles. A UAV located at a higher level needs to use lower-level UAVs to help to relay its video data until eventually reaching a ground station. Note that it will take much energy consumption and transmission delay if each UAV directly communicates with the ground station.

Now the issue is, how do we find a best routing path from any UAV at any height level to a ground station? A natural solution is to ask each UAV to use current MANET routing protocols such as AODV to randomly select a neighbor to relay its data. However, those random path discovery schemes do not utilize the special features of UAV communications: for a mission-oriented surveillance application, typically, UAVs group together if they belong to the same mission (such as monitoring a specific region). We could regard those special groups as "clusters." For each cluster, we could select one UAV as the cluster head (CH) that aggregates all cluster members' video data and forwards to the next CH. We have noticed that UAVs typically fly in certain height levels based on their camera viewing angles and resolutions. Therefore, we could regard each height level as a "ripple" and ask the upstream ripple CHs to send data to next level ripple CHs.

## 18.4 Video Analysis under Different Antennas in the UAV Systems

Video quality is defined to quantify the changing of the frames in the videos transmitted or processed by all the different systems. Usually, the quality is declined after the procedures because of the distortion of the video signals. The video analysis is important to evaluate the video quality and provide an accurate metric for future operation and system control.

Video systems have been developed from analog to digital platforms nowadays. At the beginning phase of the analog video systems, there were some traditional measurement signals transmitted beforehand for the computation of the frequency response and making sure that the system was working well in video transmissions. However, today's digital videos have used different metrics and methods for evaluation purposes. The performance of the digital video processing system is significant depending on the video's dynamic characters such as the motions or space details.

There are two types of methods to evaluate the videos. One type is objective evaluation, and the other type is based on the subjective methods. The objective video evaluation technology is to use the mathematic models to simulate the subjective views of the video to get the quantity values of the pictures. They are usually automatically finished by the computer according to the standards and indexes. The evaluation method is to compare the original and processed videos. The comparisons are divided into three categories: full reference, partial reference, and no reference. Full reference method is based on the difference of every pixel in the original and processed videos. Partial reference method focuses on the features of interest in both videos. Full reference and partial reference are both utilized under the condition when the original videos are available. In the case when the video codec is known we can use the no reference method to evaluate the videos without the original video.

The traditional methods are to calculate the SNR and PSNR differences. PSNR is the most widely used objective video evaluation metric. Whereas considering the nonlinearity of the human vision system, sometimes the PSNR value is much different from human vision. Recently, some more complex and accurate indexes are proposed, like UQI, VQM, and SSIM.

An objective VQM's performance is based on the correlation relationship of the subjective evaluation results. The most used correlation parameters are correlation coefficient, kurtosis, and Cohen's kappa coefficient.

To evaluate a video codec's quality, all kinds of metrics should be used repeatedly to evaluate the accurate video quality. Such a method will be very time consuming and complicated.

The subjective video quality evaluation methods collect the opinions from the audiences and the experts. The mean opinion score (MOS) is for the average opinions of the audiences. Different people could give opposing opinions on the same video.

In the future, video analysis will develop the technology that can be used in the computer graph vision area [2]. On the basis of the detection of the actions of objects in the video, the system could generate an alarm and trigger the programs or devices to respond and adopt proper measures.

Video analysis technology is a branch of bionics and artificial intelligence that mimics the animal functions to apply artificial intelligent algorithms that use the statistics results to update the knowledge to improve the actions for the future.

Artificial intelligent video analysis is inspired by the structure of human eyes to build a basic flow, that is, sampling, preprocessing, and processing actions. Human eyes can project realistic pictures to the brain. At this point in time the pictures generated are compound figures that are the clear focused central part and the unclear surrounding parts. Rather than processing a whole picture, the brain will remove the part that is not clear and focus on the clear part. Then more details could be extracted after the second analysis. These features will be used to make decisions and subsequent actions. Our UAV video system will only use objective metrics to evaluate the videos.

When the video stream is transmitted in the UVA network, we will focus on the following major objective video QoE metrics [3]:

## 18.4.1 Peak Signal-to-Noise Ratio

$$PSNR = 10\log_{10} \frac{\text{MaxErr}^2 wh}{\sum\limits_{i=0, j=0}^{w,h} (x_{ij} - y_{ij})^2}$$

Here, MaxErr is the maximum possible absolute value of color components difference, $w$ is the video width, $h$ is the video height, and $x$, $y$ are pixel values. Typical values for the PSNR in lossy image and video compression are between 30 and 50 dB, where the higher value is better. Acceptable values for wireless transmission quality loss are considered to be about 20–25 dB.

Figure 18.9 illustrates the PSNR concept.

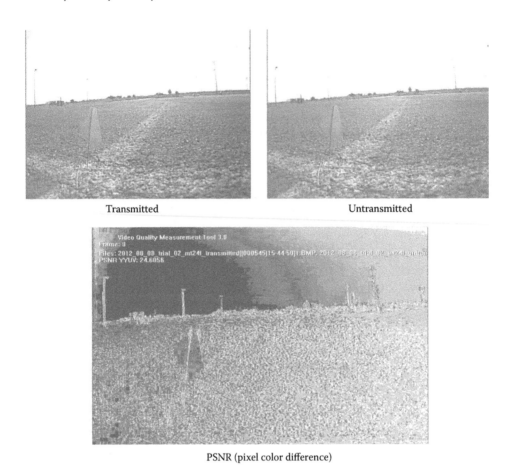

Transmitted        Untransmitted

PSNR (pixel color difference)

**Figure 18.9  PSNR concept.**

### 18.4.2 Modified Sum of Absolute Differences

$$d(x, y) = \frac{\sum_{i=1, j=1}^{w, h} | x_{ij} - y_{ij} |}{wh}$$

The value of this metric is the mean absolute difference of the color components in the correspondent points of image (Figure 18.10). Obviously, a smaller modified sum of absolute differences (MSAD) means the source and received images are more similar.

### 18.4.3 Delta

It is very similar to MSAD except for using the original pixel differences without using absolute values. It is sensitive to different colors (Figure 18.11).

**Figure 18.10    Modified sum of absolute differences.**

**Figure 18.11    Delta sample.**

## 18.4.4 *Structural Similarity Index*

If we consider one of the signals to have perfect quality, then the similarity measure can serve as a quantitative measurement of the quality of the second signal. The system separates the task of similarity measurement into three comparisons: luminance, contrast, and structure. First, the luminance of each signal is compared. Second, we remove the mean intensity from the signal. Third, the signal is normalized (divided) by its own standard deviation, so that the two signals being compared have unit standard deviation. Finally, the three components are combined to yield an overall similarity measure. Figure 18.12 shows an example.

## 18.4.5 *Video Quality Metric*

The VQM is a standardized method of objectively measuring video quality that closely predicts the subjective quality ratings that would be obtained from a panel of human viewers. It measures the perceptual effects of video impairments including blurring, jerky/unnatural motion, global noise, block distortion, and color distortion, and it combines them into a single metric. This metric uses

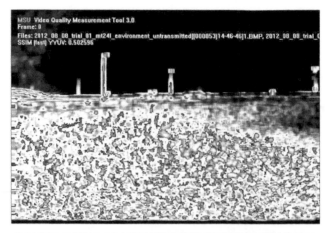

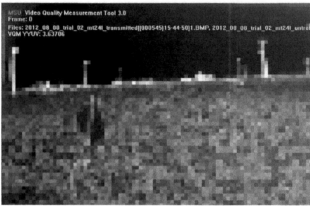

**Figure 18.12 (Top) SSIM sample and (bottom) VQM sample (luminance and contrast difference).**

the existing discrete cosine transform coefficients in image processing. VQM has a high correlation with subjective video quality assessment and has been adopted by American National Standards Institute as an objective video quality standard [4]. Figure 18.12 shows an example.

## 18.5 Field Test Setup and Measurement Results Analysis

Before the test of the UAV video system, we first use the digital ZigBee and WiFi system to test our antenna and other circuit devices indoors and in the field. For the reliable wireless communication links we install the video system on the RC car and evaluate the videos. Finally, we set up the entire video system on the UAV. On the evaluation results we can control the system performance every time.

### 18.5.1 ZigBee and WiFi Digital Wireless Test

Our testing hardware equipment consists of the following units: (1) Alfa AWUS-036H 1-W WiFi adaptor, (2) original 5-dBi antenna, (3) enhanced 9-dBi antenna, (4) TI CC2430 ZigBee EMK communication board, and (5) WLIB-BG24-1000 Amplus.

The testing environment includes an indoor hallway with WiFi signals as well as outdoor parking lot and a silent park area (Figure 18.13).

**Figure 18.13  Outdoor test environment.**

From Table 18.1 and Figure 18.14, we can see that our designed miniature antenna (UA) has better performance than another popular one, dipole, which has a bigger size than ours.

In addition, we have also done bandwidth and reliability test (see Figure 18.15). From the test we reach the goal we expected. Our system under the high interference environment can still get up to 9 Mbps to transport control protocol and stable 3 Mbps on user datagram protocol, which are more than enough for normal video transmission. For a 2.4G system the ability of overcoming noise and interference is more important than power level at our test.

**Table 18.1  Outdoor Performance of Unmanned Aerial Antenna (Gen2)**

| Distance (m) | Bandwidth (Mb) | | Antenna 2 | | | | |
|---|---|---|---|---|---|---|---|
| | | Time (s) | Throughput (Mb) | Throughput (Mb/s) | Jitter (ms) | Packet | Loss |
| 50 | 1 | 30 | 3.73 | 1.04 | 0 | 0/2662 | 0 |
| | 2 | 30 | 7.12 | 1.99 | 0.01 | 0/5082 | 0 |
| | 5 | 30 | 17.2 | 4.8 | 0 | 0/12238 | 0 |
| 100 | 1 | 30 | 3.39 | 946 | 0 | 14/2431 | 0.58 |
| | 2 | 30 | 7.14 | 2 | 4.983 | 0/5095 | 0 |
| | 5 | 30 | 17.3 | 4.83 | 0 | 13/12237 | 0.11 |
| 150 | 1 | 30 | 3.47 | 969 | 2.139 | 16/2491 | 0.64 |
| | 2 | 30 | 7.1 | 1.98 | 0.942 | 0/5062 | 0 |
| | 5 | 30 | 17.3 | 4.84 | 0.321 | 13/12354 | 0.11 |
| 200 | 1 | 30 | 3.49 | 976 | 0.001 | 11/2501 | 0.44 |
| | 2 | 30 | 7.11 | 1.99 | 0 | 0/5069 | 0 |
| | 5 | 30 | 17.9 | 5 | 0 | 0/12745 | 0 |
| 300 | 1 | 30 | 2.26 | 0.631 | 12.6 | 13/1626 | 0.008 |
| | 2 | 30 | 2.57 | 0.717 | 12.28 | 8/1841 | 0.43 |
| | 5 | 30 | 1.21 | 0.337 | 27.203 | 16/878 | 1.8 |

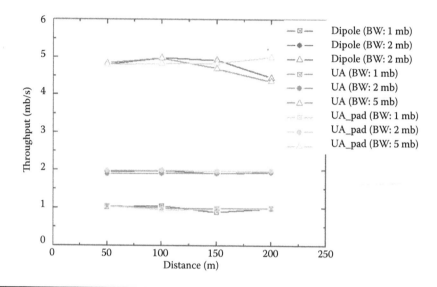

**Figure 18.14  Our antenna (unmanned aerial) performance is superior to others (dipole).**

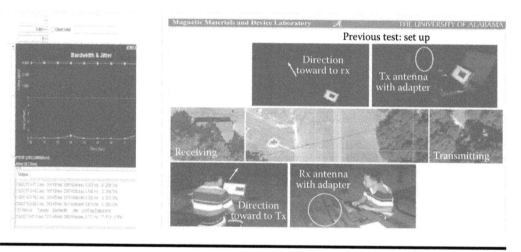

**Figure 18.15  Bandwidth and reliability test.**

## 18.5.2 RC Car Video Test

In the RC car test, our main goal is to test the video system on a RC car with the 2.45G communication board, and test the connection effect for different antennas. The results of these tests will give us a good overview of the performance of our new system in different perspectives and prepare us for the final UAV test.

Experiment setup and environment are as follows: latitude, 33.274236; longitude, -87.533899; location: Sokol Park North Parking Lot; distance: 200 m; rectangle trace; weather, clear; radio environment, low interference.

Figure 18.16 shows the PSNR trends when using our antenna in the RC car system. The average PSNR is 20 dB.

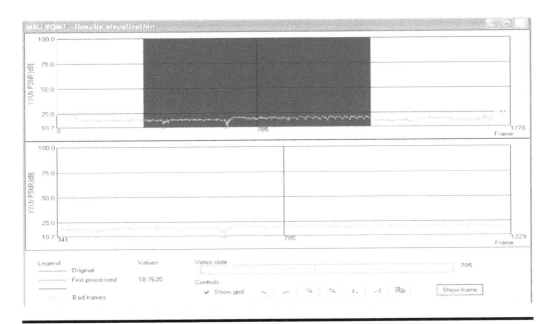

**Figure 18.16    Performance of around 20 dB PSNR in RC car test.**

**Figure 18.17    Bad frame samples.**

**Table 18.2    Video Performance in Different Trials for RC Car Case**

|       | Trail 1   | Trail 2 | Trail 3 | Trail 4 | Trail 5 | Trail 6 | Trail 7 |
|-------|-----------|---------|---------|---------|---------|---------|---------|
| PSNR  | 24.21423  | 24.0341 | 23.4017 | 23.9135 | 23.0193 | 19.732  | 23.2118 |
| SIMM  | 0.79      | 0.81    | 0.74    | 0.71    | 0.63    | 0.57    | 0.66    |
| VQM   | 3.6371    | 4.8819  | 3.1423  | 2.7998  | 3.38    | 3.6312  | 3.4238  |
| Delta | −8.3419   | 2.8893  | 0.20157 | 3.3854  | 2.8445  | −1.2782 | −1.8873 |
| MSAD  | 11.0563   | 4.9612  | 6.2898  | 15.411  | 9.3425  | 11.949  | 10.44   |

The WiFi interference could have a certain effect on the video transmission. Figure 18.17 illustrates an example.

Table 18.2 lists the video performance in different tests. We have found that some factors including the reflection of the ground, engine vibration, circuit power splitter, and the antenna directions could damage the video frames.

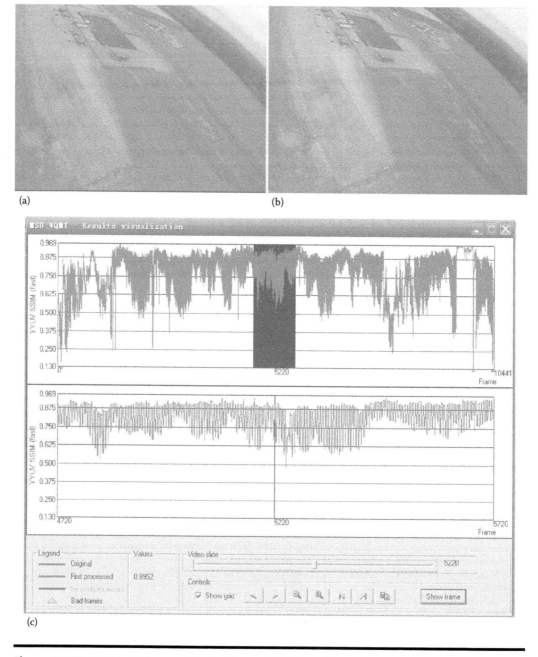

**Figure 18.18 UAV testing results (without WiFi interferences). (a) Untransmitted frame, (b) transmitted video frame, (c) UAV video performance (SSIM) without interferences.**

### *18.5.3 UAV Video Test*

The above RC car testing helps us to be prepared for the real UAV test. Figure 18.18a and b shows that the UAV with our designed ferrite antenna can well transmit the video frames after we remove some interference factors: such as making sure there is no WiFi nearby; there is a light of sight between the sender and the receiver. The video SSIM performance also verifies the good performance for the received video: the average SSIM is around 80%.

Figure 18.19 shows the delta performance. It matches with the PSNR color difference well.

Figure 18.20 shows a case with WiFi interference nearby. It shows that from time to time the PSNR could be low. The grass frame is not so clear in such a case.

Figure 18.21 shows that even with WiFi interference, most video frames have an average PSNR nearly 30 dB, which is acceptable for video applications. Generally only when PSNR <20 dB will the video become unclear.

Figure 18.22 shows a case called picture shifting. This means that different parts of a video frame could be distorted (i.e., misplaced) because of wireless sending jitters. The relative geometric positions could not be maintained in such cases. Our video analysis software can capture such a bad frame, which has low PSNR.

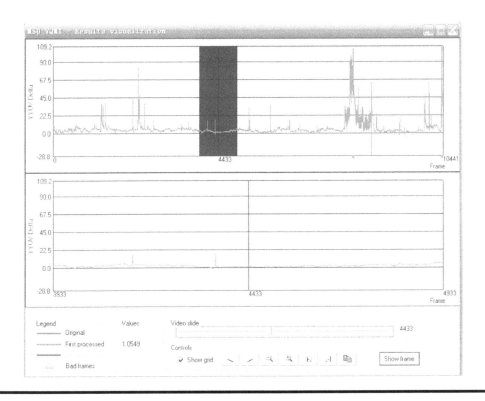

**Figure 18.19 Delta metric is matched with the PSNR color difference.**

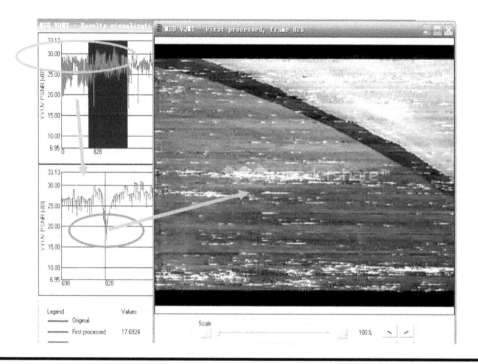

**Figure 18.20  Low-quality video frames with wireless interference (frequency hopping).**

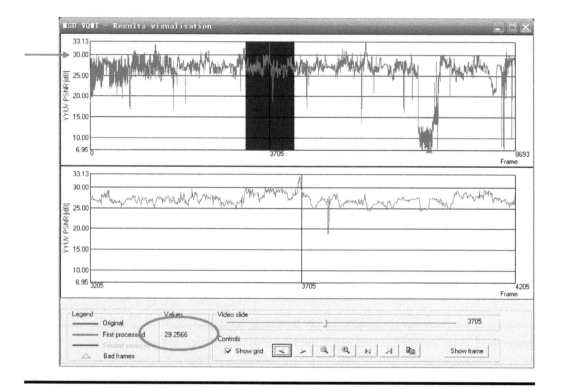

**Figure 18.21  PSNR is larger than 30 dB for most of the transmitted video frames.**

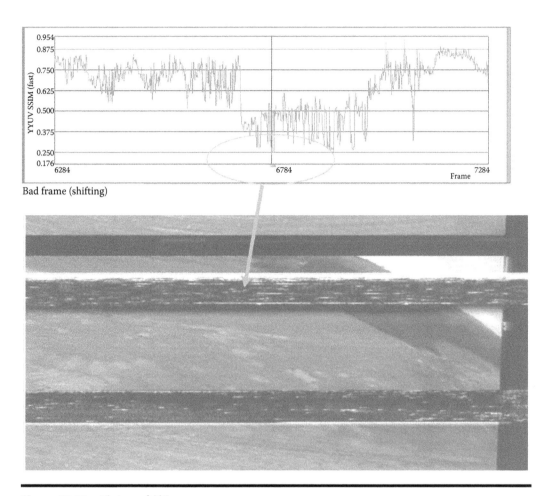

**Figure 18.22   Picture shifting case.**

Figure 18.23 shows that the video quality could be poor (PSNR < 14 dB) when the UAV consistently opens its motor, which causes much interference to the wireless video transmission system.

Table 18.3 shows that our designed antenna can ensure an average of >26 dB PSNR, which is acceptable for video applications. The SSIM is >77%, which is a good video quality maintenance. The VQM is less than 4.2, which is a small distortion, and the delta performance is lower than 7.5, which means that the color difference is small.

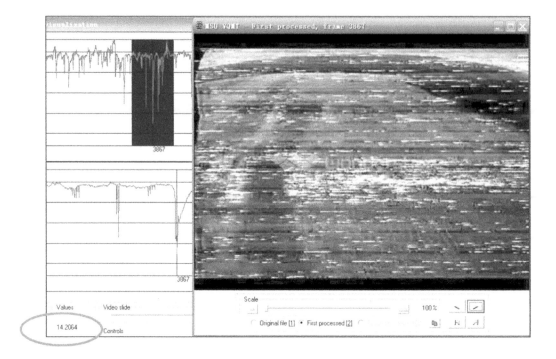

**Figure 18.23　Impacts of UAV motor.**

**Table 18.3　Antenna Field Test (Video Metric Results)**

|        | *PSNR*    | *SSIM*  | *VQM*   | *Delta* |
|--------|-----------|---------|---------|---------|
| Test 1 | 26.81791  | 0.77737 | 3.93877 | 6.77613 |
| Test 2 | 27.16919  | 0.78145 | 4.19370 | 5.62900 |
| Test 3 | 26.95242  | 0.77301 | 3.99883 | 4.89798 |
| Test 4 | 28.33111  | 0.87641 | 3.12849 | 7.49169 |

## 18.6　Conclusions

In summary, as shown in the flow chart in Figure 18.24, we have followed the CPS modeling procedure from the theory, software simulation, and video evaluation stages (in the cyber world) to hardware experiments and practice applications, that is, the UAV video system (in the physical world). Our results showed that our new designed ferrite miniature antenna has satisfactory video transmission performance.

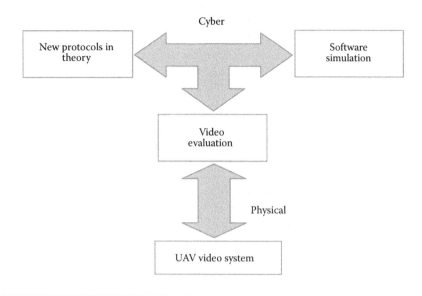

**Figure 18.24  CPS modeling process: in cognitive radio networks (CRN)-based UAV communication case.**

## Acknowledgment

This work is supported by US Army Research Laboratory, #W911QX-11-C-0017. All ideas reported here do not necessarily reflect the sponsor's opinions.

## References

1. I. J. Mitola. 1993. Software radios: Survey, critical evaluation and future directions. *IEEE Aerosp. Electron. Syst. Mag.* 8, 25–36.
2. Bellsent. Home page. ShenZhen, China. Available at http://www.bellsent.com.
3. Y. Wang. 2006. Survey of objective video quality measurements. EMC Corporation, Hopkinton, MA. Available at ftp://ftp.cs.wpi.edu/pub/techreports/pdf/06-02.pdf.
4. S. Winkler. 2005. *Digital Video Quality: Vision Models and Metrics*. John Wiley, Hoboken, NJ.

# HEALTH CARE CYBER-PHYSICAL SYSTEM APPLICATIONS

## Chapter 19

# Cyber-Physical Medication Systems and Devices to Improve Health Care

Bryant Grace, Brock Bennett, and Fei Hu

## Contents

## 19.1 Introduction

A cyber-physical medical system refers to a communication system between patients, medical devices, and caregivers. In a cyber-physical medical system, medical devices can interact with one another to better suit the patient's needs. Also, a caregiver will be able to view real-time patient information. The benefits a patient will receive from a cyber-physical medical system are obvious.

The concept of cyber-physical medical systems is close to that of sensor networks or robotics. However, each component is usually initially a stand-alone device. Although research into full-fledged cyber-physical medical systems is fairly new, many years of research has gone into embedded systems and sensor networks.

The motivation for writing about the importance of cyber-physical medical systems comes from the potential use they can have for patient care. This chapter will explain the major effects that cyber-physical medical systems will give the medical community. They will not only help patients significantly but will also help caregivers. Nurses and doctors will not need to manually adjust a medical device or constantly read a device. Instead, all of this will be done autonomously and that information will stream to a database for the caregiver to view. They will give all sorts of benefits, including interoperability and patient safety. The significance that cyber-physical medical systems will have on the medical system is great and should receive much attention from caregivers and researchers alike.

## 19.2 Cyber-Physical Medication Systems

### 19.2.1 Basics of Implementing Cyber-Physical Medication Systems

Medical devices can range anywhere from reactive and embedded systems such as pacemakers to stand-alone medical systems such as medication dispensers. These are real-time medical devices with specific requirements such as timing and safety. There are several devices in place in hospitals that are used to monitor a patient's health. To control and monitor these devices, they are often connected to remote networks so that they can be distantly controlled. However, monitoring or adjusting these devices from a remote location such as a nurse's office is typically not allowed for safety reasons. Figure 19.1 shows an example of an embedded medical system with the control component *D* and the patient being monitored.

There are a number of issues surrounding the implementation of cyber-physical medical and medication systems that allow monitoring and performing functions from a remote location. For instance, to accomplish the reliability necessary for cyber-physical medical and medication systems to be remotely controlled, there must be time syncing and confirmation technologies put in place in these devices. Although remote controlling technology does not currently exist in most cyber-physical medical and medication systems, there is a way to apply these real-time synchronous technologies to these medical systems. A prime example of a medical system that should be able to utilize technology

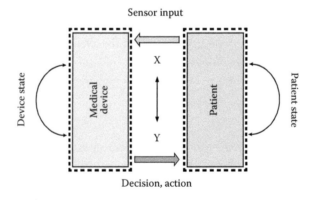

**FIGURE 19.1   Embedded medical system.**

to be remotely controlled is a breathing ventilator system. Every patient that is using a ventilator system to assist in breathing has different capabilities and conditions that change every day. A doctor should be able to monitor and adjust the ventilator according to the patients' conditions. The patient on the ventilator is the environment being controlled while the ventilator system is the embedded control component. The ventilator is connected to the patient through an endotracheal tube that connects to the patient's larynx. There is an embedded system in the ventilator machine that controls and monitors several aspects of a patient's health including airflow and lung pressure. Also, this embedded system monitors system-level parameters such as power, mechanical failure, and air leakage. As the ventilator receives sensor input from the patient, the state of the ventilator will change accordingly. Then, the ventilator will be adjusted and the action of the ventilator will satisfy the patient.

Currently, the doctor or nurse will adjust all output parameters of the ventilator system by hand. With this current system in place, the ventilator system is not able to adapt automatically to the patient. If the patient's breathing capabilities improve because of medication or healing, the ventilator will be on the same settings. The patient will simply have to wait for the nurse to change the settings of the ventilator system to adapt to the patient's health. If a cyber-physical medical system is in place, the nurse can monitor and adjust ventilator settings safely from a remote location. This way the ventilator can adapt almost immediately to the patient's health. Also, the nurse will not need to travel from room to room adjusting ventilator systems. This can all be done safely in one room from a remote location. However, in order for this to be a practical solution, we must adapt and perfect these real-time synchronous technologies into our medical systems. These technologies have great potential to improve our health care system as well as patients' health and safety [1].

High-confidence medical cyber-physical systems (MCPS) are systems that rely on software to deliver new functionality to hospitals and medical practices. Wider use of network connectivity in MCPS and the high demand for continuous patient monitoring bring great challenges into the process of MCPS development. The medical device industry is undergoing a rapid transformation, embracing the potential of embedded software and network connectivity. Instead of stand-alone devices that can be designed and used to treat patients independently of each other, new distributed systems will be used that simultaneously control multiple aspects of the patient's physiology. Just like in the example above, these embedded systems can be used to monitor heart rate, breathing patterns, and other vital regulatory processes.

Traditionally, most clinical scenarios have a caregiver, and often more than one, controlling the process. For example, an anesthesiologist monitors sedation of a patient during an operation and decides when an action needs to be taken to adjust the flow of sedative. There is a concern in the medical community that such reliance on a "human in the loop" may compromise patient safety. Caregivers, who are often overworked and operate under severe time pressure, may miss a critical warning sign. Nurses typically care for multiple patients at a time and can be distracted at the wrong moment. Using an automatic controller to provide continuous monitoring of the patient state and handling of routine situations would be a big relief to the caregiver and can improve patient care and safety. Although the computer will probably never replace the caregiver completely, it can significantly reduce the workload, demanding the caregiver's attention only when something out of the ordinary happens. Scenarios based on physiological closed-loop control have been used in the medical device industry for some time. However, their application has been mostly limited to implantable devices that cover relatively well understood body organs, such as the heart in the case of pacemakers and defibrillators. Implementing closed-loop scenarios in distributed medical device systems is a relatively new idea that has not made its way to the mainstream practice. A clinical scenario that can easily benefit from the closed-loop approach is patient-controlled analgesia (PCA) (Figure 19.2). PCA infusion pumps are commonly used to deliver opioids for pain

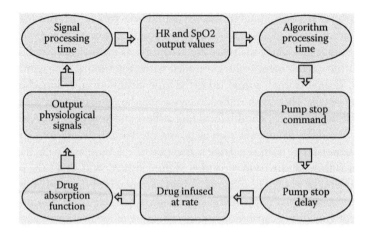

**FIGURE 19.2 System control loop for PCA.**

management, for instance, after surgery. Patients have very different reactions to medications and require very different dosages and delivery schedules. PCA pumps give the patient a button to press to request a dose when they decide they want it rather than using a schedule fixed by a caregiver. Some patients may decide they prefer a higher level of pain to the nausea the drugs may cause and can press the button less often, while patients who need a higher dose can press it more often. A major problem with opioid medications, in general, is that an excessive dose can cause respiratory failure. A properly programmed PCA system should not allow an overdose because it is programmed with limits on how many doses it will deliver, regardless of how often the button is pushed. However, this safety mechanism is not sufficient to protect all patients [2].

## 19.2.2 Research to Improve Cyber-Physical Medical Systems

The research that goes into developing medical CPSs is taking on a new direction. There are many challenges involved when building a medical CPS. The goal in mind when designing medical CPSs is to create systems that are both safe and effective. The research that goes into designing and implementing new medical CPS require meeting several important challenges that can be broken down into several important points. The first is high guarantee software. The role of the firmware inside embedded medical devices has become more important than ever. Many of the things that were previously implemented in the hardware of medical devices are now being implemented digitally. The next is state of context awareness. It is important that the nurse or doctor know immediately when the state of the device changes. This can be crucial for a patient suffering a heart attack or relying on a ventilator to breathe. The next is autonomy of medical devices. Increasing autonomous capabilities can mean that devices do not need to be manually changed but will automatically update on the basis of the current state of the patient. The security and privacy of medical CPSs is another point of high research interest. Medical records of patients should be kept confidential. Tampering with medical records can be crucial to a patient's health. And finally, verifiability is the last point of research interest. There needs to be a safe and reliable way to prove the dependence on medical devices. With all of these points of high interest of research highlighted, it can be seen that there are various challenges when implementing medical CPSs. Of the research points stated above, the most significant is the dependence of software in embedded medical CPSs. As stated previously, several features of embedded medical systems that were once implemented using hardware are now being implemented digitally. This requires major adjustments in the software of these embedded medical systems. This means that highly reliable software development for embedded medical systems is crucial for the patient's health and safety. Through various research methods, it is proven that a model-based approach to redesigning software in embedded medical systems works the best. This way, when implementing hardware digitally, it is easy to debug and make adjustments at the model level rather than at the entire system level. The most important thing is the development process. The safety and health of the patient is top priority. Because of this, the implementation of software replacing hardware must be thought out carefully. There must be something in place that will warn the user when the patient's health has fallen into a critical condition. Traditionally, nurses act as the controllers that act to determine the patients' health and make adjustments accordingly. However, with this new software implementation method in embedded medical systems, this will no longer be the case. The embedded medical system will autonomously adjust itself. Figure 19.3 illustrates how a microcontroller can be used to monitor the patient's health, make adjustments to the device, or sound an alarm if necessary. The microcontroller will be able to decide if it will be able to make adjustments to keep the patient's health in a satisfactory condition or if a nurse needs to be alerted. This software implementation will also put less strain on nurses so they can focus their time on other things rather than monitoring routine medical devices. This will allow for better overall health care [4].

Given all of the research that goes into software implementation of medical CPSs, there must be a common platform to design software for these systems. The software must go beyond the individual medical component. It must be able to interface with all other medical components on the network. It will consist of a network of medical devices and computers that interact with each other in order to treat each specific patient. This is where the "plug-and-play" method can be used to support clinical workflow. In computer systems, plug and play is a term used to describe the characteristic of a computer bus or device specification, which facilitates the discovery of a hardware component in a system, without the need for physical device configuration or user

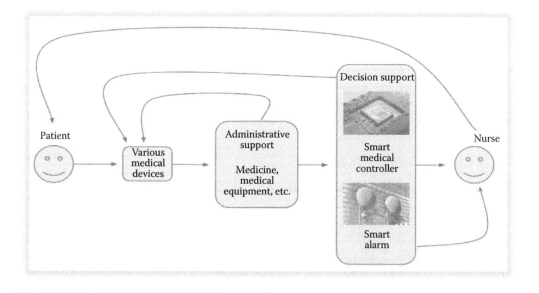

**FIGURE 19.3   Impact of smart medical controller.**

intervention in resolving resource conflicts. This is a preferred method because there is no required calibration or configuration required from the nurse. Initialization is done autonomously, so all the nurse will have to worry about is adding the patient's medical condition and things of that nature [3].

As mentioned before, the most important thing to ensure with medical CPSs is the patient's health and safety. Therefore, there must be an alarm system to alert a caregiver when the patient is in a critical condition or if the device is failing. While there are alarms already in place on most modern medical devices such as a heartbeat monitor, they usually sound one alarm when the device crosses a threshold value. This can cause false alarms and a high number of alarms. In turn, this may cause the nurses to become tired or even ignore most alarms. Also, these alarms only tell the caretaker if a threshold has been crossed. It fails to mention any medical condition that the patient may be in. Implementing a smart alarm into medical software would greatly increase the accuracy of the alarm systems associated with these medical devices. Relevant information could stream in real time from the medical device to the network system. This would not only increase the context awareness available to the nurse but also help doctors to better understand the condition of each and every patient.

# 19.3  Medical Device Coordination and Integration

## 19.3.1  Closed-Loop Control in Medical Systems

Medical devices throughout history have been developed as monolithic stand-alone units. This means they are developed, validated, and approved by regulatory authorities as stand-alone entities. The perception of a device as a unit operating without cooperation from other devices is reinforced by the fact that the US Food and Drug Administration's (FDA) regulatory regimes are designed to approve single stand-alone devices. In the past 5 to 10 years, medical devices have increasingly incorporated communication interfaces to provide connectivity including serial ports, Ethernet, 802.11, or Bluetooth wireless. Until recently, device connectivity has been used primarily to run diagnostics, dump diagnostics, or configure data to external output devices for audits, install software and firmware updates, and retrieve dosing information from Internet databases. New products are emerging, taking the form of middleware, that allow devices to be connected to a common network and allow device information to be streamed to electronic health record databases. These products are improving quality of care by reducing time to enter device readings into patient records, and also by reducing mental overhead associated with gathering information from multiple devices scattered across patient rooms [5].

### 19.3.1.1  Cardiopulmonary Bypass

Patients undergoing a cardiopulmonary bypass operation typically have their breathing supported by an anesthesia machine ventilator during preparation for surgery; then during the actual operation, their breathing is switched to a cardiopulmonary bypass machine, which oxygenates their blood directly, and then is switched back to a ventilator (after bypass). There have been instances where the anesthesiologist has forgotten to resume ventilation after separation from a cardiopulmonary bypass. It only makes sense to use a medical device coordination framework with a connected anesthesia machine and bypass machine to detect this invariant violation and to sound an appropriate alarm.

### 19.3.1.2 Laser Surgery Safety Interlock

In today's society, trachea or larynx surgery usually involves a laser to get rid of cancers or non-malignant grazes and a tracheal tube to source oxygen to the patient during the procedure. A potential hazard is the accidental slicing of the oxygen tube by the laser, which can produce a sporadic fire. There have been a number of injuries and even deaths reported due to these fires. Again, the potential system error in this scenario can be declined by coordination of the laser and ventilator system.

The Medical Device Coordination Framework (MDCF) is a software organization that includes a Medically Oriented Middleware runtime factor and associated development tools that enable researchers to rapidly implement coordination and integration systems. The MDCF architecture has been designed to support features such as dynamic information flow, heterogeneous systems, many listeners, many sources, and time-critical performance. The Java message service (JMS) standard satisfies the criteria above, while providing low-cost, open-source implementations for easy integration into research environments and ease of entry as well. A JMS endpoint is a nonconcrete entity from which a client publishes or receives a message. Destinations are retrieved and located through a Java Naming and Directory Interface (JNDI) calls as shown in Figure 19.4. The JNDI is used to locate a connection factory that summarizes a set of configuration parameters for the provider.

## 19.3.2 Open-Source Medical Device Coordination Framework

As there is an increase in demand for better health care models, it is obvious that there is a direct need for integrated medical devices that cooperate with one another. Many devices promoted today already include some type of communication port such as serial, Ethernet, RS-485, etc., that

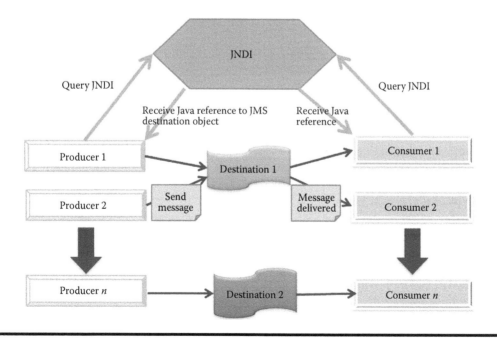

**FIGURE 19.4 JMS destinations.**

are typically used to record data or events from these devices. In the future, it is likely that medical systems will experience augmented device integration, moving beyond simple data logging to provide important functions such as device records streaming straight into patient electronic health records, integration of information from multiple devices in a hospital room into a single editable compound display, and mechanization of medical workflows through computer systems that control networks of devices as they perform compliant tasks [8].

A plain example of programming medical workflow through multiple integrated medical devices gives way to problems in obtaining precise chest x-ray images for patients on ventilators. In order to obtain a nonblurry x-ray image, the doctor must turn off the ventilator for a few seconds to keep the lungs from moving while obtaining the x-ray. However, there are risks in unintentionally leaving the ventilator off for too long. For example, one anesthesiologist stopped a ventilator that was keeping a woman alive while the x-ray technician acquired the x-ray images. However, the x-ray took longer than expected owing to some complications. While they were fixing the x-ray complications, the ventilator was still off. By the time they realized the ventilator was still off, the woman had died.

There are several research projects such as a plug and play for medical devices that specify that these hazards can be reduced by routinely organizing the actions of the x-ray imaging device and the ventilator. The ventilator can identify when the patient has fully inhaled or exhaled so that the x-ray image can be automatically taken at the best time. More generally, the technology exists to accumulate many types of medical systems that can considerably improve health care quality while lowering costs of medical care. It is very easy to create networks and integrate devices. Companies are quickly pushing integration answers into the market, and increasing numbers of clinical technicians are creating their own device networks. However, this can be dangerous because the verification and validation technology and supervisory processes to promise the safety and security of these systems are absent.

### 19.3.3 Open Test for Medical Device Integration and Coordination

Most early medical devices were operated by hand and combined electrical as well as mechanical control implementation. As the complexity of medical devices increased, analog circuitry was replaced or complemented with digital technology. This sparked an increased reliance on low- and high-level software for the implementation of serious device functionality. One example of this comes from the pacemaker. The first pacemakers on the market utilized about 10,000 lines of executable code. A pacemaker today will utilize about 100,000 lines of code.

Historically, medical devices have been developed as separate units. Current techniques of validation and verification used in business mainly target single systems. Moreover, FDA supervisory rules are designed to approve single separate devices. Currently, there are no procedures in place for how the business might bring to market an agenda that delivers new clinical functionality by exploiting an open system of collaborating medical device components from different retailers. This is driven by doubt regarding how one might regulate device collections when the full collection of device–device connections is not fully known [12]. It is expected that medical systems will experience a standard shift as device integration moves past simple connectivity to offer functionality such as device data streaming directly into patient electronic health records, integration of information from several devices in a hospital room into a single compound display, and automation of medical work through a computer system that controls systems of devices as they perform compliant tasks. Undeniably, companies are bringing to market an organization that eases streaming of device information into medical records.

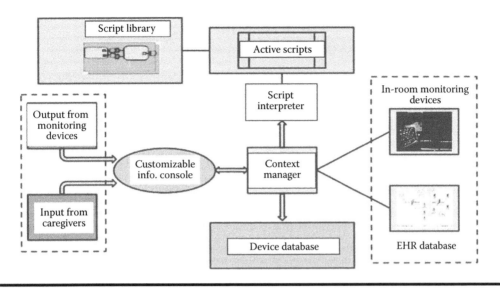

**FIGURE 19.5    In-room medical device implementation.**

A classic hospital room in the intensive care unit hosts a number of stand-alone devices. This example can be seen in Figure 19.5. Many contemporary hospital rooms are integrated with an electronic healthcare record (EHR) database to record medical events, lab statistics, treatment plans, and information for the billing of the patients. Networks to drug medicating databases may also be available to enable correct drug administration. In such settings, a number of issues lessen efficiency, reduce the superiority of the patient's meeting, and increase error probability. Each device in the room has its own graphical user interface, and these interfaces are not the same throughout. This could potentially lead to mental overload and confusion on the part of the nurse.

There is a proposed integration solution [6]. Figure 19.5 presents an idea of how devices and medical information systems can be integrated to confront these problems. Several merchants are now advertising integration structures that go over one or more features of this image. In these integrated solutions, old-fashioned medical devices are viewed as data makers that publish sporadic or flowing data to numerous types of data clients. An EHR database serving as a data customer allows information from separate devices to be integrated directly into the EHR. This stores device data into one place, which simplifies record keeping. This also enables data searches and detection and tracking of statistical trends or glitches that would otherwise go unobserved by doctors. Another type of data consumer would be a sole "heads-up" display that takes information from several devices and an EHR database and organizes it on some large monitors that are near the patient's bed.

The integration setting above must maintain lots of data quantities and amounts. For example, a pulse oximeter finger clip may harvest only heart rate and blood oxygen capacity standards that are updated as separate strictures every 10 s; yet an electronic stethoscope may need to stream information at 8 kilosamples per second over the system. The construction must allow simple addition of "data transformation" components that total, screen, and transform data. The organization must be supported by a software design and authentication environment that allows easy definition and composition of data producers, consumers, and transformers as well as provides facilities for thorough validation and easy auditing so as to support regulatory oversight.

## 19.4 Medical Device Integration Options

### 19.4.1 Medical Devices "Plug and Play"

Medical devices have traditionally been designed to operate independently, unlike the connected environment of networked computers and other electronics. Up to this point, most medical devices have been designed as stand-alone entities, but with the increasing complexity of the health care environment, monolithic devices and systems no longer provide an acceptable solution. These systems must easily integrate with other vendors' equipment, software, and systems in order to improve health care quality, reduce health care costs, and provide for a more comprehensive and secure management of health information.

Standards-based medical device interoperability can provide real-time comprehensive population of the electronic medical record and lay a foundation for the more comprehensive improvements in patient safety and quality that can arise from the integration of medical devices. Interoperability will enable the creation of integrated medical systems to support highly developed capabilities such as automated system readiness assessment; physiologic closed-loop control of medication delivery, ventilation, and fluid delivery; decision support; safety interlocks; smart alarms; monitoring of device performance; plug-and-play modularity to support "hot swapping" of replacement devices and selection of "best of breed" components from competitive sources; comprehensive data collection for the analysis of near misses and undesirable events; enhanced disaster preparedness and response capabilities; and other innovations to improve patient safety, treatment efficiency, and workflow effectiveness.

Interoperability of medical devices requires many elements to be aligned. There must be clinical based regulatory protocols, liability concerns must be met, and widely adopted standards should be used.

There is an almost unlimited opportunity when it comes to the future of cyber-physical medical systems. The future, however, is leaning toward a plug-and-play system. There is a huge demand for these in the medical field. Some of the needs for the near future include connecting ventilators, monitors, pacemakers, and things of that nature to local networks. This will enable local care of patients, decision support, alarms, etc. There is also a need for resource management. This refers to the amount of data that gets streamed from the patient back to a caregiver or database. There needs to be some way to manage all of this information that comes through. Finally, there is a need to develop an open platform approach. This could help support or automate some of these needs. With an open platform, it will be much simpler to manage all of the resources in a controlled manner.

There are multiple health care organizations and medical representatives that are making it understood that they demand to accept developing interoperability principles for medical device connectivity. For example, the health care company, Kaiser Permanente, began to contain restricted necessities for medical device interoperability in the contracts with their clients in 2006 [7]. As an outcome of association with the Plug and Play program, MGH/Partners, Kaiser, HealthCare, and Johns Hopkins Medicine became dynamically involved in this determination with the objective of increasing and firming up the original language to make it certain that vendors know that customers want interoperability. These organizations have issued a National Call to Action to advance the safety of the patients by endorsing that medical device interoperability requirements should be included as a crucial component in retailer selection standards and attaining procedures. This relationship has shaped sample RFP and tightening language that is being shared with device producers as well as other medical institutions. The FDA now approves interoperability and says that it improves patient safety. In March 2007, the first endorsement from a clinical society occurred. Since then, there have been numerous other endorsements by other medical companies.

There are several crucial projects to look for in the future. The successful completion of these projects will be dependent on the amount of research and significance that cyber-physical medical systems, plug and play in particular, gain. One project includes provoking clinical situations to advise interoperability resolutions. This also includes classifying hostile events and things that could have been avoided through the integration of stand-alone medical devices and CPSs. The next project includes developing some way of analyzing the methodology that enables the use of engineering systems to improve these medical systems. This includes, but is not limited to, functional and physical requirements. The final project would be to develop an open IDE platform to develop and perfect the software that goes into the plug-and-play systems. This could also be used to deploy and evaluate important reference implementations of proposed standards, technologies, and products.

Anyone can participate in the movement and research to help improve the quality of health care through the use of cyber-physical medical systems. Caregivers can replicate and contribute clinical scenarios to ensure the new interoperability standards are up to par. Engineers can provide real-time solutions and guidance in the creation of these systems. They can also analyze these scenarios and develop a mathematical and scientific solution. Health care organizations, regulatory agencies such as the FDA, medical device manufacturers, as well as promoting organizations can all support and help raise the bar for the new standard of cyber-physical medical systems.

## 19.4.2 Safe Interoperability of Medical Devices in the Event of Failure

There is a large need for automatic interoperability among medical devices in modern health care systems. The need for automatic interoperability is not just for handiness but also to avoid the likelihood of mistakes due to the complication of communications between the devices and their operator [9]. There are several systems that allow for this to happen. One system, for example, is the Network-Aware Supervisory System [10]. This system integrates medical devices into a medical interoperability system that uses real networks already in place. In many cases, it is dangerous to operate certain medical devices manually. For example, an unintentional burn caused during an airway surgery occurs when an airway surgery laser is started before the oxygen concentration is turned down from the oxygen supply. In turn, this will burn the patient. If these devices were on a communications network, the laser could be programmed to remain off until the oxygen concentration was turned down. If this was the case, accidents like this would not happen. However, this is not the first case of this happening nor will it be the last. The American Society of Anesthesiologists predicted that between the years 2003 and 2005, there were 247,662 patient safety-related incidents that happened that could have possibly been prevented.

One model that has been established is Network-Aware Safety Supervision (NASS). This is the first time that a communication network has been established for medical devices to operate and interact with each other in the company of communication failures. The basic model for the NASS contains one central supervisor that connects all medical devices together through a star network. There are certain supervisory modes that declare devices safe to use. When each device receives its supervisory mode, it will reply back to the supervisor with its "state." By using the different states given by the devices, the supervisor will determine the state of the patient based on the states of the devices. The formalism of the basic models is as follows. First, let $D = \{d_1, d_2, d_3, ..., d_m\}$ represent the set of devices in the system. Now, the supervisor of the system is defined by the tuple $S = (P, \ Modes^*, \ States^*, \ \xrightarrow[p]{}, \ \xrightarrow[spv.]{}, \ Safe^*)$. $P$ is the set of patient states. *Modes\** is the set of global supervisory modes for all devices combined. *States\** is the set of all possible states from each medical device. $\xrightarrow[p]{}$ represents the transition function of the patient state based on the state of the devices. The $\xrightarrow[spv.]{}$ symbol is used for the transition of modes. *Safe\** is used to represent the compositional state transition function. With the set of devices and the supervisor of the system being defined, the supervisor then updates its patient state of cycle $k$ as follows:

$$P_{k-1} \xrightarrow[\quad P \quad]{\left(state\frac{(d_1)}{k-1}, \ state\frac{(d_2)}{k-1}, \ ..., \ state\frac{(d_n)}{k-1}\right)} P_k,$$

where $P_k \in P$ is the patient cycle at state $k$. This representation is much like a finite-state machine [11]. When each device is in a certain state, it will advise the supervisor to update the patient status in one way or another. Figure 19.6 gives an example of how this NASS system will work. As you can see, it functions like a finite-state machine. It reads an input and changes states based on that input. Figure 19.6 gives a real-life example of how the system would work for an airway laser, similar to the one that burned if the oxygen concentration is too high.

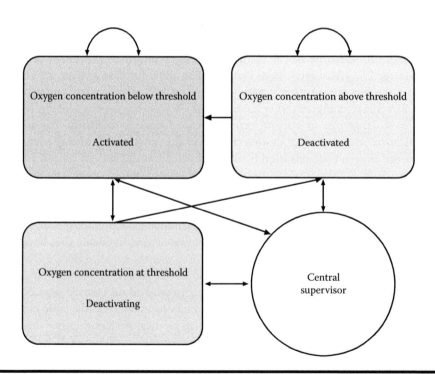

**FIGURE 19.6** NASS state machine airway laser.

## 19.5 Conclusions

The duty of designing a system that is considered very safe on top of unpredictable communication networks is highly challenging. Although medical devices exist as stand-alone units, there is a great need for them to join together and communicate with one another. This can be made possible through implementing a cyber-physical medical system. There have been countless examples of medical devices failing and hurting patients simply from human error. The American Society of Anesthesiologists predicted almost 250,000 medical injuries and deaths resulting from medical devices failing patients due to human error [12]. This is a number that should be close to zero. If these medical devices were not stand-alone units, this number would decrease significantly.

The impact that cyber-physical medical systems would have on a global scale would be enormous. Alarms would alert caregivers. Medical devices would autonomously adjust themselves and update the state of the patient. All of these things could be made possible through the implementation of cyber-physical medical systems. However, if we are to use these systems, they must be completely reliable. Although the technology exists, the communication network structure has not proven to be flawless. In order for a full scale implementation of a cyber-physical medical system to be successful, the communication framework must be impenetrable. Table 19.1 shows the differences between CPS (for a medical system) and a general stand-alone unit.

While the present version of the current cyber-physical medical system communication framework has not been scrupulously analyzed and examined for dependability and consistency, it is conceivable to build a closed-loop safety control system that is safe and reliable to use globally. This sort of safe fallback behavior may not be appropriate or even possible for other coordination scenarios. One example would be a closed-loop blood sugar management system for people with diabetes. Another example would be someone on a breathing ventilator that is recovering from a collapsed lung. In order for the cyber-physical medical communication systems to be widely accepted in global trials, energy must be focused toward constructing a safety case of the communication framework reliability, safety, and security. There are several areas where safety and reliability properties must be properly established and validated. The first would be timing constraints within a cyber-physical medical system inside of each stand-alone medical unit. There must be some sort of timing guarantee that requires on time deliverance of critical messages between units and databases. The next key area to improve is message

**TABLE 19.1  Comparison of CPS and Stand-Alone Unit**

|  | Cyber-Physical Medical System | Stand-Alone Unit |
|---|---|---|
| Update automatically | Yes | No |
| Reduce caregiver staff | Yes | No |
| Provide feedback | Yes | No |
| Update patient state | Yes | No |
| Increase patient safety | Yes | Yes |
| Provide 100% reliability | No | Yes |
| Provide single point of failure | Yes | No |

deliverance reliability. If the deliverance reliability is not 100%, then there is no way of trusting a cyber-physical medical system. Finally, the next area that needs improvement is the security of communication network systems. Because cyber-physical medical systems are managing multiple physiological data streams from many patients and caregivers, there should be secure access to these data streams. If there is not, that could violate doctor–patient confidentiality laws and lawsuits could occur.

There is a huge need for research into cyber-physical medical systems. However, more attention needs to be drawn to this area of the medical industry. Also, medical devices are typically very expensive. Therefore, it is hard for researchers to develop systems that are integrated into cyber-physical medical networks because they cannot afford to carry out research on so many devices. There are thousands of different medical devices on the market, and it is very expensive to carry out or implement full scale research on all of these devices in order to integrate them into cyber-physical medical systems. Researchers simply do not have the resources to do this. Therefore, there needs to be much more research funding in the area of cyber-physical medical systems. Also, there should be certain standards for faux medical devices that will work just as well during research periods. The safety and care of patients will be held to a much higher standard with the implementation of cyber-physical medical systems into the medical industries [4].

There are many challenges that come along with the implementation of cyber-physical medical systems. As mentioned earlier, the first challenge would be research funding. The second implementation challenge is the safety of the patients. Obviously, when each medical device becomes part of a communication network, it must be 100% reliable. This is especially important for life-saving devices such as pacemakers or ventilators. There are also a number of software challenges that will occur during the implementation phase. The security of the software must be top-notch. This is because multiple patients' health and safety records will be streamed across communication networks and could be made public if they fell into the wrong hands. Some of these challenges that are presented are similar challenges that are faced by other industries. However, most of these challenges are strictly tailored to the medical industry. Regulatory agencies are well aware of these tasks and are interested in finding precise and sound solutions to them. This gives an opportunity to the research community to get involved in this academically challenging and significant problem.

# References

1. A. M. K. Cheng. 2008. Cyber-physical medical and medication systems. *IEEE Comput. Soc.* 67, 529–532.
2. O. S. Insup Lee. 2010. Medical cyber physical systems. In *Design Automation Conference (DAC)*, ACM, New York, pp. 743–748.
3. I. Lee, O. Sokolsky, S. Chen, J. Hatcliff, E. Jee, B. Kim, A. King et al. 2001. Challenges and research directions in medical cyber-physical systems. *Proc. IEEE*, 100:1, 75–90.
4. O. Sokolsky, I. Lee, and M. Heimdahl. 2011. Challenges in the regulatory approval of medical cyber-physical systems. In *Proceedings of the International Conference on Embedded Software*. ACM, New York, pp. 227–232.
5. A. King, D. Arney, I. Lee, O. Sokolsky, J. Hatcliff, and S. Proctor. 2010. Prototyping closed loop physiologic control with the medical device coordination framework. In *Proceedings of the 2010 ICSE Workshop on Software Engineering in Health Care*, ACM, New York, pp. 1–11.
6. A. King, S. Procter, D. Andersen, J. Hatcliff, S. Warren, W. Spees, R. Jetley, P. Jones, and S. Weininger. 2008. An open test bed for medical device integration and coordination. SAnToS technical report, Air Force Office of Scientific Research.

7. M. Julian and M. Goldman. 2009. *Medical Device Safety and Innovation*. National Science Foundation, Boston, MA.
8. A. King and J. Hatcliff. 2009. An open-source medical device coordination framework. Kansas State University, Manhattan.
9. J. M. Goldman. 2009. *Medical Device Interoperability*. Boston, MA.
10. D. Arney, I. Lee, M. Pajic, Mangharam, J. M. Goldman, and O. Sokolsky. 2010. Toward patient safety in closed-loop medical device systems. In *Proceedings of the 1st ACM/IEEE International Conference on Cyber-Physical Systems*, ACM, New York, pp. 139–148.
11. C. Kim, H. Yun, M. Sun, L. Sha, S. Mohan, and T. F. Abdelzaher. 2010. A framework for the safe interoperability of medical devices in the presence of network failures. In *Proceedings of the 1st ACM/IEEE International Conference on Cyber-Physical Systems*. ACM, New York, pp. 149–158.
12. A. King, S. Procter, D. Andresen, J. Hatcliff, S. Warren, W. Spees, R. Jetley, P. Jones, and S. Weininger. 2009. An open test bed for medical device integration and coordination. In *31st International Conference of ICSE-Companion 2009*, IEEE, New York, pp. 141–151.

## Chapter 20

# Augmented Cognition for Intelligent Rehabilitation

Chad Buckallew, Sarah Duncan, Walter Hudgens,
Matt Smith, Hoyun Won, and Fei Hu

## Contents

## 20.1 Introduction

Stroke is the third leading cause of death in the United States, and it is also the leading cause of adult disability in the United States. Rehabilitation of brain-damaged patients involves a combined use of medical, educational, social, and vocational techniques. Typical stroke rehabilitation programs are utilized to provide the optimal neurological recovery, to allow patients to overcome residual deficits, as well as to teach the activities of daily living (ADLs) and skills required for community living [1]. The body recovery of stroke patients is based on the level of intensity of the rehabilitation services. Using coordinated multidisciplinary stroke rehabilitation teams has assisted in the lowering of the mortality rates for stroke patients. Patients receiving care at an inpatient rehabilitation facility have a higher percentage of recovery. A step has been taken to lower these mortality rates in constructing more stroke systems throughout the United States.

In the early twentieth century, rehabilitation of stroke patients was limited to five different methods [1]: Repetitive drills such as reading and verbally saying words were used for aphasia. Exercise programs were used to battle moderate paralysis. Splints and support braces were used as aids for patients to maneuver. In order to prevent muscle wasting, electrical stimulation of the muscles was implemented. Other common methods of stroke rehabilitation include the various surgical procedures.

The technological advances of today have allowed for the stroke rehabilitation systems to be safe as well as cost efficient. The first 36 weeks of at-home physical therapy with an appointed aid boasts an average cost of $20,000 [1]. By using an augmented cognition such as a virtual reality (VR) environment, which can manipulate the data received from sensors and integrate those data with video games, the stroke systems constructed would be more interesting and efficient to use.

## 20.2 VR SYSTEMS

### 20.2.1 Basic Concept

Stroke survivors usually require a long therapy process to regain some sensor and motor skills, which include physiological recovery, body function recovery, and activity recovery. Improvements in these motor skills can be achieved through the following therapies: constraint-induced therapy, robotic-assisted therapy, repetitive movement therapy, and VR therapy. Successful stroke rehabilitation helps the patient regain the ability to complete functional tasks and relearn premorbid movement patterns. Feedback to the patient is important because it can encourage and motivate the patient. Digital computing systems can be used to supply this feedback either visually or through audio. The feedback and the tasks can be tailored to each individual patient in order to promote learning while engaging the patient mentally and physically. Detailed kinematic parameters can be used to tell whether or not the patient is improving because the therapy can be tailored to each individual.

When starting VR-based therapy, the patient will first focus on just making general movements so as to have a baseline of where to begin. As the therapy progresses, the readings will be able to tell if the kinematic gestures have or have not improved. Later, the patient will then begin to actually focus on completing the tasks that he or she is prompted. Customized computational algorithms were developed to customize the sessions for each individual because the kinematic readings drive the feedback for each individual patient. Not all patients progress the same, so this helps patients to improve to the best of their abilities. This is better than the physical therapy done in the hospital, which is basically just a checklist of items to complete.

In [2] a 10-camera three-dimensional (3D) infrared passive motion capture system is used to capture the kinematic motions of the patient. Reflective markers are placed on various places on the body. The data are then collected at 100 Hz while using a low pass filter to suppress the noise. Data from various sensors are used to determine the different "grasp and reach" stages of the movements. The patient is asked to reach in four different directions. When this happens, the patient is being assessed on several different items including goal completion, trajectory, accuracy, speed, velocity profile, joint coordination, arm and torso movement, and ranges of the angles of the joints. Average values are obtained for all of these measurements and are used to map feedback to the patient. The measurements are kept in an archive in order to compare with later data to see how the patient is improving. The feedback will also offer the patient helpful advice about how to improve the next-step training based on the data obtained. Visual and audio feedback is also used to help the patient. For visual feedback, a picture appears on the screen for the patient to reach toward. As the patient comes, the picture disappears until the patient puts his/her hand where the picture previously was; when they do, the picture reappears. The audio feedback mostly plays music while this therapy is going on in order to help the patient connect his/her mind with his/her body.

Feedback is important for life. The feedback sensitivity uses hulls in order to determine the error from the patient's movement. If the error is large enough, the patient will receive feedback to tell ways to correct the mistakes. The therapy is broken into three major steps: In step 1, the patient relearns a simple reaching activity like moving the hand to a certain location. Step 2 focuses on the recovery of the patient's body function like improving the coordination between the patient's arm, torso, and shoulder. When these steps are completed during the therapy, step 3 can be initiated. Step 3 is the integration of steps 1 and 2. It helps the patient develop more complex movement and obtain some level of accuracy. Three patients were chosen to test this research performance in [2]. All three of the subjects showed improvement in their motor functions.

## 20.2.2 Pneumatic Glove and Immersive VR

One of the most common motor deficits reported by stroke victims is the extension of the fingers. The design in [3] allows for independent movement of each digit while still allowing the patient to have full arm movement. This research uses a pneumatically actuated glove for the training of the hand for stroke victims. It focuses more on the extension and flexing of the fingers in order to regain the use of the hand. The pneumatic glove actually assists the patients in the extension of the fingers by using air pressure. It also uses a wireless system called shadow monitor. The shadow monitor uses bend sensors located around the fingers to read the kinematic measurements for each patient.

The glove (Figure 20.1) has an air bladder located on the palm side of the glove and a Lycra backing, which is located on the "top of the hand" side of the glove. Also, on the palm side of the glove, polyester pockets are sewn into the glove near each of the joints of the fingers. These polyester pockets also hold the flexing sensors. Each of these polyester pockets is isolated from all others in order to work on one finger at a time. Air pressure is used down a channel from the palm to the polyester pocket being targeted at the time to force the finger into further extension. When the patient is ready to grab a certain object, the air pressure is removed, and the patient can grab the object. This pneumatic glove is designed to either interface with a designed VR system or be used on its own. One of the fundamental tasks of the hand is the grasp and release. The glove was designed to help improve this task on stroke survivors by breaking it down into smaller subtasks. The glove is designed to break the grasp and release task into the following subtasks: successfully

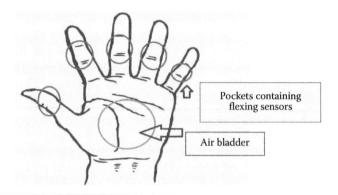

Pockets containing flexing sensors

Air bladder

**Figure 20.1  Layout of the pneumatic glove.**

open the hand, position the hand, grasp the object, and let go of the object. This technology is designed so that the extension and contraction of each finger can be exercised in order to achieve these subtasks.

A pilot study was held to determine if the technology would actually improve the patients' lives and also to determine how it could one day be introduced into the rehabilitation of the patients. The subjects for the study were recruited from the Rehabilitation Institution of Chicago Clinical Neuroscience Research Registry and also from the Rehabilitation Institute of Chicago. The subjects were required to arrive at the clinic for 18 training sessions. Each session consisted of two 30-min sessions. The first session was with a VR environment in which the subject practiced the grasp and release task. In the second 30-min session, the subject focused on hand therapy. This consisted of the subject practicing the "grasp and release" task with real objects. The subjects involved with this study had to meet several requirements including the following: at least 18 years old, at least 6 months poststroke, at least 6 months since the last Botox injection in the upper arm, and able to follow auditory commands without seeing the instructor. Sixteen individuals participated in this study.

Of these 16 individuals, 8 were in a group that used the VR training only with a Shadow Glove. The other eight subjects wore the PneuGlove for the VR and the intensive hand therapy, which focused on the grasp and release of real objects. Fourteen subjects completed the study. In the VR environment, 60 grasp and release tasks were performed during each 30-min session. In the intensive hand therapy environment, each subject was allowed to choose which real-life activity to practice such as opening a medicine bottle or setting a table with dishes. For the group wearing the PneuGlove, the therapist directly controlled the air pressure in the glove to help the subject with hand opening and closing. Although both groups showed significant improvement during the study, the group wearing the PneuGlove improved more than the group wearing the Shadow Glove. One follow-up was done by testing patients directly after the study was completed, and another follow-up was done after the postfollow-up. The group wearing the PneuGlove was seen to improve more significantly than the group that wore the Shadow Glove.

## 20.3  Augmented Cognition

Augmented cognition refers to the use of information technology (IT) to augment human cognition. Augmented cognition (often abbreviated as AugCog) research has been on the rise in recent years as computers have become small and low power enough that they can be easily and unobtrusively portable. There are several augmented cognition systems in development today. Some of these could possibly be adapted to aid stroke patient rehabilitation.

## 20.3.1 Augmented Cognition Concept in Web Design

In the last decade, this world has experienced the rapid expansion of IT. Despite its prevalence, IT failed not only to produce any technologies for people who have visual or other impairment but also to give proper guidelines to IT users, who do not have any impairment, for developing an effective Web site or software. Despite the fact that there are luxuriant Web sites in the Web, most Web sites lack the effective resources and designs that help the users to obtain the resources that one needs. The problem was the designers. To discover more specific problems, four examinations were taken to discover the weaknesses of each 169 Web practitioners. These examinations revealed that 96% of designers use guidelines to inform their work but not effective guidelines to produce an indispensable Web site. The results of the four examinations laid out two crucial weaknesses that designers lack: evaluating and generating effective Web sites [4].

For evaluation, a Web-based and desktop application were developed to assess many aspects of Web interfaces, including the amount of text on a page, color usage, and consistency by using 157 adequate measures, the empirical data to develop guidelines, and profiles as a comparison basis [4]. After evaluating the submitted Web site, the application sends the feedback including the rate of the site and the comments from the Internet professionals.

Even though there are many different types of Web sites, the software can only generate the design for instructor to use in a classroom. First, a database is made for evaluating the Web site; the database consists of an expert's reviews and other toolkits. After applying this database to 79 sites, three problems arose. These are accessibility, organization, and navigation. The solution was making a simple text-based mechanism that does not require knowing HTML and software that supports the automated transformation of documents and site's design (e.g., navigation, layout, and consistency) except the content organization. The latter version would include text-based content entry, perform automated content mapping to layouts, or provide feedback to instructors.

The next topic is a help system. Like the above topic, an online assessment was performed to 107 IT users and discovered two breakdowns: help content and pertinent resources. To solve the first bottleneck, the structured and unstructured help content was made to garner for three email applications that are used within three different computing domains: Web-based help content, Windows help resource, and transformation of jargon-heavy help content to everyday language help content. To address the second bottleneck, the help matching system was constructed to allocate the help resource to user's experience levels or help-seeking preferences, to offer different types, and to organize help content by the user's level. However, the difficulty is that there is no way to know whether the user is really novice, intermediate, or expert. One of the solutions that the paper mentions is making a wizard that asks users a few questions and recommends the appropriate resources, especially for novice.

The last topic is about tactile graphics for blind students. Long ago, blind students had limited opportunities to experience an important graphical image that required sight. The only opportunity to touch and feel the image is through tactile version (making palpable image) made by a tactile graphics specialist. However, tragically, even if a blind student has financial leeway as well as passion, there are limited resources for the student to touch and experience because a tactile graphics specialist has limited tools as well as resource to produce advanced and adept drawings.

Moreover, after conducting an online questionnaire to 51 people, only half of specialists use a computer program while the other half draw by hand only. More shockingly, the specialists who use a program mostly relied on simple drawing kits like Microsoft Word because advanced drawing tools like Adobe Photoshop are hard to learn and operate, even though it is effective when the specialist knows the commands. In order to help the specialist produce complex and advanced pictures, the software called tactile graphics assistant (TGA) is built to translate graphical images because there are no machines that can print out the tactile graphics for blind students automatically. First, TGA identifies the type of the image after converting a hard copy image to a digital file. Then it goes through the processing steps like color reduction and edge detection and assembles graphical and textual image components into an effective layout that can help the tactile specialist draw effective images for blind students.

Even though the technology itself advances very rapidly, most users could not follow up well. This causes users to produce less effective Web sites and it is hard for tactile graphic specialists to adapt to the advanced program. To address the first issue, the automated evaluation and Web site generation tool were developed to help users design the Web site efficiently. Further, a help portal system and help matching system are constructed to assist the users to find the solution for the problem and match their problem to adequate Web sites. For the last breakdown, TGA is built to help tactile graphics specialists produce effective images for blind students [4].

## 20.3.2 Augmented-Reality Visualization Guided by Cognition

Mixed and augmented reality (MR/AR) systems [5] help the user to see through the wall or any obstacle that hinders the user's view by using the combination of real and virtual images. These systems are very useful for medical issues. Examples include watching a baby in a womb, military issues, and spotting hostile criminal(s) behind a wall or injured soldiers. To achieve the solution for this bottleneck, obscured information visualization (OIV) is applied and has accomplished three goals: identifying the main problem in the system, providing general theoretical guidelines for each problem, and demonstrating sample results from an experiment concerning AR-OIV systems.

After surveying the problems in OIV, they condensed into two main parts: depth ambiguity and visual complexity. Figure 20.2 shows the overall tree diagram for the problems and solutions for an OIV system.

Depth ambiguity conveys the difference between obscured (not visible) and plain sight (visible in a naked eye). It is important for AR/VR displays because it converts 3D environment into 2D, a planar projection. It means that the depth will be gone because there is no way to distinguish the distance from displays to desired location in 2D. To solve the problem, two methods are applied. One is for static while the other for mobile.

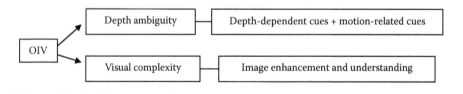

**Figure 20.2   Tree diagram.**

The first applicable solution is depth-dependent perceptual cues. One of the popular cues with depth dependence is monocular depth cues. Monocular depth cues are just as informative when presented to one eye as two and are efficient for static image, as well as dynamic images. Moreover, within monocular depth cues, transparency is one of the common approaches to solve the depth issue. Simply, overlapping regions of transparent surfaces that overlap become darker than non-overlapping regions. Another common approach is called size-scaling gradients and texture, which varies the size of the object as a function of distance.

The second applicable solution is motion-related percepts. This method is very effective when there are moving creatures as obstacles. Two of the common motion-related cues are motion parallax and structure from motion (SFM). In motion parallax, objects that are closer move farther and faster than objects that are farther. In SFM, a complicated situation, where there are many moving obstacles, converts into a simple situation, where all the obstacles transform into small bits or structure.

Within the depth ambiguity, the visual complexity occurs when the displays convert 3D into 2D, where all the contrast, brightness, and transparency of the view rapidly change. There are two ways to solve this bottleneck. The first solution is using image enhancement, which manipulates contrast, brightness, and transparency to reinforce the image within 2D environment by changing each pixel in the displays. Traditionally, histogram equalization is applied to solve the bottleneck by changing contrast. Nowadays, homomorphic filtering is used because it combines multiple images of difference exposure. In this application, regions with more homometric certainty are accentuated than regions that are midtones. The second solution is using image understanding, which accentuates the specific building and others. To distinguish the things that are to be accentuated or not, a variety of techniques is applied. Among the techniques, information theoretic scale space approaches, bottom-up feature-based approaches, and saliency networks for extracting salient curves are most common approaches.

In order to enhance the effectiveness of an OIV system, there are a few guidelines to follow. Table 20.1 shows the four major guidelines to follow in order to achieve the maximum effectiveness from an OIV system.

For a test plan, a camera HiBall 3000 optical tracker is used for AV systems under dual processor Xeon 1.7 GHz PC running windows XP, with an NVIDIA Quadro 2 graphics board. The output of the test is recorded on video tape and edited with the software DirectShow. There are three goals to accomplish with this test: (1) validate how much people suffer from depth ambiguities, (2) select the most appropriate cues for the test, and (3) test the OIV system.

**Table 20.1  Guidelines and Description for OIV System**

| Guidelines | Description |
|---|---|
| Distance conveyance | Distance has to be told before. |
| Proper motion physics | For motion parallax, the system has to have proper defined geometries and metrically accurate models of the environment. |
| Eliminate unneeded AR motion | Minimize unneeded motion of rendered material practically. |
| Selective cues | One specific cue has to be chosen before using it. |

## 20.4 Examples of Augmented Cognition Designs

### 20.4.1 Augmented Cognition for Fire Emergency Response

Researchers at the University of California, Berkeley, proposed an advanced IT network/augmented cognition system for use by firefighters [6]. The inspiration for this system began in the wake of the September 11, 2001, attacks on the World Trade Center. In the course of the New York Fire Departments' response to the plane crashes, there were multiple communications and management failures that resulted in many unnecessary deaths. Most of the communication failures involved failures with hand-held voice radios. The resulting lack of communication between firefighters and Incident Command led to more firefighters entering the World Trade Center than necessary, as they tried to reach their comrades. As a result, 350 firefighters perished in the attack.

The Berkeley team decided that one way to address these issues was to develop a robust, reliable IT network to take some of the strain off fire departments' radio networks. Additionally, the team decided to take advantage of augmented cognition (AugCog) technology to improve firefighters' situational awareness [7].

The system was developed following interviews with firefighters and fire chiefs in San Francisco and Chicago. In addition, these fire departments assisted in testing and evaluating the system over the course of several iterations.

Following the interviews, the team came up with a guiding philosophy for the project. The fire departments all shared three common concerns: budget, technology dependence, and reliability of equipment. Fire department budgets typically do not allow for equipment that is not an absolute necessity, so the IT/AugCog systems would need to be somewhat minimalist. The firefighters were concerned about technology dependence because if a firefighter were too reliant on a system and it failed, that firefighter could easily become lost or get hurt. This was also part of the concern with reliability. With all these factors in mind, the team decided that their IT/AugCog system would need to have the minimum number of necessary features. There should be nothing distracting, and "feature creep" should absolutely be avoided.

The system the Berkeley team came up with consists of three components: (1) a head-mounted display called FireEye, (2) a wireless network called SmokeNet, and (3) an electronic Incident Command System.

The FireEye system consists of a small display mounted to the firefighter's mask. Its purpose is to provide a graphical user interface to the firefighters that can show both text and images in real time. This has several uses. For example, a floor map can be displayed. Combined with data from the SmokeNet network, this map could show the location of the fire and of other firefighters in the building. This would help firefighters get to the fire quickly and, if necessary, know exactly where to go to aid fallen or unresponsive comrades. FireEye can also be used to show text messages from the incident commander. Such messages would consist mostly of routine checks from incident command. In future versions, it may be possible to incorporate a Global Positioning System-like navigation system. This is considered with caution, however, as there may be danger along the path that the system chooses. Overreliance on the technology could cause firefighters to implicitly trust the navigation and blunder into harm's way.

The first version was mounted outside the mask, near the top. It made use of holographic technology to project an image onto the mask. In testing, this was found to be somewhat distracting. Having the projected image between the eye and the outside of the mask was confusing and made it difficult to focus on anything outside the mask. In addition, the bright image was often indistinguishable against a bright background. Later versions have all featured a display that projects the image to the eye rather than on the mask. Another problem with the first version was the image location. Firefighters often have to crawl to navigate burning buildings, which necessitates looking "up" for navigation. Having the display at the top of the mask interfered with their ability to see in this case. For later versions, the display was moved to the bottom of the mask. One additional problem with the first version was it being mounted outside the mask. Firefighters did not believe an outside-the-mask mounting system could be reliable, so the display was moved inside the mask. After all of this, the current version of FireEye is an inside-the-mask, bottom-mounted display that must be looked into for the image to be visible.

The SmokeNet system consists of wireless transceivers, called Motes, and sensors mounted inside smoke detectors. Firefighters also have Motes on their person. This system has two purposes. First, it allows firefighters to know where they are within a building. The Motes carried by the firefighters can communicate with Motes in the smoke detectors to tell which smoke detector is closest. This gives room-accurate location information to incident command and other firefighters in the building. SmokeNet's second purpose is to help pinpoint the location of a fire. It has two sensors: a smoke detector and a temperature sensor. Smoke detectors sample these data every 10 s and transmit them to a centralized node. The incident commander can connect a computer to this central node and use the information from SmokeNet to narrow down the likely location of a fire. (A room with fire will be hotter and smokier than other rooms.) All of this information can also be shared with FireEye [6].

The electronic Incident Command System is a Visual Basic program that displays a floor plan and data from SmokeNet on a laptop or table PC. It currently only has support for one team of firefighters at a time, so it features a toggle function to switch between teams. It also has the ability to send text messages to FireEye. The complete system is shown for one firefighter and incident commander in Figure 20.3.

Text messages are used to reduce the load on the firefighters' voice radios. Currently, most communications conducted over radio are routine check-in questions with simple yes/no answers. Using text messages for these routine questions frees up the radio bandwidth for more

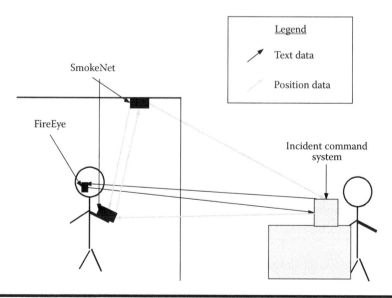

**Figure 20.3 Complete firefighter AugCog system for one firefighter and incident commander.**

important/detailed communications. Currently, FireEye does not have the ability to respond to these text messages, but future versions will add the ability to provide yes/no answers.

The response of firefighters to the current versions of the IT/AugCog system has been mostly positive. Future work will consist of allowing individual departments to customize the system for their district and to have the system be cleaned and serviced by nontechnical personnel.

Some aspects of this system could be modified to aid stroke victims. For example, a modification of SmokeNet could detect that a patient has not moved in some time. The system could then alert caregivers that the patient may have suffered some medical emergency. Patients who have not regained their ability to speak or use a phone could use a system like FireEye to communicate using simple, prewritten messages.

## 20.4.2 Rapid Decision-Making in Hostile Environments

In support of the US Army's Future Force Warrior project, Honeywell Laboratories developed a cognitive augmentation system for use in hostile environments. This system, called the Joint Human–Automation Augmented Cognition System (JHAACS), aims to improve decision-making ability and vigilance and to reduce or automate workload under stress [8].

JHAACS consists of five main components: the cognitive state assessor (CSA), augmentation manager (AM), human–machine interface, virtual environment, and automation. In addition, there is an experimenter's console for the purposes of testing. The main components of JHAACS are shown in Figure 20.4.

The CSA's function is to monitor the subject's cognitive state. This information is used to determine when, and how much, to activate the cognitive augmentation technology. The CSA assesses cognitive state through the use of five "gauges." The first gauge is the engagement index. The engagement index uses electroencephalographic (EEG) power bands to measure alertness and vigilance. The arousal meter determines cognitive arousal by measuring cardiac interbeat intervals. The stress gauge measures stress levels by measuring pupil diameter, heart rate, and cardiac waveform amplitude. The P300 Novelty Detector uses an auditory alert tone to provoke an EEG response; the strength of this response is used to determine remaining attentional resources. The XLI gauge measures EEG power to determine comprehension levels. Together, all of this information is used to determine the cognitive state of the user. Information on the cognitive state is then passed on to the AM.

The AM uses four mitigation strategies to augment user cognition. The first strategy is that the Communications Scheduler automatically prioritizes incoming messages as they are received. High-priority messages are immediately displayed for the user to read. Low-priority messages are

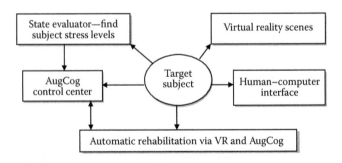

**Figure 20.4   Main components of the JHAACS system. Arrows indicate direction of data flow.**

deferred to a tablet PC to be read later. This system is triggered if the user's workload is detected to be too high or too low. When activated, it will remain active until all messages have been read. The second strategy is the Tactile Navigation Cueing Interface System. This system guides the soldier in the best direction to his/her destination. The third strategy is the Task Offloading Negotiation Agent, which loads information gathered from communications into forms. These forms are used to reduce verbal communications necessary to trigger a Medevac. The final mitigation strategy is the Mixed Initiative Target Identification Agent. This system searches the soldier's field of view to aid in detecting targets, friendliness, or other items of interest.

The JHAACS was tested in four stages. The first stage of evaluation was to determine the effectiveness of the system at high workloads. For this stage, test subjects were tasked with navigating an urban environment in a virtual desktop environment. In addition to the navigation, subjects were tasked with identifying friend from foe and with monitoring, and responding to, communications. During the test, the CSA would monitor the subjects' cognitive state and activate the Communications Scheduler as necessary. Trials were conducted with both high task load (i.e., subjects received many messages) and low task load. As a control, trials were also conducted without any mitigation system. Four trials were conducted in each combination of these variables for a total of 16 runs. Subjects' performance was measured via message content probes, responses to messages, and posttest questions. The results showed the Communications Scheduler caused "significant improvement" [9] in situational awareness in the high workload scenarios [9].

The second stage of testing examined the performance of JHAACS at both extreme high and extreme low workloads. Performance at both extremes was measured by introducing longer duration into the tests and by varying the task load during testing. This stage was also conducted in a virtual desktop environment. In addition to the Communications Scheduler, this stage also included the other mitigation systems: the Tactile Navigation Cueing Interface System, the Task Offloading Negotiation Agent, and the Mixed Initiative Target Identification Agent. The tasks subjects were required to perform were similar to those in the first test, but now the situations focused more on multitasking, vigilance, and divided attention. As before, control tests were conducted without any JHAACS mitigation systems. The results greatly increase subjects' performance. The Communications Scheduler increased message comprehension by 100% and situational awareness by 125%. The Tactile Navigation Cueing Interface System reduced evacuation time by 20%, and enemy encounters were reduced by 350%. The Task Offloading Negotiation Agent improved Medevac negotiation time by 300%. The Mixed Initiative Target Identification Agent improved target identification by 30%. Subjects reported no negative impacts on their ability to engage enemies. Subjects also reported that all tasks were easier with augmentation activated [9].

The third stage of testing was conducted in a mobile environment, a motion-capture laboratory at Carnegie Mellon University. Subjects stood in a space of fixed area and engaged targets with motion-tracked prop rifles. Subjects were tasked with monitoring a building, engaging enemies, monitoring radio traffic, and maintaining counts. The Communication Scheduler was compared to a random scheduling scheme. The results showed a 150% improvement in working memory tasks versus the random scheduler [9].

The fourth and final stage of testing was a combination of desktop and mobile trials. The Communications Scheduler and the Mixed Initiative Target Identification Agent were the augmentations used. The desktop portion of the test had subjects monitor radio communications on the number of enemies, friendlies, and civilians in the environment, guide three squads through the environment with radio, and monitor images for signs of enemies and report their position. The mobile portion was conducted in an outside, wooded environment. The radio monitoring tasks were the same as in the desktop case. Rather than examining images, however, subjects scanned the woods for concealed targets. Monitoring tasks saw a 94% improvement. Count recall had a 36% improvement. The visual search task saw a 40% improvement when augmentation was introduced [9].

The augmented cognition system designed by Honeywell could be a fantastic addition to the US Army's arsenal. It is speculated that this system could allow a squad to cover the same amount of area that would have previously required an entire platoon.

Similar technology could be used to aid stroke victims. For example, the cognitive gauges could detect if a patient is becoming frustrated with their training and reduce workload appropriately so the patient will not become discouraged. If the gauges detected levels of stress consistent with a medical emergency, the Mixed Initiative Target Identification Agent (or something similar) could help a patient locate a phone, emergency exits, or other possible forms of aid.

### 20.4.3 Augmenting Cognition in Complex Situation

Complex decisions are required for complicated areas in life. However, complex decisions can overwhelm the decision maker and lead to tunnel vision, which is defined as insufficient problem formulation and situation assessment, producing narrow minded interventions that may lead to failure or even to the emergence of new unintended problems. In these types of complex situations, studies have shown that it can be very important for the decision maker to focus instead on short-term goals instead of trying to tackle the bigger problem at once. These short-term gains help the decision maker to reach the long-term goal without so many unintended problems cropping up along the way. This study tries to provide cognitive support for the individuals in order to help the decision maker to reach the long-term goal. The cognitive support given to the decision maker is obtained by using approximate models of the system in order to find reasonable outcomes to supply the decision maker with [10,11].

Decision support technology can possibly have the potential to extend human reasoning capabilities from just the outcomes of long-term goals to hypothetical assumptions of what could possibly happen. Although this type of research sounds good, studies have shown that decision support technology does not actually extend human reasoning capabilities or further goal attainment for each individual. This study uses a game called Ecopolicy to investigate how using a decision aid game impacts the information acquisition and decision-making behaviors of the participants involved in the study. Ecopolicy is a simulated society management game that is designed to inform the participants about the importance of "networked thinking." Networked thinking is teaching the participants how to understand how different variables relate to one another when looking at the whole of a complex system.

In this game, the causes and effects were clearly shown for the participants, but it was seen that the participants still had a difficult time trying to bring the system to the long-term goal. The participants failed to recognize the implications of each of the decisions he or she made during the course of the game. This game showed the participants how bad it can be to focus on an isolated problem within the system without stopping to think what the implications could be for the entire long-term goal experiment. Forty subjects from a wide range of backgrounds participated in two sessions lasting 2 h a piece. The experiment used the game Ecopolicy and was run on a standard personal computer having two flat screen monitors.

The results showed that with the decision-making aid, the participants improved mapping out for their long-term goals. However, discouraging results were also recorded for this group of individuals. Because the individuals were giving a decision-making aid to help with making decisions toward the long-term goal, the individuals decreased when it came to information acquisition. Because they were being helped, they stopped trying to learn on their own or gather more information about it [10].

# 20.5 Design Principles of Augmented Cognition

## 20.5.1 Adaptive Multiagent Framework

This section will discuss the development of an Adaptive Multiagent Integration (AMI) framework and its application for the use of augmented cognition systems (AugCog). Researchers in the field of augmented cognition seek to develop a closed-loop human–computer interaction system. This would allow the state of the human's functions to be analyzed, measured, and then automatically updated to improve on the human's performance. The developers were faced with two main tasks for the development of this system. Because many different highly specialized sensors were used, the different sensors were limited to different platforms, and each of the sensors had different limitations. The first task was for the components to be able to communicate with each other using the same language. The second task was for the components to be able to find each other. In order for the AMI to be useful for an augmented cognition system, it had to be flexible and useful to the researcher, so the researcher would be able to run the equipment and make successful evaluations.

The developers of the AMI framework used the augmented cognition system developed by DARPA Improving Warfighter Information Intake under Stress program. This AugCog system consists of three main parts: the application, the Augmentation Manager, and the Cognitive State Assessor. The Cognitive State Assessor is composed of all of the computer algorithms and the sensors for which the system uses. The application component is just the application that the user of the system is using at the time, and the Augmentation Manager just manipulates the application in order to improve the user's performance based on the components of the Cognitive State Assessor. The Cognitive State Assessor was able to use different algorithms for the different cognitive state detection techniques.

The main goal of the AugCog system was to read the physiological data of the user, and then it needed to interpret those data in order to determine the cognitive state of the user. The goal of the AMI was to interpret the data from the AugCog system if multiple types of different sensors and algorithms were used. The different sensors all read in different data for different types of information including gaze, stress, executive load, novelty, engagement, and arousal of the system. The task of the AMI was to be able to read in the data from these different types of sensors asynchronously and process the data accordingly. These data could then be used to help the human. The algorithms for the different types of sensors were created by individuals of the Honeywell team. However, each algorithm was done by a different researcher, and each sensor had different requirements it needed in order for it to function properly. A design problem was that these algorithms needed to have a way to interface easily. The AMI fulfilled this task by creating an extremely simple and flexible interface that could handle the different requirements and algorithms from the different types of sensors.

## 20.5.2 Platform-Based Design

A main goal of the field of augmented cognition is to research and develop technologies capable of extending the information management capacity of individuals working with current computing technologies. AugCog science and technology research and development is therefore focused on accelerating the production of novel concepts in human–system integration and includes the study of methods for addressing cognitive bottlenecks [12].

Augmented cognition systems are complex and resemble embedded and reactive systems, driven by temporal and spatial logic. These components of the system contribute the most advanced technology available in cognitive sensing and information processing bottleneck mitigation [13]. This advanced nature often proves to be unstable. This increases the need of platform-based principles during design.

Component- and platform-based design makes use of modularity. Through small modular systems, the design process is made more manageable instead of a large complex system. This allows many small groups of researchers to focus on specialized functionalities of the system. The platform-based design helps shorten development, deliver customer tailored products, and improve profit [13]. However, platform-based design also makes use of top-down and bottom-up design principles. The top-down approach focuses on the high level view of the system. In turn, the bottom-up approach focuses on individual components, while the system remains constrained by the rest of the system. Other benefits of platform-based design are reuse of design components, shortened design time, and early system verification and validation. By emphasizing the reuse of key components, much of the possible redundant design is avoided.

The typical augmented cognition system is separated into two generalized systems: sensors and interfaces. The sensor system (see Table 20.2 [13]) includes equipment used to monitor and track conditions of the user and surrounding environment. This monitoring provides valuable information on the status of the user's health as well as any possible threat that may be encountered.

**Table 20.2  Sensor System**

| Types | Measures | Monitoring Equipment |
|---|---|---|
| Cognitive | Direct brain | EEG |
| | | fNIR |
| | Psychophysiological | HR, EKG |
| | | Temperature |
| | | EOG |
| | | Pupilometry |
| | | Gaze |
| | | Tracking |
| | | Pulse ox |
| | | Posture |
| | | GSR |
| Environmental | Platform | Fuel |
| | | Location |
| | | Internal conditions |
| | | Weapons |
| | External | Weather |
| | | Hostility |
| | | Chemical/biological agents |
| | | Obstacles |
| | | Situational awareness |

*Note:* List of sensors used in [13].

**Table 20.3   Interface System**

| Type of Interface | Characteristics |
|---|---|
| Visual | Heads up display (HUD) |
| | Traditional display |
| | Alert |
| | Warning |
| | Picture |
| | Text |
| Auditory | Voice |
| | Warning |
| | Spatially locatable |
| Tactile | Warning |
| | Directional cue |

*Note:* All interface units used in [13].

An AugCog system would collect, sort, and prioritize the information for use with an effective interface system (see Table 20.3 [13]). Any combination of visual, auditory, or tactile interfacing equipment relays refined information back to the user for assistance. The most important aspect of augmented cognition systems is the closed-loop nature (see Table 20.2). This closed-loop functionality allows for the human inclusion into this type of system. The human interface acts as the real-time sensor of the system by continuously gathering information.

## 20.5.3 *Mitigating Cognitive Bottlenecks*

Augmented cognition is an experimental technology that aims to bridge the gap between human–computer interactions. This technology seeks to increase the ability of a user via computational processes. These processes focus on bottlenecks, limitations, and biases in cognition that, in turn, would improve decision-making abilities of the end user. The obvious challenges stem from the human–computer interaction and the associated information processing bottlenecks [14].

The rise of this technology derives from the excess of information that people encounter everyday. The concept is to separate valuable information from the useless data and process it at a higher level. Through sorting of the information streams, an AugCog program would drastically increase the computational capability of the user.

In order to increase the capability of the human–computer cognitive system, the nature of the task and the aspects for improvement must be explored from the cognitive perspective. To improve the flow of information between humans and the system, the area where the majority of constrain must be addressed is cognitive bottlenecks. These bottlenecks tend to ultimately compromise the total system performance but allow for the largest gains in improvement. Fundamentally, the source of these bottlenecks lies with the human limitations for gathering, processing, and completing tasks. The inability to multitask and poor short-term memory are only two examples of human-related limitations to the flow of information. Such a wide gap between the abilities of a user and computer leads to designing function schemes that allocate tasks separately between the two. In this case, the result is a fundamentally flawed AugCog system.

The Cognitive Bottleneck Framework (CBF) identifies the most constraining areas in the human–computer cognitive system and draws on research in cognitive psychology, cognitive neuroscience, human–computer interaction design, and computer science [14]. This framework addresses four primary cognitive bottlenecks that could adversely affect the throughput of decision making. These include address sensing and environmental assessment, information processing, information exchange between the human and automated processing system, and acting based on the combined human–automation interpretation of the information.

The first bottleneck is information overload that deals with the inability of a human user to manage the large amounts of information afforded by sensor, database, communication, and display advances. Next, is the sequential cognitive processing. This addresses the information being presented to a user in parallel. However, humans by nature are serial processors and only able to address single threads or information at the time. While possible, most humans do not possess the ability to parallel processes information. The narrow user input capabilities allow the telling of the system of their knowledge, questions, and tasks and are overly constrained. Thus, the human has a lower level of communication with the system than the system has with the human user. Finally, there is the misallocation of functions. This evolutionary development of automation defines the user's role by default as opposed to by design. This leaves the user with tasks that they are not meant to complete.

In order to mitigate these bottlenecks, knowledge of human processing and adaptive automation can be exploited [15]. To mitigate the neuroergonomic adaptive automation, an unobtrusive measure of a user's cognitive capacity has the potential to relieve each of the four cognitive bottlenecks. The approach of adjustable autonomy calls for moving the human–automation interaction, which the user assumes more or fewer duties, depending on the situation as well as the level of interaction of the user. Next, Wickens' Multiple Resource Theory contends that the dynamic allocation of incoming information to the most readily available attention resource pool will avoid overtaxing the operator. In an attempt to reduce the resume time of suspended tasks, an automated system would retain memory of the task status to assist with multitasking support. The automated system could update the user on any changes to the task status that may occur during the suspension period as well as remind the user to complete the task when appropriate. Mitigation of the abstraction of information presentation allows an overloaded user the option to store, reason, and present information at different levels of abstraction to support different users that are completing disparate tasks. Finally, the cross-modality functional logic allows for a single consistent set of functional logic, defined to have equal counterparts in all modalities, to be mixed and matched by their input methods to accommodate the current task demands. The user is not required to finish an operation the way it was started.

The CBF identifies the most crucial constraints in a human–computer system. Without properly addressing these cognitive bottlenecks the information flow in an augmented cognition system will not be effective in the increase in performance in the human–automation system.

## 20.6 Advanced Sensing Technologies for Augmented Cognition

### 20.6.1 Accelerometers [16]

Sensors allow for the recording and analysis of various physical phenomena that are otherwise rather difficult to measure. In order to make a certain phenomenon measurable, a sensor will convert it into a more useful signal. Today's sensors are capable of recording measurements of displacement, velocity, acceleration, force, pressure, chemical concentration, or flow into electrical signals. From the electrical signal produced by the sensor, the original physical parameters can be back-calculated.

Sensor manufacturing technology has recently taken several strides forward. By using applications of semiconductor fabrication technology and applying them to construction of sensors, companies are able to produce a higher-quality product at a lower manufacturing cost. By etching both electrically conductive and nonconductive layers on the silicon wafers, sensors can be created with the elements needed for electrical sensing built in. These new sensors are microelectromechanical systems (MEMS).

Accelerometer sensors are types of sensors that are capable of measuring acceleration values experienced by the sensor, and anything that is attached to the sensor. In order to record the various measurements of acceleration, the accelerometer sensor will split the signal into two distinct signals. The first is the acceleration due to gravity, which can be used to determine the measurement of tilt of the sensor by recognizing which direction is down. The second signal is designated for external acceleration. By allowing the external acceleration to be filtered out, the orientation of the accelerometer sensor can be calculated from the accelerations on various axes [17].

Accelerometer sensors are used to convert linear or angular accelerations into an output signal. The most common way accelerometer sensors obtain this output signal is by an application of Newton's second law of motion. The accelerometer sensor measures the force from acceleration of an object whose mass is known. The most common way found in accelerometers to obtain the force exerted on the mass is by measuring the displacement of the mass when it is suspended by springs. A diagram of this accelerator spring system can be found in Figure 20.5.

The forces that are acting on the proof mass are depicted by

$$\bar{F} = M\bar{a}_{\text{external}} = M\frac{\mathrm{d}^2\bar{x}}{\mathrm{d}t^2} + B(\bar{x})\frac{\mathrm{d}\bar{x}}{\mathrm{d}t} + K(\bar{x})\bar{x}$$

where $K(\bar{x})\bar{x}$ is the restorative force of the spring relative to position, $B(\bar{x})\dfrac{\mathrm{d}\bar{x}}{\mathrm{d}t}$ is the dampening force proportional to velocity, and $M\dfrac{\mathrm{d}^2\bar{x}}{\mathrm{d}t^2}$ is the force from external acceleration. When in equilibrium, the restorative force is equal to the force from acceleration on the proof mass. The displacement of the spring, $x$, can be converted into an electrical signal by various methods. The two most common methods of conversion of the spring displacement are by measuring a change

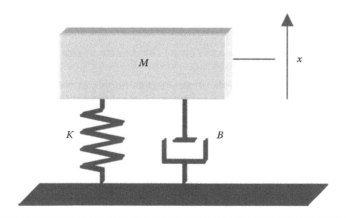

**Figure 20.5** **Diagram of accelerometer spring system.**

**Table 20.4   Types of Involuntary Hand Tremors**

| Cause | Frequency (Hz) | Notes |
|---|---|---|
| Normal hand tremor | 9–25 | Small amplitude |
| Essential tremor | 4–12 | May worsen with position |
| Parkinson's disease | 3–8 | Resting tremor |
| Cerebellar lesions | 1.5–4 | |

in resistance of a piezoresistive material and by measuring a change in capacitance between moving and fixed electrical elements [18].

Typical hand motions can be placed into one of two different categories: voluntary and involuntary. Voluntary hand motion is intended, such as writing on a piece of paper or making a fist. Involuntary hand motion is unintentional. An example of this is small vibrations of a person's hand as he or she is attempting to keep it still. A few types of involuntary hand motions along with their frequency ranges are listed in Table 20.4.

The maximum amplitude of hand motion during a movement is 5 g. Higher accelerations, around 20 g, arise from shocks. These shocks can be caused by various things, such as tapping a metal rod on the ground, allowing it to vibrate the hand [18].

### 20.6.2 Sensor Design Examples

There are various types of accelerometer sensors. The differences between them are the types of sensing element used and the principle of their operation.

Capacitive accelerometers sense changes in electrical capacitance with respect to acceleration to vary the output of an energized circuit. The sensing element of a capacitive accelerometer consists of two parallel plate capacitors acting in differential modes. The capacitors operate in a bridge circuit and alter the peak voltages generated by an oscillator when the sensor is under some acceleration.

In a piezoelectric accelerometer, the sensing element used is a crystal that is able to emit a charge when it is subjected to a compressive force. The crystal is bonded to a mass so that when the accelerometer sensor is subjected to a g-force, the crystal is compressed by the mass. The crystal being compressed emits a signal that can be related in value to the imposed g-force. Piezoresistive accelerometer sensors experience a change in resistance of the material rather than a change in charge or voltage.

In Hall effect accelerometers, the Hall element is attached to a spring with a seismic mass deflecting because of the forces due to acceleration. The element moves in a nonuniform magnetic field. The transverse Hall voltage is proportional to the measured acceleration.

Magnetoresistive accelerometers work by measuring changes in resistance due to a magnetic field. The structure and function is similar to the Hall effect accelerometers, except the magnetoresistive accelerometer measures resistance [18].

### 20.6.3 Distributed Intelligent Sensor Network

Parkinson's disease (PD) occurs mostly in senior people, especially 70 to 90 years of age. Further, PD attenuates the coordination between locomotion and respiratory system. In order to improve the coordination between respiration and locomotion and measure the health status of Parkinson's patients, a mobile sensor network system that concedes long-term data monitoring of steps and respiration and real-time analysis of the locomotion–respiration coordination during rehabilitation, is needed. In mobile sensor network system, there are three subparts: a step detector, a communication coordinator, and a breathing sensor. In this chapter, measuring simple gait analysis, instead of complex, is the goal. Further, because the system aims to be usable during the daily life of a patient, there are five restrictions within the system: (1) patient comfort, (2) low power consumption for long-term operation, (3) continuous monitoring of vital signals, lossless continuous data acquisition (DAQ), and storage for data, (4) signal processing, and (5) communication between three subsystems [19].

The system includes two sensors: "control tower" nodes, named coordinating network operating device (cNODE), and "scout" nodes, called intelligent network operating device (iNODE) [20]. Two iNODE-based sensors are located on both feet to detect the step of a patient, while two other iNODE-based sensors are located on the chest to measure the respiratory data. One cNODE-based sensor is placed on the abdomen. There are four important operations that iNODEs have to handle: (1) DAQ, (2) data storage, (3) signal processing, and (4) communicating with cNODE. For patients' comfort, the iNODE is implemented with flexible PCB, which can be bent and folded into 20 × 20 × 20 mm sizes. The microcontroller (μC) is the MSP430F161 from TI, incorporating a 16-bit reduced instruction set computer CPU and connected with 4-GB memory (SanDisk iNAND) [17]. An operating system is called FRABOX, written in C++ in "IAR Embedded Workbench development environment." This software can implement an optimized direct memory access (DMA) based DAQ system. Figure 20.6 shows the function diagram of a DMA-based DAQ system [19].

After receiving analog acceleration data from an accelerometer, the data pass through an analog to digital converter. The transformed data are stored in an internal register. This event triggers data streaming into one of the buffers by the double-buffer concept of DMA. Each buffer can handle 256 samples of 16-bit each. If buffer 0 is filled, DMA 0 halts its transmission to buffer 0 and transmits the data to buffer 1. While buffer 1 is receiving the data, the data in buffer 0 now move DMA 1 to SPI to SD card. If buffer 1 is filled, buffer 0 is now opened to receive the new data.

A coordinating node called cNODE is made to control the iNODEs. Tasks that cNODE has to function are (1) updating the list of a network member, (2) receiving iNODE messages, (3) synchronizing network time, and (4) providing interface for external users. There are no significant differences in specifications between iNODEs and cNODE, but μC is different. The cNODE is using MSP430 F2618, where it can be driven up to 16 MHz and can store 116-kB flash memory.

Among many different wireless communications, ZigBee, IEEE 802.15.4, has been used for wireless communication system because it is very suitable for low-power and short-distance devices. In addition, within the cNODE, a network layer protocol (iTalk) is implemented to locate iNODEs and enable data exchange between iNODEs.

In step detection, a force sensitive resistor (FSR) sensor is used to get data for step detection. In this chapter, the FSR sensor is designed to identify two simple actions: foot off and foot strike.

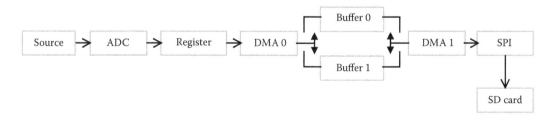

**Figure 20.6  DMA-based DAQ system.**

The output signal is in square waves. Furthermore, only a simple system is applied to handle the locomotion problem: output rectangular signal, whose falling and rising edges arise from the foot-lifting and foot-striking events, respectively. For any discrepancies, the system sets two thresholds, Thresh+ and Thresh-, to strongly detect the data needed.

For respiration, the respiratory inductive plethysmography (RIP), which is noninvasive quantitative assessment of respiratory effort, is implemented. It consists of insulated coils with expandable band that wraps around the rip and abdomen and an electrical device that converts the "variation in self-inductance into a proportional voltage change." The output of this sensor is a low-level ac signal.

Unlike the simple procedure for step detection, a complicated procedure is applied to get the respiratory data. Figure 20.7 shows the overall processes for a respiratory signal. The first thing that the source signal encounters is the analog-to-digital converter (ADC), which converts the 12-bit digital representation outputs into a common fixed-point data-type format. In order to estimate current dominant frequency, a 256-point fast Fournier transform is computed on the respiratory data at a sample rate of 16 Hz. Then, the band-pass filter filters the input into a bandwidth of 0.2 Hz around the spectral peak. Next, by using Hilbert transform we can shift the spectrum $X(f)$ by 90°. The transfer function of Hilbert transform is given below as well as its impulse response [19]:

$$
G_Q(t) \begin{cases} -j & \text{for } f > 0 \\ 0 & \text{for } f = 0 \\ j & \text{for } f < 0 \end{cases} \tag{20.1}
$$

$$
g_Q(t) \begin{cases} \dfrac{1}{\pi t} & \text{for } t > 0 \\ 0 & \text{for } t = 0 \end{cases} \tag{20.2}
$$

Even though the above Hilbert transform function can be used, certain approximations require to be implemented for possible consideration to make the filter causal, as given below:

$$
g_Q(n) \begin{cases} \dfrac{1}{A(n)} \sin^2(A(n)), & 0 < n \leq M \quad n \neq \dfrac{M+1}{2} \\ 0, & n < 0, \quad n = \dfrac{M+1}{2}, \quad n > M \end{cases}
$$

$$
A(n) = \frac{\pi(n - (M+1)/2)}{2} \tag{20.3}
$$

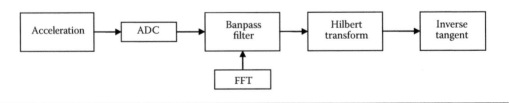

Figure 20.7   Block diagram of a respiratory signal.

where $M$ is the length of the filter. Finally, the convolution of a real signal $x(n)$ and impulse function is given below:

$$x_{ht}(n) = g_Q(n) * x(n) = \sum_{k=0}^{M-1} g_Q(n)\, x(n-k) \tag{20.4}$$

The final process was taking the inverse tangent of the original signal, $x(n)$ and $x_{ht}(n)$.

FSR and RIP sensors were connected with DAQ systems and iNODEs for the test. The average current of the FSR system is 27 μA at 3.6 V operating voltage, while the average current of the RIP system is 120 μA at 3.6 V operating voltage. For the battery, one that has 3.7 V and 200 mAh LiPo is used to supply the voltage. The cNODE consumed more power than iNODE because cNODE has more tasks to accomplish.

The experiment was completed in the Center for Sleep and Rehabilitation Research of the Clinic for Neurology in Hagen, Germany. Two FSR sensors were attached on both feet of the patients when they were walking on the treadmill. The experiment began by collecting the healthy gait from a 28-year-old client. Then the data of PD patients were collected. As a result, there was no significant difference between the two gaits.

To evaluate the respiratory system, the experiment recruited seven PD patients and one healthy 29-year-old. Patients' age ranged from 42 to 84 years old. Like the experiment in the step detection, this experiment was conducted on a treadmill. Even though the main purpose of this section is to get the respiration data, FSRs were applied for the latter experiment. Consequently, PD patients wore the thoracic and abdominal respiratory bands and FSR at the same time. Like step detection, there was no significant difference between PD patients and a healthy person.

To combine the results of step detection and respiratory detection, the common logarithm of the Kuiper Index log10(P) is applied to test the coordination between locomotion and respiratory. Younger patients' Kuiper Index was –11, while older patients' was –7.

### 20.6.4 Human Motion Reconstruction Using Wearable Accelerometers

Traditional optical motion capture system could capture only where there is less interference between sensor and receiver and lighting changes. In order to ameliorate the capability of the motion capture system, Kelly et al. built body-worn wearable wireless accelerometers. Even though this portable motion capture can be used for other purposes, like capturing inconsistent motions, it is most suitable for athletic purposes, like capturing consistent motions [21].

The body-worn wearable wireless motion capture system has three stages within the system: an offline stage, a precapture stage, and the motion reconstruction stage. In the offline stage, the system creates a database that contains a sample set of motion that a player should act. For example, if a sport is tennis, capturing a different stage in serve motion would be a desired task in the offline stage. In this chapter, eight different motions were captured. In the precapture stage, the system assigns a specific acceleration to the sample set of motion from above. In the motion reconstruction stage, the system gathers the player's motion during games and compares with the sample set of motion from the offline stage. The results are computed by using a dynamic programming solution to get a most adequate motion through the graph. Then, the system tells the player which motions have to be improved.

To evaluate this system, the two search algorithms including 16 tennis motion sequences of six different types have been applied. The result of the system was improved compared to Slyper system, a traditional optical motion capture system, in most cases.

The two main objectives of this paper [21] are constructing a system that can capture the motion by using the data from off-the-shelf accelerometers, and producing a comfortable and innocuous outer garment, with five accelerometers attached, for a user to wear. This system is very effective for repeated actions rather than irregular motions. Furthermore, there is some background information mentioned in this paper. For example, Vlasic et al. created a portable system with off-the-shelf accelerometers as well as gyroscopes and acoustic sensors attached on different locations of the human body to capture the motion. The main difference between two systems is that a system made by Vlasic utilizes double integration of accelerations to prevent the drift, while a system from this paper only uses accelerations.

Arduino is used for the platform, because the system has to be constructed with easily achievable hardware from the store. Moreover, the system also consists of a μC containing input and output node and is written in C++ programming language. For the garment, five sewable Lilypad accelerometers (built on triple-axis ADXL330), μC, and conductive thread are sewn onto the shirt, and each accelerometer is attached on each forearm, upper arm, and on the chest. Threads are sewn to a standard snap and an accelerometer. Figure 20.8 shows the locations of the accelerometers. The last component that the garment consists of is "hard" hardware, where the current best hardware is implemented. The hard hardware contains an Arduino USB board with 16:1 mux, where the output is connected with 10-bit ADC of Arduino board. The sampling rate of this device is 125 Hz [8].

The software that is implemented in this paper is written in C++, is run on any Unix platform, and has three threads: serial, search, and display. Before any calibrations, the locations of accelerometers of garments have to be closely matched with the location of accelerometers of a motion capture clip. After matching the locations, the calibration takes place. In this step, the user, wearing the garments with accelerometers, mimics a sample motion from database while watching a motion capture clip. The next step is then run. The user can manipulate the location

**Figure 20.8   Five accelerometers in different body locations.**

of the virtual sensors, software-based, and achieve the polished and optimal acceleration data for specific locations.

The following is the procedure that the software uses for searching. First, the system scans all data set of acceleration, which has 128 frames—a total of 30,000 records. For an individual set, a Haar wavelet transform is computed and the first 15 coefficients retained. These coefficients now combine into a group of vectors along with the joint positions of the pose in the 128th frame. By the way, "15" coefficients and "128" frames are used because the system operates effectively at those coefficients and frames. The system finishes its search in roughly .060 s, even though the larger data sets take more time. After the search is completed, the system sleeps for a few microseconds in order to reduce the errors in transitions between current search and upcoming search. Finally, if the users want to search more after the first search, the new motion clip starts at 10 frames less from the last previous frame. For instance, if the last frame was the 128th frame, the new motion clip starts at the 118th frame.

## 20.7 Conclusions

The technology presented in this chapter elaborates on ways to optimize the outcome of stroke rehabilitation patients. This technology combines technological, social, vocational, and medical techniques that help the patients to recover better. This research uses new technology to teach the ADLs and skills required for community living. The possibilities that can be achieved with this technology have several other uses other than just for stroke rehabilitation. The technology elaborated on in this chapter can allow for increased productivity, safer working environments, as well as the possibility of assisting in the medical fields for rehabilitation other than just for stroke rehabilitation.

The VR section of this chapter includes different types of systems being used or being researched on today. The pneumatic glove allows the doctor to improve the stroke patient's hand movements by controlling the extending and bending of each individual finger. This is done by inflating and deflating an air bladder contained on the palm of the glove. Although this research allows the doctor to control the patient's hand, it allows the patient to become in better control of his/her hand by committing the movements to muscle memory. Another type of research focuses on increasing the patient's skills through "mixed reality" therapy. This is achieved by allowing the patient initially to perform hand movements and receive feedback for these. As the therapy increases, the kinematic process will be able to tell the patient if the gestures are or are not improving. This type of technology is virtual so it can be performed with little manpower and still achieves the common goal of rehabilitation.

Augmented cognition technologies [13,14,22,23] can improve people's lives in a variety of ways. One system tested looked at the complex decision-making skills of individuals. This research tried to eliminate the "tunnel vision" effect when it came to making complex decisions. The researchers gave the subjects decision-making aids to help and teach about improving long- and short-term decision-making skills. Although the complex decision-making skills of the subjects improved, the subjects declined in their information–acquisition skills. Another study tried to develop a computer program in order to study gaze, stress, executive load, novelty, engagement, and arousal of subjects. Measuring these variables can be useful because they can determine a subject's cognitive state. This can be used to activate other augmentations as needed, as in the case of Berkeley's firefighter program and Honeywell's military program. These AugCog technologies, which were not originally developed for stroke rehabilitation, could be adapted for that purpose.

Mainly, these technologies can be adapted by changing the way patients interact with IT and by being repurposed to improve patient safety.

Implementing accelerometers into digital systems in order to record human motion has helped in making the rehabilitation more effective. Using an accelerometer to obtain data of a human's movement, and implementing it into a VR system utilizing augmented cognition, decreases the cost of the rehabilitation while increasing its effectiveness. Utilizing the accelerometer sensors in sync with the augmented VR systems increases the number of actions that can be simulated and provides more accurate acceleration data. By using the acceleration data obtained, it can be implemented into various types of programs that can manipulate it as the user sees fit.

# References

1. L. M. Reeves and D. D. Schmorrow. 2007. Augmented cognition foundations and future directions: Enabling "anyone, anytime, anywhere" applications. In *Proceedings of the 4th International Conference on Universal Access in Human Computer Interaction: Coping with Diversity* (UAHCI'07), Constantine Stephanidis (Ed.). Springer-Verlag, Berlin, pp. 263–272.
2. Fifth Dimension Technologies. 2011. Virtual Reality Pain Distraction System (VRPDS). Available at http://www.5dt.com/products/pvrpds.html.
3. AnthroTronix, Inc. 2012. Acceleglove control in hand. Available at http://www.acceleglove.com/.
4. A. Akl et al. 2011. A novel accelerometer-based gesture recognition system, *IEEE Trans. Signal Process.* 59:12, 6197–6205.
5. M. Johnson, S. Kulkarni, A. Raj, R. Carff, and J. M. Bradshaw. 2005. AMI: An adaptive multi-agent framework for Augmented Cognition. In *Proceedings of the 11th International Conference on Human-Computer Interaction*, Las Vegas, NV.
6. D. Lafond, M. B. DuCharme, M.-È. St-Louis, and S. Tremblay. 2011. Augmenting cognition in complex situation management: Projection of outcomes improves strategy efficiency. In *Proceedings of the IEEE Conference on Cognitive Methods in Situation Awareness and Decision Support*, pp. 296–300.
7. D. Steingart et al. Augmented cognition for fire emergency response: An iterative user study, Technical report. University of California, Berkeley. Available at http://vertex.berkeley.edu/fire/Misc/AugCogFinal.pdf.
8. M. C. Dorneich, P. M. Ververs, S. Mathan, and S. D. Whitlow. 2005. A joint human-automation cognitive system to support rapid decision-making in hostile environments. In *Proceedings of the International Conference on Systems, Man and Cybernetics*, vol. 3, IEEE, Piscataway, NJ, pp. 2390–2395.
9. L. Connelly et al. 2010. A pneumatic glove and immersive virtual reality environment for hand rehabilitative training after stroke. *IEEE Trans. Neural Syst. Rehab. Eng.* 18:5, 551–559.
10. M. Duff et al. 2010. An adaptive mixed reality training system for stroke rehabilitation. *IEEE Trans. Neural Syst. Rehab. Eng.* 18:5, 531–541.
11. M. Ingebretsen and G. Goth. 2008. Augmented-cognition research on the rise. *IEEE Intell. Syst.* 23:1, 4–7.
12. R. Slyper and J. K. Hodgins. 2008. Action capture with accelerometers. In *Proceedings of the 2008 ACM SIGGRAPH/Eurographics Symposium on Computer Animation* (SCA '08). Eurographics Association, Aire-la-Ville, Switzerland, 193–199.
13. L. Marshal and C. Raley. 2004. Platform-based design of augmented cognition systems, ENSE 623, Class project report. Available at http://www.eng.umd.edu/~austin/ense623.d/projects04.d/Marshall-Raley-Paper.pdf.
14. L. M. Reeves, D. D. Schmorrow, and K. M. Stanney. 2007. Augmented cognition and cognitive state assessment technology: near-term, mid-term, and long-term research objectives. In *Proceedings of the 3rd International Conference on Foundations of Augmented Cognition* (FAC '07), D. D. Schmorrow, L. M. Reeves, M. B. Russo, and M. Crosby (Eds.). Springer-Verlag, Berlin, pp. 220–228.

15. P. E. Taylor et al. 2010. Classifying human motion quality for knee osteoarthritis using accelerometers. *Conf. Proc. IEEE Eng. Med. Biol. Soc.* 339–343.

16. P. Kelly, C. Ó. Conaire, and N. E. O'Connor. 2010. Human motion reconstruction using wearable accelerometers. In *SCA 2010—ACM SIGGRAPH/Eurographics Symposium on Computer Animation*, Madrid, Spain.

17. B. B. Graham. 2000. Using an accelerometer sensor to measure human hand motion, MS thesis. Department of Electrical Engineering and Computer Science, Massachusetts Institute of Technology, Cambridge. Available at http://www-mtl.mit.edu/researchgroups/MEngTP/Graham_Thesis.pdf.

18. T. R. Burchfield and S. Venkatesan. 2007. Accelerometer-based human abnormal movement detection in wireless sensor networks. In *Proceedings of the 1st ACM SIGMOBILE International Workshop on Systems and Networking Support for Healthcare and Assisted Living Environments* (HealthNet '07). ACM, New York, 67–69.

19. A. A. Gopalai and S. M. N. A. Senanayke. 2008. 2D human motion regeneration with stick figure animation using accelerometers. *Int. J. Electr. Comput. Eng.* 3:7, 55–60.

20. H. Ying, M. Schlösser, A. Schnitzer, T. Schäfer, M. E. Schläfke, S. Leonhardt, and M. Schiek. 2011. Distributed intelligent sensor network for the rehabilitation of Parkinson's patients. *Trans. Info. Tech. Biomed.* 15:2, 268–276.

21. S. Yang and M. Gerla. 2010. Energy-efficient accelerometer data transfer for human body movement studies. In *Proceedings of the 2010 IEEE International Conference on Sensor Networks, Ubiquitous, and Trustworthy Computing* (SUTC '10). IEEE Computer Society, Washington, DC, pp. 304–311.

22. M. Y. Ivory, A. P. Martin, R. Megraw, and B. Slabosky. 2005. Augmented cognition: An approach to increasing universal benefit from information technology. In *Proceedings of the 1st International Conference on Augmented Cognition*, Las Vegas, NV.

23. F. Hu, X. Cao, D. Brown, J. Park, M. G. Q. Sun, and Y. Wu. 2013. Tele-rehabilitation computing: From an cyber-physical perspective. In *Tele-Healthcare Computing and Engineering: Principles and Design*, Chapter 4. CRC Press, Boca Raton, FL.

*Chapter 21*

# Using Wiimote and Kinect for Cognitive Rehabilitation: Toward an Intelligent Sensor/ Image Signal Processing

Fei Hu, Yufan Wang, Xiaojun Cao, Qi Hao, and David Brown

## Contents

## 21.1 Introduction

Disability is a critical issue in the well-being of any country. For example, around 1.9% of the US population reported some form of paralysis in 2009 [1], which affects the life quality of about 5,596,000 people [1]. The socioeconomic burden imposed on the patients, relatives, and caregivers is enormous. For example, since 1994, disability-related costs for medical care and lost productivity have exceeded an estimated $300 billion annually in the United States [2]. Efficient rehab training could greatly help to recover functional body *coordination and flexibility* (C&F) capabilities. Efficient rehab training is especially important to the prevention of disabilities for stroke survivors because stroke is the leading cause of disability among adults, and more than 4 million people in the United States have suffered a stroke and are living with poststroke disabilities [3]. Of the 730,000 individuals who will survive a stroke each year, 73% will have residual disability [4]. Stroke is also the third leading cause of death in the United States [5]. Other neurodisorder diseases such as Parkinson's disease and cerebral palsy are also accompanied by disabilities. Therefore, there is an urgent need to investigate new, intelligent rehab methods to efficiently restore C&F of people with various disabilities.

### 21.1.1 Importance of Robot-Aided Rehab Training

Today, many in-clinic rehab training methods are based on labor-intensive assistance from clinicians such as physical/occupational therapists. Because the patient needs to practice various hazard-based training (e.g., intentionally using slippery surface to walk) [6], it is difficult and dangerous for a clinician to hold the patient all the time to protect them from training hazards. To overcome such an issue, we have used our preinvented robot-aided rehab training system, the KineAssist [7], to achieve complex training tasks within the context of hazard environments, such as perturbation recovery training and gait training in hazardous conditions [8].

### 21.1.2 Importance of Virtual Reality-Based Rehab System

A virtual reality (VR)-based rehab system allows the trainee to recover his or her body C&F through various virtual scenes without the need of going through complex real environments. In our research, the digital signals collected from devices, including the neuroimages from the functional near-infrared (fNIR)/electroencephalogram (EEG) devices, trajectories from the digital glove, eye focus data from the eye tracker, and other signals, can be processed to *quantitatively* analyze the rehab training effects. Moreover, we extend the VR system to a *mixed reality* (MR) platform, which can embed a reconstructed 3D body motion graphics into the VR scenes. The MR facilitates the trainee to be more interactive with the scenes.

In this chapter, we will first introduce our robot-aided rehab training platform. We will discuss its robot, programmable treadmill, and VR/MR system. Then, we will provide our research results on the use of Wiimotes and Microsoft Kinect for gait training. The correct gaits are important to the disability recovery for poststroke patients.

## 21.2 Cognitive Rehab Research Platform

We are enhancing our current rehab research platform [9,10] through the integration of the corobot (i.e., KineAssist), treadmill, mixed reality, and wireless gait/medical sensor networks:

1. *Robot-aided training*: It has been found that body weight supported treadmill training is effective in improving locomotor flexibility after a stroke [11–13]. Hyndman et al. [14] found that repeat fallers had significantly reduced walking ability compared with those who did not fall. Our platform integrates our preinvented KineAssist [7] (Figure 21.1a) with the VR system. KineAssist has a cognitive training feature called intent-based holding: when a trainee moves normally, it operates in a transparent way; that is, it can move quickly in the intended direction in response to a small force from the trainee. When he/she loses his/her balance, it smoothly catches him/her through the dynamic model of the stiff spring coupled to a damper (Figure 21.1b).

2. *Treadmill–computer interface*: We adopt a new, powerful treadmill system (model: Desmo) from Woodway [15], which supports treadmill-on-demand, real-time belt control through

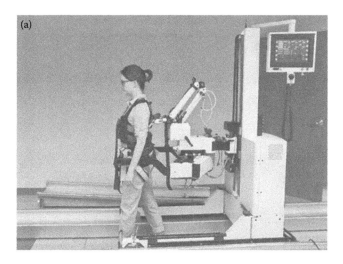

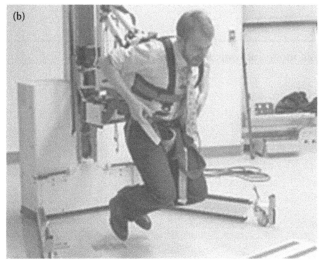

**Figure 21.1** (a) Robot-aided rehab and (b) intent-based holding.

the RS-232 interface to a computer (Figure 21.2). The treadmill control programs can be embedded into our VR animations.

3. *Mixed reality*: Our platform is able to reconstruct a trainee's 3D body gesture and motion and overlay them with the VR interactive scenes. Such a design is called mixed reality (MR). We adopt inexpensive depth cameras (Microsoft Kinect, <$200 each) to capture 3D human motions at video rate. We can avoid the interference between cameras through only three Kinects deployed as shown in Figure 21.2: two Kinects capture the upper and lower parts of a human body, respectively, and a third Kinect is used to capture the middle body part. We will use the free software, KinectFusion [16], for accurate 3D body motion reconstruction. It can even capture the dress wrinkles (Figure 21.3). We use 3DVIA Virtools and other VR tools to embed the Kinect-based 3D motions into the rehab-oriented VR animation scenes.

4. *Wireless gait/medical sensor networks*: Our rehab platform integrates our previously designed medical sensors such as electrocardiogram (ECG) and EEG [17–21] with the VR system.

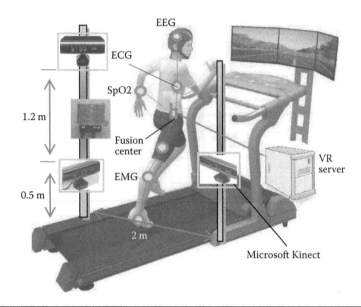

**Figure 21.2   Treadmill mode with mixed reality.**

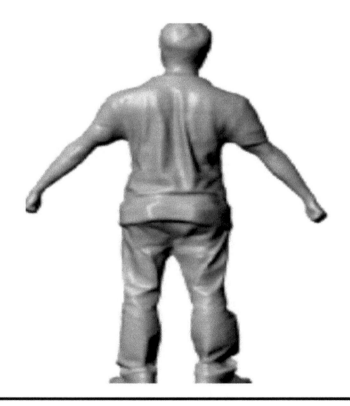

**Figure 21.3   3D reconstruction.**

There exists a wireless interface between the medical fusion center and the VR server (Figure 21.2). The treadmill automatically stops if any sensor detects health issues (such as premature heart beats). In addition, we have built a low-cost gait network based on pyroelectric sensors for accurate gait pattern capture [22–27]. Such a gait network can be deployed around the trainee to recognize various walking/running gaits (more details in the next section).

Our above proposed cyber-physical system uses a *heterogeneous* network that consists of digital glove [28], acoustic sensors [29], Wiimote chips (accelerometers) [30], etc. (see Figure 21.4) to recognize various abnormal gaits that have only minor differences between them (thus needs much more accurate pattern extraction than our previous system). More importantly, it could accurately recognize patient *motor disorders* through integrated hand, arm, and leg motion analysis. We achieve the high-fidelity recognition of various gait anomalies and motor disorders through *information geometry* (IG)-based multimanifold fusion and Bayesian learning.

Besides poststroke training, our research also has significant social impacts once successful. On the basis of a World Health Organization report [31], over 1 billion people worldwide suffer from neurodisorder diseases (NDDs), ranging from epilepsy to Parkinson's disease. Every year over 7 million people die as a result of NDDs [12]. As an example, in the United States alone, NDDs cost over $148 billion per year, and the annual cost of care for each NDD patient is over $64,000 per year [32]. Before determining the medication or surgery treatments for many different types of NDDs (>50 [33]), a neurology doctor needs to analyze the patient's concrete NDD symptoms, which include abnormal gaits or motor disorders.

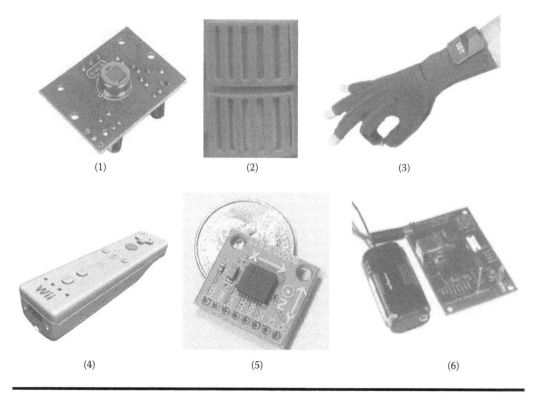

(1)  (2)  (3)

(4)  (5)  (6)

**Figure 21.4** **(1) Pyroelectric sensor, (2) lens, (3) digital glove, (4) Wiimote, (5) accelerometer, and (6) our designed wireless board.**

## 21.3 Abnormal Gait Capture via Geometry-Preserved Non-Negative Matrix Factorization (NMF)

It is important to capture different abnormal gaits during rehab training. By comparing them with a prestored gait pattern database, we will know whether or not the detected abnormal gait corresponds to a certain neurodisease. Our designed pyroelectric sensors (Figure 21.3) could detect the gait of an approaching walker. We have used machine learning algorithms (such as Bayesian patient controlled analgesia and hidden Markov model [HMM]) to recognize the *normal* gaits of different walkers [22–27]. In this research, we extend our previous work on the recognition of different patients' normal gaits to the same patient's various abnormal gaits. Such an extension is not trivial for two reasons:

1. Higher gait recognition accuracy is needed: While it is relatively easier to detect different walkers' normal gaits because of people's obvious walking habit differences, it is difficult to distinguish among various abnormal gaits because of their minor differences. For example, the borderland between two types of motor disorders (epileptic seizures and paroxysmal movement disorders) is not obvious [34]. It needs high-fidelity gait recognition schemes to distinguish among the following daytime motor disorders: episodic ataxia, stereotypies, drop attacks, paroxysmal kinesigenic dyskinesia, and other symptoms.

   In order to achieve higher resolution gait recognition, we have redesigned our pyroelectric sensors with a richer gait sensing mode. Inspired by insect compound eyes (Figure 21.5), we have designed the Fresnel lens with interleaved, small thermal signal filters. Such a special lens architecture could segment the surrounding thermal detection space into many delicate regions and thus improve the gait detection sensitivity by checking which regions have signals.

2. The temporal and spatial correlation between different gait sensors' signals needs to be maintained during gait signal processing: In order to accurately record the gait features from different angles, we propose to use pyroelectric sensor array in each deployment position (Figure 21.6). Data correlation exists between different array signals. One is intraarray geometry: in the same sensor array, their data have high similarity since they sense the same body part's gait. The other is interarray geometry: between different sensor arrays, A and B (Figure 21.6) should have stronger spatial correlation than A and C because A and B catch gaits in the same vertical line. Additionally, A and C should have stronger spatial/temporal

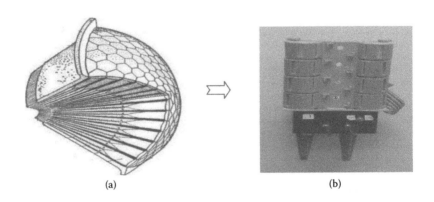

(a)  (b)

**Figure 21.5** **(a) Insect compound eyes and (b) Fresnel lens.**

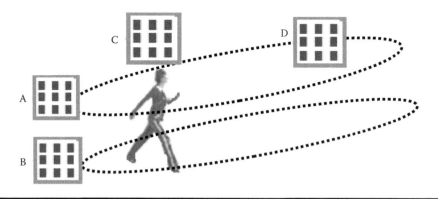

**Figure 21.6 Geometry-aware gait sensing.**

correlation than A and D due to two reasons: First, A and C are closer to the patient while D is far away. Second, A and C capture signals with closer temporal correlation when the patient walks from A to C.

Our research seeks a pattern extraction solution for high-dimensional array signals. Such a solution can maintain sensor *array geometry structure* (*AGS*) for delicate gait discrimination. For example, since sensor arrays A and B record body motions in the same location (but deployed in different heights to capture arms' and legs' motions, respectively), by maintaining their AGS we can describe the entire body's gait features. By maintaining A and C, and A and D pairs' AGSs, we guarantee the repeatability and consistency of all extracted gait patterns because we can analyze the signals along the patient's walking path (in this example, it is A→C→D). In order to retain such intraarray and interarray AGS information, we extend our previous general NMF-based gait recognition model [22] to a *geometry-preserved NMF* (denoted as *gNMF*) based on *graph embedding* models [35]. The gNMF model can maintain the geometric structure among neighboring arrays' signals even after we map the original high-dimensional sensor array signals to a low-dimensional gNMF feature subspace.

Because NMF uses iterative $W$ (basis matrix) and $H$ (feature matrix) updates [36], it is time consuming to analyze a *large* observation time window. Therefore, before gNMF analysis, we first use the online signal segmentation (described in section 5.1) to limit our gNMF analysis within a window with proper size. For a window of sensor array data, we can use a *weighted graph* $G = \{X, S\}$ to represent the geometric relation between all data points. Here, $X = [x_1, x_2,..., x_n]$, and $n$ is the window size (i.e., how many data points). $S = \{S_{ij}\}$ is the *graph similarity* matrix, which can be formed through a *Gaussian kernel* denoted as $\mathrm{Exp}(\|x_i - x_j\|^2/t)$, or using other kernel methods [37]. The diagonal matrix $E$ of the graph is $E = \{S_{ii}\}$, and Laplacian matrix $L = E - S$. For each original point $x_i$, we map it to gNMF low-dimensional subspace through $\tilde{x}_i = W^T x_i$. All mapped points form a data matrix: $\tilde{X} = [\tilde{x}_1, \tilde{x}_2,..., \tilde{x}_n]$.

In order to retain the AGS information of the original weighted graph, we could add a new constraint to the original NMF's cost function as follows:

$$\mathrm{Cost}(X \parallel WH) = KL(X \parallel WH) + \zeta \left( \sum_{ij} \| \tilde{x}_i - \tilde{x}_j \|^2 S_{ij} \right)$$

Here $KL$ (.||.) is K-L divergence. Obviously, the second item of the above function penalizes the graph distance of two data points that are far away from each other (thus, with less geometric similarity).

Besides the reservation of AGS information between neighboring sensor array readings, another important advantage of gNMF is its capability of suppressing redundant dimensionality from high-dimensional signals. This is due to its matrix factorization nature and the use of kernel mapping. We have verified gNMF's signal suppression characteristics through the example of fence data analysis (Figure 21.7a). While general NMF generates basis vectors with detailed feature points (Figure 21.7b), gNMF generates very *sparse* feature basis vectors (Figure 21.7c).

The complete solution to gNMF (i.e., solving $W$ and $H$ through a gradient optimization) involves a *stationarity-guaranteed conditional optimization* problem. Conventional NMF uses simple multiplicative update through an *auxiliary function*. However, it cannot guarantee the algorithm to converge to a stationary point. We can form a conditional optimization problem in order to find such a stationary convergence point (here we use the basis matrix $W$'s update rule as an example):

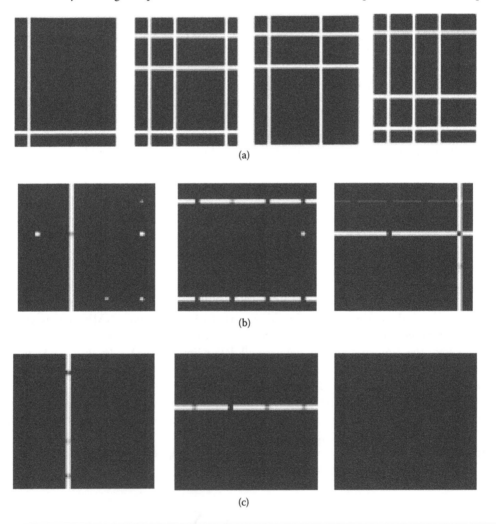

Figure 21.7    (a) Fence data, (b) feature basis obtained from general NMF, and (c) basis from gNMF.

$$W^{(t+1)} = [W^{(t)} - \beta^\gamma \nabla KL(X \| WH)^{(t)}]^+$$

Here $[.]^+$ means max $[., 0]$ (thus, it keeps nonnegative entries) and $KL$ $(.\|.)$ is the K-L divergence between two distributions. Here, $\beta$ can be set up to a constant based on some empirical analysis. The value of $\gamma$ is the first nonnegative integer that meets the following inequality:

$$KL(X \| WH)^{(t+1)} - KL(X \| WH)^{(t)} = c < \nabla KL(X \| WH)^{(t)}, (W^{(t)}(\beta^\gamma) - W^{(t)}) >$$

Here $c$ is a constant and $<.,..>$ is the *Frobenius inner product* between two matrices.

It is also important to see if the trainee can achieve a specified static pose such as raising the leg to a certain height. We use a low-cost depth camera, Microsoft Kinect (Figure 21.3), to capture the 3D pose depth data, and then we perform normalization to remove the noise. A kinematic chain [38] is used to model the body pose. Finally, we compare the captured pose data to the prestored pose template database to see how well the trainee can achieve a specified pose. The difference between two poses can be measured by the average Euclidean distance of the corresponding kinematic joint positions. The static pose measurement procedure is illustrated in Figure 21.8.

## 21.4 Experimental Results

Here, we report part of our experimental results on Wiimote-related training (other results will be published in a journal paper). Our purpose is to use the accelerometers of the Wiimotes to automatically recognize the patient's arm gestures, and match with a pretrained template database to see whether the patient's arm gestures are correct or not.

We have defined 14 complex arm gestures as shown in Table 21.1.

For all the 14 gestures, two data sets have been generated.

Data set 1: 1 subject, 14 gestures, 25 repeats of each gesture, with 1 Wiimote in hand
Data set 2: 1 subject, 14 gestures, 25 repeats of each gesture, with 2 Wiimotes (one is in hand, the other is bound with the upper arm)

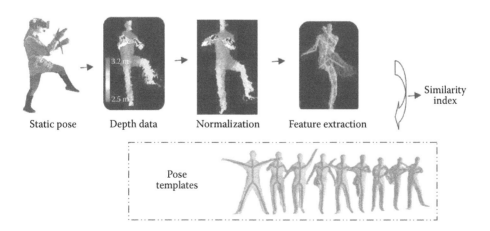

**Figure 21.8 Static pose training effect measurement.**

**Table 21.1  Complex Arm Gestures**

| No. | Gesture Terms | Gesture Descriptions |
|-----|---------------|----------------------|
| 1 | Forward reach | Start with the elbow at the side of the body and extend arm straight out slightly lower than shoulder height; return to starting position. |
| 2 | Upward reach | Start with the elbow at the side of the body with the wrist near the shoulder and extend arm straight up; return to starting position. |
| 3 | Wiping motion | With the elbow near the side of the body and the arm bent, rotate forearm from roughly 45° inside the frame of the body to 45° pointing away from the body; return to starting position. |
| 4 | Sawing motion | Start with the elbow bent and near the side of the body; move arm forward and back roughly 6 in. |
| 5 | Touching face | Start with the elbow near the side of the body, arm bent, palm up, then touch the palm to the forehead; return to starting position. |
| 6 | Lifting using forearm (dumbbell curl) | Keeping the elbow stationary on a table, lift the forearm from parallel to the table to a 90° angle; return to starting position. |
| 7 | Rotating elbow | With the arm bent and keeping the wrist at a stationary point, the elbow was rotated from straight down to 90° pointing away from the body; return to starting position. |
| 8 | Lifting barbell | Start with the arm straight down near the side of the body, bend arm upward to the front of the breast, like lifting a barbell. |
| 9 | Stretching to level 1 | Start with the arm straight down near the side of the body; stretch the arm forward to level in front of body. |
| 10 | Stretching to level 2 | Start with the arm straight down near the side of the body; stretch arm out forward to level to the side of body. |
| 11 | Coving mouth | Start with the arm straight down near the side of the body; bend arm upward to 4–6 in. in front of the mouth. |
| 12 | Stretching upward 1 | Start with the arm straight down near the side of the body; stretch arm upward to the highest point while keeping arm in front of the body. |
| 13 | Stretching upward 2 | Start with the arm straight down near the side of the body; stretch arm upward to the highest point while keeping arm in the side of the body. |
| 14 | Dumbbell flying | Start with the arm bent in front of the breast; move the arm levelly to the back of the body while keeping bent. |

*Note:* First 7 gestures are referred from a B.S. paper from Massachusetts Institute of Technology (http://www.nlm.nih.gov/medlineplus/tutorials/strokerehabilitation/htm/index.htm).  Others are designed according to stroke rehabilitation tutorial materials (http://groups.csail.mit.edu/lbr/hrg/2001/learn_gest.pdf.)

The basic NMF method has been implemented. The whole system is depicted in Figure 21.9 and described step by step in the following:

Step 0. Data preprocess: The training data are preprocessed with moving slide window and interpolated to the same length as the longest gesture.

Step 1. Training stage (V-construction): Two different ways have been implemented to construct the nonnegative matrix $V_{n \times m}$, which is to be factorized to $W_{n \times r}$ and $H_{r \times m}$.

The first way is to directly use the data of each gesture as a row of $V$, as shown in the following graph:

$$\begin{bmatrix} x_1 & y_1 & z_1 \\ x_2 & y_2 & z_2 \\ \cdots & \cdots & \cdots \\ x_k & y_k & z_k \end{bmatrix} => \begin{bmatrix} x_1 & x_2 & \cdots & x_k & y_1 & y_2 & \cdots & y_k & z_1 & z_2 & \cdots & z_k \end{bmatrix}$$

The highest average recognition rate of this method for different basis factor $r$ is only 44.1% on data set B.

The second way is to take each gesture data as an image, which can be easily understood with the graph in Figure 21.10.

The second $V$-construction method will generate a very big matrix $Vn$, for example, for data set 1, $n$ = range * max_length = 150 * 727 = 109,050; here, 150 is the range of the raw acceleration data and 727 is the size of the longest repeat in the data set (727 sampling point, about 7.27 s). $m$ = 14 * 10 = 140, and 10 of 25 repeats of each gesture are used as training data. A total of 14 gestures are processed, and $r$ = 19 is selected as the optimal basis factor.

According to the trial on data set 1 (20 iterations), the average recognition rate is 91%, which is much better than the first $V$-construction method but is not as good as the dynamic time warping (DTW), affliation propagation (AP), compressive sensing (CS) method [39],

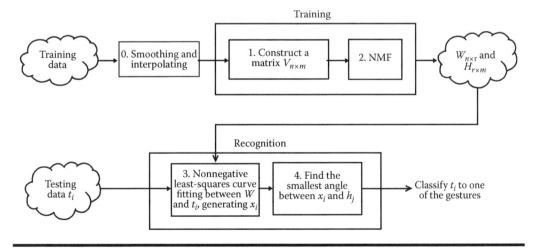

**Figure 21.9  NMF-based arm gesture recognition.**

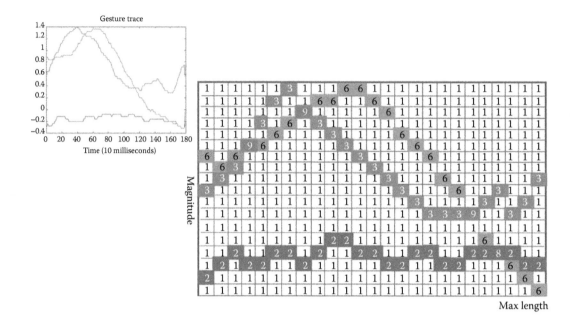

**Figure 21.10  Image-based gesture recognition.**

which is 96.8% in the same trial. The NMF method with the second *V*-construction method consumes less computation time (about half in total time) than the DTW-AP-CS method. The average cost of time (seconds per iteration) is shown in Table 21.2 (in the same trial on data set 1).

Step 2. Training stage (NMF): The basic NMF method is employed.

Step 3. Testing stage: Calculate the coefficients of test gesture regarding to *W*.

For each test gesture $t_{n \times 1}$ (after smoothing and interpolating), which also has *n* columns, solve the *nonnegative least squares curve fitting problems* between *W* and *t* (with a MATLAB function *lsqnonneg*).

$$\min_{x} \left\| Wx - t \right\|_2^2$$

The result $x_{r \times 1}$ includes the coefficients of the basis images regarding to *t*.

Step 4. Testing stage: Find the most similar coefficients in *H*.

**Table 21.2  Average Cost of Time per Training Phase**

|  | DTW-AP-CS | NMF |
|---|---|---|
| Training phase | 325.4 | 135.7 |
| Testing phase | 101.9 | 79.9 |
| Total | 427.3 | 215.6 |

For each of the $m$ column vector in $H$, calculate its *cosine similarity* with $x$.

$$\theta_j = \arccos\left(\frac{x'h_j}{\|x\| \times \|h_j\|}\right)$$

The smallest $\theta_j$ should generate from the most similar gesture in the training set and the test gesture $t_{n\times1}$.

Overall, for data set 1, our above algorithm has reached 96.8% recognition rate. For data set 2, we have reached 97.1% recognition rate.

# References

1. Reeve-Irvine Research Center. 2009. Approximately 1.9% of the U.S. population, or some 5,596,000 people, report some form of paralysis, among whom, some 1,275,000 are paralyzed as the result of a spinal-cord injury. Christopher & Dana Reeve Foundation, University of California, Irvine 2009. Available at http://www.reeve.uci.edu/.
2. US Department of Health and Human Services. The Surgeon General's call to action to improve the health and wellness of persons with disabilities. Office of the Surgeon General, Washington, DC. Available at http://www.surgeongeneral.gov/library/calls/disabilities/index.html.
3. University Hospital. 2010. Stroke statistics. Newark, NJ. Available at http://www.theuniversityhospital.com/stroke/stats.htm (Accessed on May 2012).
4. G. E. Gresham et al. 1995. Post-stroke rehabilitation clinical practice guidelines. No. 16. Agency for Health Care Policy and Research, Public Health Service, US Department of Health and Human Services, Rockville, MD.
5. Centers for Disease Control and Prevention. 2009. Leading causes of death. Available at http://www.cdc.gov/nchs/fastats/lcod.htm (Accessed on May 2012).
6. P. Y.-C. and T. S. Bhatt. 2007. Repeated-slip training: An emerging paradigm for prevention of slip-related falls among older adults. *Physical Therapy* 87, 1478–1491.
7. HDT Robotics—KineAssist: http://devwww.hdtglobal.com/services/robotics/portfolio/KineAssist/. For more detailed materials on KineAssist, see http://www.kineadesign.com/portfolio/kineassist/. Note that Kinea Design Inc. is now renamed as HDT Robotics Inc., a division of HDT Global.
8. J. Patton, E. Lewis, G. Crombie, M. Peshkin, E. Colgate, J. Santos, A. Makhlin, and D. A. Brown. 2008. A novel robotic device to enhance balance and mobility training post-stroke. *Topics in Stroke Rehabilitation* 15.2, 131–139.
9. F. Hu et al. 2013. Tele-rehabilitation computing: From a cyber-physical perspective. In *Tele-Healthcare Computing and Engineering: Principles and Design*. CRC Press, Boca Raton, FL, pp. 85–99.
10. F. Hu, X. Hu, Q. Sun, and M. Guo. 2013. Neuro-disorder patient monitoring via gait sensor networks: Towards an intelligent signal processing. In *Intelligent Sensor Networks: The Integration of Sensor Networks, Signal Processing, and Machine Learning*. CRC Press, Boca Raton, FL, pp. 181–204.
11. K. J. Sullivan, B. J. Knowlton, and B. H. Dobkin. 2002. Step training with body weight support: Effect of treadmill speed and practice paradigms on poststroke locomotor recovery. *Arch. Phys. Med. Rehabil.* 83, 683–691.
12. S. Hesse, M. Malezic, A. Schaffrin, and K. H. Mauritz. 1995. Restoration of gait by combined treadmill training and multichannel electrical stimulation in non-ambulatory hemiparetic patients. *Scand. J. Rehabil. Med.* 27, 199–204.
13. M. C. Kosak and M. J. Reding. 2000. Comparison of partial body weight-supported treadmill gait training versus aggressive bracing assisted walking post stroke. *Neurorehabil. Neural Repair* 14, 13–19.

14. D. Hyndman, A. Ashburn, and E. Stack. 2002. Fall events among people with stroke living in the community: circumstances of falls and characteristics of fallers. *Arch. Phys. Med. Rehabil.* 83, 165–170.

15. Woodway treadmill (Model: Desmo). Available at http://medical.woodway.com/sports_medicine/ sports_medicine_desmo.html.

16. R. A. Newcombe, S. Izadi, O. Hilliges, D. Molyneaux, D. Kim, A. J. Davison, P. Kohli, J. Shotton, S. Hodges, and A. Fitzgibbon. 2011. KinectFusion: Real-time dense surface mapping and tracking. In IEEE ISMAR, October 2011: General introduction on Open Source Kinect Fusion—Instant Interactive 3D Models. Available at http://www.i-programmer.info/news/144-graphics-and-games/3616-open-souce-kinect-fusion-instant-interactive-3d-models.html.

17. F. Hu, Y. Xiao, and Q. Hao. 2009. Congestion-aware, loss-resilient bio-monitoring sensor networking. *IEEE J. Selected Areas Commun.* 27(4), 450–465.

18. F. Hu, L. Celentano, and Y. Xiao. 2009. Error-resistant RFID-assisted wireless sensor networks for cardiac tele-healthcare. *Wireless Commun. Mobile Comput.* 9, 85–101.

19. F. Hu, Y. Wang, and H. Wu. 2006. Mobile telemedicine sensor networks with low-energy data query and network lifetime considerations. *IEEE Trans. Mobile Comput.* 5(4), 404–417.

20. F. Hu, M. Jiang, M. Wagner, and D. Dong. 2007. Privacy-preserving tele-cardiology sensor networks: Towards a low-cost, portable wireless hardware/software co-design. *IEEE Trans. Inf. Tech. Biomed.* 11(6), 617–627.

21. F. Hu, Q. Hao, M. Qiu, and Y. Wu. 2009. Low-power electroencephalography (EEG) sensing data RF transmission: Hardware architecture and test. In *Proceedings of the First ACM International Workshop on Medical-Grade Wireless Networks*. ACM, New York, pp. 57–62.

22. Q. Hao, F. Hu, and J. Lu. 2010. Distributed multiple human tracking with wireless binary pyroelectric infrared (PIR) sensor networks. In *2010 IEEE Sensors*. IEEE, New Brunswick, NJ, pp. 946–950.

23. X. Zhou, Q. Hao, and F. Hu. 2010. 1-bit walker recognition with distributed binary pyroelectric sensors. In *2010 IEEE Conference on MultiSensor Fusion and Integration for Intelligent Systems*. IEEE, New Brunswick, NJ, pp. 168–173.

24. F. Hu, Q. Sun, and Q. Hao. 2010. Mobile targets region-of-interest via distributed pyroelectric sensor network: Towards a robust, real-time context reasoning. In *2010 IEEE Sensors*. IEEE, New Brunswick, NJ, pp. 1832–1836.

25. Q. Sun, F. Hu, and Q. Hao. 2010. Context awareness emergence for distributed binary pyroelectric sensors. *2010 IEEE Conference on MultiSensor Fusion and Integration for Intelligent Systems*. IEEE, New Brunswick, NJ, pp. 162–167.

26. Q. Sun, F. Hu, and Q. Hao. 2012. Mobile targets scenario recognition via low-cost pyroelectric sensing system: Towards an accurate context identification. *IEEE Trans. Syst., Man Cybernetics: Systems*.

27. Q. Hao, F. Hu, and Y. Xiao. 2009. Multiple human tracking and recognition with wireless distributed pyroelectric sensor systems. *IEEE Syst. J.* 3(4), 428–439.

28. Reprogrammable wired data glove: http://www.technologyreview.com/computing/22838/. Also see 5DT wireless glove: http://www.5dt.com/products/pdataglove_wirelesskit.html (Accessed on October 2011).

29. An example high-sensitive acoustic sensor: See the following web site for some specification data: http://www.bksv.com/Products/TransducersConditioning/AcousticTransducers/Microphones/4953.aspx (Accessed on November 2011).

30. 3-Axis accelerometers (TG series from Memsic): http://www.memsic.com. It can connect any Memsic wireless motes (such as Iris) to achieve wireless sensing data collection.

31. Neuro-disorder data (2007) from World Health Organization (WHO) report (visited in November 2011): http://www.who.int/mediacentre/news/releases/2007/pr04/en/index.html.

32. Neurodisorder cost in the United States. Please refer to the following site for detailed statistics: http://www.silverbook.org/browse.php?id=52.

33. Wikipedia Foundation, Inc. 2012. Types of neuro-disorders symptoms. Available at http://en.wikipedia.org/wiki/List_of_neurological_disorders.

34. C. W. Huang and W. J. Hwang. 2009. The borderland between epilepsy and movement disorders. *Acta Neurol. Taiwan* 18(1), 42–55.

35. X. Liu, S. Yan, and H. Jin. 2010. Projective nonnegative graph embedding. *IEEE Trans. Image Process.* 19(5), 1126–1137.
36. D. D. Lee and H. Sebastian Seung. 2001. Algorithms for non-negative matrix factorization. In *Advances in Neural Information Processing Systems 13: Proceedings of the 2000 Conference.* MIT Press, Cambridge, MA, pp. 556–562.
37. S. V. N. Vishwanathan, N. N. Schraudolph, I. R. Kondor, and K. M. Borgwardt. 2010. Graph kernels. *J. Machine Learning Res.* 11, 1201–1242.
38. R. M. Murray, Z. Li, and S. S. Sastry. 1994. *A Mathematical Introduction to Robotic Manipulation.* CRC Press, Boca Raton, FL.
39. A. Akl. 2011. A novel accelerometer-based gesture recognition system. *IEEE Trans. Signal Process.* 59(12), 6197–6205.

*Chapter 22*

# Functional Near-Infrared Spectroscopy for Autorehabilitation Cyber-Physical Systems: Toward Intelligent Signal Processing for Rehab Progress Indication

Fei Hu and David Brown

## Contents

## 22.1 Autorehabilitation System

Most disabilities are from neurodiseases (such as stroke and brain trauma or other nonnatural reasons such as arthritis). Many in-clinic rehab-training methods are still based on labor-intensive assistance from physical therapists (PTs). Because the patient needs hazard-based training (e.g., intentionally using slippery surface to walk or using intervention to reduce fall risk) [1,2], it is very difficult and tiring for a PT to hold the patient all the time to overcome these training hazards, especially when the patient has heavy body weight. To overcome these issues, we have invented a corobotic system called KineAssist [3] to achieve complex rehab-training tasks with the support of hazard environment, such as progressive gait training, perturbation recovery training, and functional balance training [4]. While KineAssist greatly relieves PTs from intensive involvement, it still needs the continuous training progress monitoring and manual setup of each patient–environment interaction scene (e.g., holding a ball to ask the patient to grab it). Moreover, it lacks *quantitative, automatic* mechanisms to measure the patient's body coordination and flexibility (C&F) levels. The PT needs to estimate the C&F levels based on subjective judgment.

We have performed a project that crosses multiple disciplines including computing, medical science, and robotics. Our project goal is to build an innovative autorehabilitation platform through a tightly coupled cyber-physical system (CPS) (as shown in Figure 22.1); we call it RAPID (robot-aided virtual reality [VR] for cyber-controlled physical training of individuals with disabilities). We are significantly improving our invented KineAssist [3] through three synergic designs:

(1) KineAssist-treadmill integration: RAPID will seamlessly interface KineAssist to a treadmill by controlling its belt velocity. Such a robot (KineAssist + treadmill) design is innovative because it enables fully expressed *stumbling corrective responses* [5,6].
(2) VR integration: The robot connects to a VR-based game interaction system to help the patient recover normal hand gestures (e.g., grabbing a pot in a virtual kitchen) and body motions (e.g., crossing a virtual bridge).
(3) Gait/medical sensor integration: RAPID achieves intelligent robot control based on the pattern recognition from heterogeneous sensor signals. For example, it uses a gait sensor network (Figure 22.1) to quantitatively measure the gait balance improvement in rehab training. It also allows a patient to drive the speed of the treadmill based on his/her intent through the worn force sensors. Medical sensors will be used to monitor the patient status (e.g., ECG sensor can measure the heart beat rhythm).

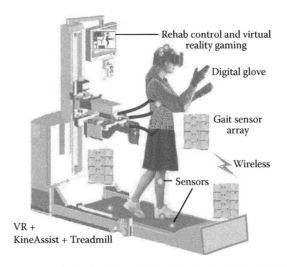

**Figure 22.1  Proposed RAPID platform.**

In order to achieve this goal, we are focusing on the following five topics:

1. *Body coherence measurement during CPS interactions:* In the VR-based CPS interactions, we aim to achieve a quantitative measurement of arm–leg and eye–hand coordination levels:
    (i) Arm–leg coherence analysis. RAPID uses three-axis accelerometers (attached to the joints of arms/legs) and VR games (e.g., kicking a ball) to achieve automatic, accurate analysis of body coherence level. The issues we focus on include the following: how do we design pattern learning to capture the intrinsic motion patterns of arms and legs, and how do we design a multisensing flow correlation scheme to define an arm–leg coherence level?
    (ii) Eye–hand coherence analysis. Disability from neurodisorders often makes it difficult for a patient to concentrate on a life activity. RAPID uses eye tracker (ET) to trace the eye focus movement and then compare it with hand trajectory to check eye–hand coherency. However, ET and hand movement data have both spatial deviation (an ET provides 2D eye focus spots, while a hand path is a 1D curve) and temporal deviation (a patient's hand movement lags behind eye change). The issue is, how do we find an efficient spatial-cum-temporal warping scheme to automatically measure the alignment level between ET and hand trajectory?

2. *Learning of complex hand gestures during CPS interactions:* Our VR-based CPS interactions have an accurate hand gesture learning from high-dimensional, nested digital glove (DG) motion signals—RAPID uses different VR games (e.g., cup grabbing, boxing, and ball catching) to train hand flexibility. A nested hand motion requires the recognition of two-level actions (i.e., entire hand event and smaller actions). For example, a cup-grabbing event consists of segmented actions including going up and left, slow-to-fast movements, flattening the palm, and grabbing the cup handle. Moreover, there exists a Markov transition state explosion issue due to the high-dimensional signal processing from dozens of DG sensors. The issue is, how do we design an intelligent hand gesture learning scheme that can handle the uncertain, high-dimensional, and nested DG signals?

3. *Cyber-based physical balance analysis via gait sensor arrays:* We use distributed sensor arrays to recognize various gait disorder patterns for physical balance analysis—while it is relatively easy to recognize different people's normal gaits, it is much more difficult to distinguish among various abnormal gaits related to stroke disability due to the minor gait differences. We are enhancing our previous pyroelectric sensor-based gait recognition [7–12] through the design of a higher-resolution gait sensor circuit with a special Fresnel lens for the detection of minor gait differences. We also build a distributed gait sensor array network in order to comprehensively capture the entire body's gesture. The issue is, how do we design such a high-fidelity sensor array network with a gait pattern learning scheme that is able to recognize the signal geometry features from different sensor arrays?

4. *Cyber-physical coupled design for an integrated RAPID test bed:* We target a seamless hardware/software integration among the robot, VR gaming, and sensor systems. First, we interface our previously invented KineAssist with an off-the-shelf treadmill through a high-performance servo motor. To make the treadmill run with the patient's intent, a closed-loop impedance controller utilizes force feedback from load cells on the pelvic mechanism to compute the desired velocity for the treadmill belt. Second, we use a mobile agent-based approach to manage the CPS interactions and the interoperations among RAPID units (including KineAssist, treadmill, gait/medical sensors, and VR subsystem). Our invented KineAssist has passed the US Food and Drug Approval (FDA) clearance [13].

5. *Practical clinical test to validate CPS usability:* Stroke patients will be enrolled for a comprehensive evaluation of RAPID rehab-training performance. Because all devices are harmless to humans' bodies, we will first perform a cyber-physical coupling test in our research labs. After such proof of concept is verified, Dr. Brown will lead the practical clinical test in the Physical Therapy Department at the University of Alabama–Birmingham Health Center. Over 30 individuals (no children will be tested) with chronic stroke (>6 months post ictus), after passing Institutional Review Board approval, will be recruited for the clinical study. We will use our RAPID platform to perform VR-based body weight support treadmill training [6] for disability recovery. These testing results will be used to further improve our CPS design.

## 22.2 CPS Characteristics of the Proposed Rapid Design

The proposed RAPID design has the intrinsic characteristics of a CPS, that is, it has the tight coupling of computational objects (including programmable sensors, VR gaming system, digital inputs such as digital glove) and physical objects including patient body and the robot (i.e., the integrated KineAssist and treadmill). As shown in Figure 22.2, RAPID has the following three cyber-physical interactions:

1. Interaction between the intelligent sensors (cyber objects) and the robot (physical object): The sensing signal analysis module (Figure 22.2(1)) collects signals from medical, gait, and motion sensors and then uses machine learning algorithms to extract the intrinsic patterns such as gait anomaly level. Then, the module (Figure 22.2(1)) will use these patterns to control the operations of the robot.
2. Interaction between head mounted display (HMD)/ET/digital glove (cyber objects) and the patient body (physical object): The patient wears a digital glove and HMD to interact with a virtual cyber world through VR interfacing software module (Figure 22.2(2)). Another cyber object, ET, is used to capture the patient's eye pupil movement for the analysis of patient brain concentration level (if the patient has mind distractions, the eye movement will not synchronize with his/her hand movement).

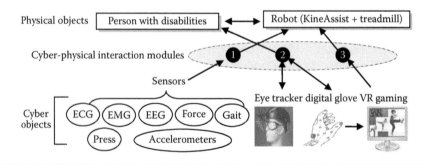

**Figure 22.2 Cyber-physical coupled design: (1) sensing signal analysis module, (2) VR interfacing module, and (3) gaming-output-triggered control module.**

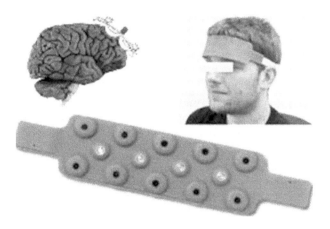

**Figure 22.3   fNIR-based neuroimaging.**

3. Interaction between VR virtual scenes (cyber object) and the robot (physical object): The VR gaming system can control the robot's behaviors. For example, if the patient has successfully achieved a low-speed walking training, the VR system can use the control module (Figure 22.2(3)) to speed up the belt.

Note that there also exist interactions between the two physical objects (i.e., patient and robot) as follows:

(1) Patient → robot: the treadmill belt can move in response to the forces applied by the patient to the pelvic mechanism. Such user intent-based robot control makes RAPID achieve intelligent rehab training.
(2) Robot → patient: KineAssist uses a few bands with controllable holding strength in case the patient loses body balance and falls down. This is achieved through signal analysis of the force sensor attached to the pelvis. Such a feature does not disturb the patient's natural gait pattern (e.g., it does not turn the body into a pendulum). Such a feature helps the patient to overcome the fear of falling down. Such fear could occur in traditional rehab training.

## 22.3  Functional Near-Infrared Spectroscopy Basics

In this chapter we will focus on the use of a critical neuroimaging tool, functional near-infrared spectroscopy (fNIR), which can detect brain activity signals. One of the critical tasks in neuroscience is to study how the brain structure, such as the cortical activation modes, changes with the rehab training process. Some important questions remain unanswered: Does a stroke survivor show certain consistent spatiotemporal patterns in his or her neuroimage signals when the body training has a satisfactory progress? Does the EEG spectrum show stable, dominant signal features when the patient can successfully grab any object after hand flexibility training? Do the fNIR statistics change when the patient shows improved arm and leg coherence during body training? Can one person's neuroimaging dynamics model be generalized to a population of individuals?

Neuroscientists are currently attempting to characterize the complex relationship between brain activity signals (measured by neuroimaging) and human motions. Although central pattern generators for locomotion are important in the control of walking, supraspinal networks including the brainstem, cerebellum, and cortex must also be critical as the result of the additional demands imposed by bipedal walking in humans [14–18]. This view has been supported by neuroimaging studies showing that rhythmic foot or leg movements recruit primary motor cortex especially in hazardous conditions [19–22].

Unlike the above studies, our research will not just simply link neuroimage to limb movements. Instead, we will establish an accurate and valid mapping model from (1) *quantitative* neuroimage pattern descriptions obtained through the latest manifold learning theory to (2) *quantitative* rehab progress measurements obtained through the mathematical analysis of the signals generated from the rehab devices.

Figure 22.3 shows a typical fNIR headband used in our system, which can be worn during rehab training. Figure 22.4 shows our proposed research methodology from neuroimage signals to rehab progress mapping.

The *fNIR principle*: The human brain can have different states (such as drowsiness and sleepiness). Those functional states correspond to certain blood properties. The fNIR can detect the optical properties of the brain tissues through the use of light in the near-infrared range (700–900 nm). Especially, the oxygenated and deoxygenated hemoglobin (briefly called oxy-Hb and deoxy-Hb) can be well detected by fNIR.

A fNIR device typically includes light sources that issue lights and light sensors that receive the light after it has interacted with the brain tissue. Because the light can scatter after entering the brain, a photodetector that is 2–7 cm away from the optode can sensor the reflected light. Especially when the distance between the source and the photodetector is set to 4 cm, the fNIR signals are very sensitive to hemodynamic changes. Therefore, fNIR can be used to assess brain activities [23].

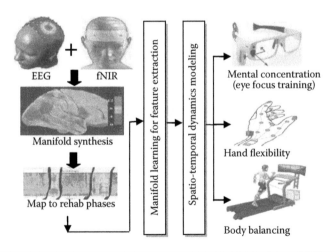

**Figure 22.4   Research methodology.**

## 22.4 Utilize the Synthesized Neuroimaging Signals between fNIR and EEG

Our neuroimaging tools adopt two noninvasive, low-cost methods, that is, EEG and fNIR, for the temporal and spatial brain activity measurements during rehab training. Other neuroimaging methods are either invasive (such as intracranial recordings) or require expensive machines (such as fMRI). Both EEG and fNIR are portable; that is, they do not require fixed head positions (this is unlike fMRI) and therefore can be used during tasks where a person is interacting with the VR environment. This enhances the ecological validity of the measurement system. Additionally, they can be worn simultaneously: EEG signals can be measured on the scalp surface through electrodes, while fNIR can be worn as a headband.

More importantly, *they can generate complementary imaging data*: EEG can capture the event-related potentials [24] through the measurement of electrical amplitudes in different scalp positions. Especially, it has different features in each frequency band. The fNIR uses lights in the near-infrared range (700–900 nm) to monitor changes in the concentrations of oxygenated or deoxygenated hemoglobin (i.e., oxy-Hb or deoxy-Hb) [25–27].

While using only EEG or only fNIR may not provide high-resolution, pattern-rich neuroimage data, synthesizing them together into a new spectrum through a data-geometry-preserved scheme allows us to exploit their complementary patterns to yield a view on brain activities with a higher spatiotemporal neuroimage resolution. Such a (EEG + fNIR) synthesis can help us better understand the relationship between neural and hemodynamic activities, that is, neurovascular coupling.

To more clearly explain the advantage of (EEG + fNIR) synthesis, here we use an analogy from a face recognition task. As shown in Figure 22.5, it is difficult to recognize a person's identity by just looking at the eyes' or the nose's image *alone*. However, by synthesizing all images through a geometry-preserved scheme (such as ensuring the nose is below two eyes), facial identity is accomplished more successfully.

Therefore, by synthesizing EEG and fNIR through special methods such as manifold learning, we can generate a more comprehensive neurovascular dynamic model for each rehab-training exercise. We can then further build the neuroimage-to-rehab mapping model.

Just as when we cannot accurately recognize the entire face's features without the correct geometry preservation among the eyes and nose (Figure 22.5), we cannot generate a correct spatial

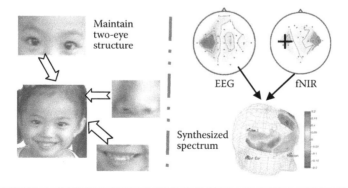

**Figure 22.5    Manifold synthesis of (left) an analogy and (right) our case.**

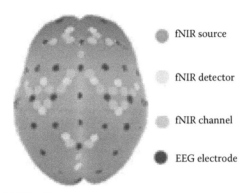

fNIR source

fNIR detector

fNIR channel

EEG electrode

**Figure 22.6  Geometry preservation.**

spectrum by simply mixing the EEG and fNIR data together without maintaining their data geometry relationship. As shown in Figure 22.6, it is important to maintain the *interposition* sensor data geometry among all EEG or fNIR channels since each EEG electrode or fNIR light source collects the brain activity signals *in certain scalp locations*. Conventional multisource fusion methods [28–32] can "flatten" the data sources, that is, losing data geometry information. A more powerful data synthesis tool is needed to handle multiformat, *high-dimensional* signals (EEG may be generated from >64 channels, and fNIR data can come from >10 light detectors). In this project, we will utilize information geometry and manifold learning to achieve a geometry-preserved EEG and fNIR synthesis.

## 22.5  Intelligent Signal Processing of fNIR and EEG

Our research uses two portable, complementary neuroimaging methods: EEG and fNIR, each of which has important *spatial* spectrum features:

(1) EEG: It has been found that different *scalp regions* in which the brain activities can be captured by the EEG electrodes make different contributions to the human's body movements [33]. Therefore, it is important to maintain the interelectrode data geometry structure when performing EEG processing.

(2) fNIR: A fNIR headband could have over 10 optical detectors to measure the oxy-Hb levels from the blood of the brain. The fNIR signals can be processed for the purpose of registering the voxel locations onto the frontal lobe brain surface image and visualize them from front, right, and left views [34]. Such *location information* is important to understand the cognition contribution of each brain region.

Therefore, when we use a data fusion method to synthesize EEG and fNIR into a comprehensive neuroimage spectrum, such a method should *preserve the geometry information* among EEG and fNIR signals. Otherwise, the synthesized image will lose scalp position-related patterns.

A challenge we face here is that both EEG and fNIR are not simple 1D, single-format time series. In other words, they have different data formats as well as high dimensionalities. For example, the EEG sensor could have over 64 electrodes (i.e., its dimensionality may be >64) in order to capture high-resolution brain activities, and the EEG data format is the electrical signal amplitude. While the fNIR headband can collect oxy-Hb through over 10 optical detectors (i.e., its dimensionality may be >10), and its data format is the energy absorption factor of oxy-Hb. Thus, the conventional data fusion methods [28–32] cannot be applied here. Then the issues are, how do we design an efficient EEG and fNIR synthesis mechanism that can preserve the intrinsic data geometry information among all neuroimage sources and in the meantime efficiently merge high-dimensional, multiformat neuroimage signals?

In this research, we propose to synthesize EEG and fNIR neuroimages through our recently designed machine learning scheme called manifold mergence (MM) that can fuse high-dimensional, multiple complex signal sources while in the meantime maintaining the geometric structure in each manifold. Our MM scheme first uses a weighted graph $G^{(k)}$, $k = 1, 2, \ldots, M$, to represent each manifold. In order to compute the distance matrix of the merged manifold, we first transform the *distance* between any two signal points in the $k$th manifold, denoted as $D_{ij}^{(k)}$, into a probability $P_{ij}^{(k)}$, which represents a transition probability from the $i$th to the $j$th point. Without loss of generality, we use a Gaussian kernel as follows:

$$P_{ij}^{(k)} = \Omega_i^{(k)} EXP[-(D_{ij}^{(k)})^2/(\sigma^{(k)})^2]$$

Here $\sigma^{(k)}$ is the standard deviation and $\Omega$ is for normalization purpose so that all probabilities initiated from point $i$ are summed to 1. Now we define an important concept for MM implementation:

*Definition 1 ($\alpha$-Integration):* Consider positive measures of random variable $x$, denoted as $m(x)$. Let us define a differentiable monotone function, $f_\alpha(x)$, as

$$f_\alpha(x) = x^{(1-\alpha)/2} \text{ if } \alpha \neq 1; \quad \log(x) \text{ if } \alpha = 1$$

Then, $\alpha$-*integration* is defined as a weighted mean of $m_1(x)$, $m_2(x)$, …, $m_J(x)$ with weights $\omega_i$:

$$\tilde{m}(x) = f_\alpha^{-1}\left\{\sum_{i=1}^{J} \omega_i f_\alpha[m_i(x)]\right\}$$

where $\omega_i > 0$ and $\sum_1^J \omega_i = 1$.

After we obtain all manifolds' transition probability matrices, that is, $P^{(1)}$, $P^{(2)}$, …, $P^{(M)}$, we can apply the above $\alpha$-*integration* concept to fuse those $M$ manifolds into one manifold with a compromised probability matrix $P_\alpha$ with each of its element as follows (from the $i$th to the $j$th point):

$$P_{\alpha,ij} = \Omega_{\alpha,ij} f_\alpha^{-1}\left[\sum_{i=1}^{M} \omega_k f_\alpha(P_{ij}^{(k)})\right]$$

where $\Omega_\alpha$ again is for probability normalization purposes.

The synthesized (EEG + fNIR) stream is still a high-dimensional, time-continuous signal that cannot be processed without data segmentation, since any pattern learning scheme can only process one window of data (i.e., a matrix) each time. The window size needs to be determined during the segmentation such that we do not miss any "atom" events, that is, the basic brain activity patterns extracted from a data window. However, it is challenging to determine the window boundaries in real time due to the lack of proper information analysis tools that could quickly detect the information statistics jump.

To capture the information jump in the synthesized neuroimage signal $X$, we notice that a probability distribution of $X$ has a tight coupling with its information "surprise" level. This can be seen from the definition of information entropy: $H(X) = -\Sigma P(X_i)\log_b P(X_i)$. Here $P(.)$ is the probability mass function of $X$. For $\log_b(.)$, we typically use natural logarithm. Therefore, in order to find the information jump of $X$, we propose to use a concept related to probability distribution, called Bregman ball, which belongs to *information geometry* theory [35,36]. Here, we define a Bregman ball as the minimum manifold with a central $\mu_k$ and a radius $R_k$. For any data point $X_t$ at time $t$, if it is inside this ball, it should have a strong statistical similarity (or, small signal distortion) between itself and the central $\mu_k$. That is,

$$B(\mu_k, R_k) = \left\{ X_t \in X : D_\Phi(X_t, \mu_k) \le R_k \right\}$$

where $D_\Phi(p, q)$ is the well-known *Bregman divergence* [36], which is defined as the manifold distance between two signal points $p$ and $q$. Both points are probability distribution mass values in a manifold space of $X$. We can use Fisher information, which can be approximated by the Bregman divergence, to measure the geodesic walking distance, which is actually the shortest path distance between any two manifold points (Figure 22.7). The geodesic distance reflects two distributions' signal similarity that can be defined as follows:

*Definition 2 (Signal similarity)* (see Figure 22.7): For any two manifold points P and Q, we define a distance called symmetric Bregman divergence (SBD): *Dis* (P, Q) = [$D_{\Phi-L}$(P, Q) + $D_{\Phi-R}$Q, P)]/2. Here $D_{\Phi-L}$ is the *left-type* Bregman divergence and $D_{\Phi-R}$ is the *right-type* Bregman divergence [37]. If SBD is less than a predefined threshold, we say P and Q are similar to each other.

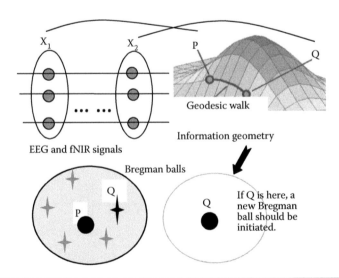

**Figure 22.7  Information geometry-based segmentation.**

On the basis of the signal similarity level between a new point (say, Q), and a central point (say, P), we can regard Q as part of the Bregman ball of P (if they are similar), or we should use Q as a center and initiate a new Bregman ball (if not similar).

## 22.6 Conclusions

This chapter has introduced our innovative research methodology in cognitive neuroscience, that is, using the latest manifold learning and information geometry theories to seek the mapping relationship between (1) the synthesized spectrum of electroencephalogram (EEG) and functional near-infrared (fNIR) images, and (2) the progress of rehabilitation (briefly called *rehab*) training. Specifically, we have performed manifold synthesis of EEG and fNIR, that is, a geometry-preserved neuroimage synthesis via manifold merging of two complementary, noninvasive neuroimaging methods, EEG and fNIR. Such a multimodal synthesis not only maintains the cortical activation geometry features but also provides richer neuroimage patterns than unimodal neuroimaging (such as using EEG alone).

## References

1. Centers for Disease Control and Prevention. 2010. Leading causes of death. Available at http://www.cdc.gov/nchs/fastats/lcod.htm (accessed on May 1, 2011).
2. P. Y.-C. Bhatt and T. S. Bhatt. 2007. Repeated-slip training: An emerging paradigm for prevention of slip-related falls among older adults. *Physical Therapy* 87, 1478–1491.
3. HDT Robotics—KineAssist: http://devwww.hdtglobal.com/services/robotics/portfolio/KineAssist/. For more detailed materials on KineAssist, see http://www.kineadesign.com/portfolio/kineassist/. Note that Kinea Design Inc. is now renamed as HDT Robotics Inc., a division of HDT Global.
4. J. Patton, E. Lewis, G. Crombie, M. Peshkin, E. Colgate, J. Santos, A. Makhlin, and D. A. Brown. 2008. A novel robotic device to enhance balance and mobility training post-stroke. *Topics Stroke Rehabil.* 15(2), 131–139.
5. C. L. Richards, F. Malouin, S. Wood-Dauphinee, J. I. Williams, J. P. Bouchard, and D. Brunet. 1993. Task-specific physical therapy for optimization of gait recovery in acute stroke patients. *Arch. Phys. Med. Rehabil.* 74, 612–620.
6. K. J. Sullivan, B. J. Knowlton, and B. H. Dobkin. Step training with body weight support: Effect of treadmill speed and practice paradigms on poststroke locomotor recovery. *Arch. Phys. Med. Rehabil.* 83, 683–691.
7. Q. Hao, F. Hu, and J. Lu. 2010. Distributed multiple human tracking with wireless binary pyroelectric infrared (PIR) sensor networks. In *2010 IEEE Sensors*. IEEE, New Brunswick, NJ, pp. 946–950.
8. X. Zhou, Q. Hao, and F. Hu. 2010. 1-bit walker recognition with distributed binary pyroelectric sensors. In *IEEE Conference on MultiSensor Fusion and Integration*. IEEE, New Brunswick, NJ, pp. 168–173.
9. F. Hu, Q. Sun, and Q. Hao. 2010. Mobile targets region-of-interest via distributed pyroelectric sensor network: Towards a robust, real-time context reasoning. In *IEEE Conference on Sensors*. IEEE, New Brunswick, NJ, pp. 1832–1836.
10. Q. Sun, F. Hu, and Q. Hao. 2010. Context awareness emergence for distributed binary pyroelectric sensors. In *IEEE Conference on MultiSensor Fusion and Integration*. IEEE, New Brunswick, NJ, pp. 162–167.
11. Q. Sun, F. Hu, and Q. Hao. 2012. Mobile targets scenario recognition via low-cost pyroelectric sensing system: Towards an accurate context identification. *IEEE Trans. Syst. Man Cybernetics: Sytems*, in press.
12. Q. Hao, F. Hu, and Y. Xiao. 2009. Multiple human tracking and recognition with wireless distributed pyroelectric sensor systems. *IEEE Syst. J.* 3(4), 428–439.

13. Please see our published article posted in U.S. National Institutes of Health resources Web site: http://www.ncbi.nlm.nih.gov/pubmed/18430678. It points out that a prototype KineAssist has been constructed and has received U.S. Food and Drug Administration (FDA) classification and Institutional Review Board clearance for initial human studies.

14. J. T. Choi and A. J. Bastian. 2007. Adaptation reveals independent control networks for human walking. *Nature Neurosci.* 10, 1055–1062.

15. S. Grillner, P. Wallen, K. Saitoh, A. Kozlov, and B. Robertson. 2007. Neural bases of goal-directed locomotion in vertebrates—An overview. *Brain Res. Rev.* 57, 2–12.

16. J. B. Nielsen. 2003. How we walk: Central control of muscle activity during human walking. *Neuroscientist* 9, 195–204.

17. S. Rossignol, R. J. Dubuc, and J. P. Gossard. 2006. Dynamic sensorimotor interactions in locomotion. *Physiol. Rev.* 86, 89–154.

18. L. O. D. Christensen, P. Johannsen, T. Sinkjaer, N. Petersen, H. S. Pyndt, and J. B. Nielsen. 2000. Cerebral activation during bicycle movements in man. *Exp. Brain Res.* 135, 66–72.

19. B. H. Dobkin, A. Firestine, M. West, K. Saremi, and R. Woods. 2004. Ankle dorsiflexion as an fMRI paradigm to assay motor control for walking during rehabilitation. *Neuroimage* 23, 370–381.

20. A. R. Luft, G. V. Smith, L. Forrester, J. Whitall, R. F. Macko, T. K. Hauser, A. P. Goldberg, and D. F. Hanley. 2002. Comparing brain activation associated with isolated upper and lower limb movement across corresponding joints. *Human Brain Mapp.* 17, 131–140.

21. C. Sahyoun, A. Floyer-Lea, H. Johansen-Berg, and P. M. Matthews. 2004. Towards an understanding of gait control: Brain activation during the anticipation, preparation and execution of foot movements. *Neuroimage* 21, 568–575.

22. M. Wieser, J. Haefeli, L. Butler, L. Jancke, R. Riener, and S. Koeneke. 2010. Temporal and spatial patterns of cortical activation during assisted lower limb movement. *Exp. Brain Res.* 203, 181–191.

23. M. Cope. 1991. *The Development of a Near-Infrared Spectroscopy System and Its Application for Noninvasive Monitoring of Cerebral Blood and Tissue Oxygenation in the Newborn Infant.* University College London, London.

24. M. Liu, H. Ji, and C. Zhao. 2008. Event related potentials extraction from EEG using artificial neural network. In *Proceedings of the 2008 Congress on Image and Signal Processing, Vol. 1—Volume 01* (CISP '08). IEEE Computer Society, Washington, DC, pp. 213–215.

25. A. Villringer, J. Planck, C. Hock, L. Schleinkofer, and U. Dirnagl. 1993. Near infrared spectroscopy (NIRS): A new tool to study hemodynamic changes during activation of brain function in human adults. *Neurosci. Lett.* 154, 101–104.

26. B. Chance, E. Anday, S. Nioka, S. Zhou, L. Hong, K. Worden, C. Li, T. Murray, Y. Ovetsky, D. Pidikiti, and R. Thomas. 1998. A novel method for fast imaging of brain function, non-invasively, with light. *Opt. Express* 2(10), 411–423.

27. A. Villringer and B. Chance. 1997. Non-invasive optical spectroscopy and imaging of human brain function. *Trends Neurosci.* 20, 435–442.

28. J. Bleiholder and F. Naumann. 2009. Data fusion. *ACM Comput. Surv.* 41(1), Article 1.

29. C. Siaterlis and B. Maglaris. 2004. Towards multisensor data fusion for DoS detection. In *Proceedings of the 2004 ACM Symposium on Applied Computing* (SAC '04). ACM, New York, pp. 439–446.

30. U. Ramachandran, R. Kumar, M. Wolenetz, B. Cooper, B. Agarwalla, J. Shin, P. Hutto, and A. Paul. 2006. Dynamic data fusion for future sensor networks. *ACM Trans. Sensor Networks* 2(3), 404–443.

31. L. Lemieux, A. Salek-Haddadi, O. Josephs, P. J. Allen, N. Toms, C. Scott, K. Krakow, R. Turner, and D. Fish. 2001. Event-related fMRI with simultaneous and continuous EEG: Description of the method and initial case report. *Neuroimage* 14(3), 780–787.

32. J. R. Ives, S. Warach, F. Schmitt, R. R. Edelman, and D. L. Schomer. 1993. Monitoring the patient's EEG during echo planar MRI. *Electroencephalogr. Clin. Neurophysiol.* 87(6), 417–420.

33. A. Presacco, L. W. Forrester, and J. L. Contreras-Vidal. 2012. Decoding intra-limb and inter-limb kinematics during treadmill walking from scalp electroencephalographic (EEG) signals. *IEEE Trans. Neural Syst. Rehabil. Eng.* 20(2), 212–219.

34. H. Ayaz, M. Izzetoglu, S. M. Platek, S. Bunce, K. Izzetoglu, K. Pourrezaei, and B. Onaral. 2006. Registering fNIR data to brain surface image using MRI templates. *Conf. Proc. IEEE Eng. Med. Biol. Soc.* 1, 2671–2674.

35. S. Amari and H. Nagaoka. 2000. *Methods of Information Geometry*, vol. 191. Oxford University Press, New York.

36. A. Banerjee, S. Merugu, I. S. Dhillon, and J. Ghosh. 2005. Clustering with Bregman divergences. *J. Mach. Learning Res.* 6, 1705–1749.

37. F. Nielsen and R. Nock. 2009. Sided and symmetrized Bregman centroids. *IEEE Trans. Inf. Theor.* 55(6), 2882–2904.

# Index

Page numbers f and t indicate figures and tables, respectively.